NEMATODE MANAGEMENT IN PLANTS

NEMATODE MANAGEMENT IN PLANTS

Editor

Pravin Chandra Trivedi

Ph.D., Post Doct. (U.S.A.), F.L.S. (London), F.B.S.,
F.P.S.I., F.N.S.I., F.B.R.S., F.M.A., F.E.S.
Professor of Botany,
University of Rajasthan,
Jaipur - 302 004, India

SCIENTIFIC PUBLISHERS (INDIA)

P.O. BOX 91 JODHPUR

Published by
PAWAN KUMAR
Scientific Publishers (India)
5-A, New Pali Road, P.O. Box 91
JODHPUR - 342 001
Tel.: +91-291-433323
Fax.: +91-291-512580
E-mail: artrans@tantramail.com
artrans@satyam.net.in
website : www.scientificpub.com

ISBN: 81-7233-333-1

Laser typeset : Rajesh Ojha
Printed in India

Dedicated to
Prof. B.N. Mathur

PREFACE

Plant parasitic nematodes are hidden enemies of crops and represent a unique challenge to agricultural research, in that they combine their potential for serious reduction in growth and yield in a wide range of crop plants, often with, rather, non-specific and easily mis-diagnosed symptoms. Equally, nematode populations themselves are comparatively difficult to identify and quantify and their relationship between nematode population size and crop damage are subject to many environmental and crop specific factors. Finally, nematodes are system pests, rather than single crop pest-particularly in the tropics, where cropping sequence may support similar species when cropping intensity is high and conditions favourable for large population build- up. Control or management systems are therefore, likely to be much more complex in tropical agriculture than monocrop temperate agriculture.

The science of nematology has developed only during the last 40 years in India. The importance of nematodes as a constraint to successful crop production was first realized with the prevalence of the *cyst nematodes* on wheat in 1958 and on potato in 1961. Root-knot nematodes are prevalent in 90% of agricultural crops and are considered to be the number one problem. Cyst-forming nematodes are restricted to wheat, barley, rice, maize, sorghum, potato and pulses. Reniform and lesion nematodes are associated with a wide range of agricultural crops. These nematode pests have assumed a highly damaging role on a variety of crops, comprising cereals, vegetables, fruits, pulses, oilseeds, fiber, plantation, spices, horticultural, medicinal, aromatic, ornamental and forest crops.

Development of the concept of pest-management and their implementation have led to a greater appreciation of the need for a wide range of tactics for nematode control. The objective of nematode control is to improve growth and yield of plant without adversely affecting the environment. The increasing environmental, economic and human health problems, resulting from traditional control of nematodes is a clear signal for serious examination of practices regarded as biological control.

The present book "Nematode Management in Plants" incorporates critical review articles on nematode pest-management with special emphasis on biocontrol management practices over other control practices in the present scenario. The volume contains 18 articles by leading Plant Nematologist who have contributed immensely in the area of their research. Articles on application of cropping sequences, plant products and botanicals, latex, bioagents and biological control practices for the management of nematode pests have been included. Information has been compiled on the use of *Azotobacter, Bacillus thuringiensis* and VAM Fungi for reducing nematode pests. Detailed account on the integrated nematode management practice has been incorporated for the economical management of the pest.

This Festschrift — "Nematode Management in Plants" is a humble dedication to Prof. B.N. Mathur and to commemorate his outstanding contributions in the field of Plant Nematology. Dr. Mathur worked extensively on the "Molya" disease of wheat and barley in Rajasthan and suggested economical management practices for *Heterodera avenae*. He has published more than hundred research papers, completed many research projects and participated in many national and international Conferences. He established a Nematology laboratory at Durgapura Agricultural Station, Jaipur. For his excellent contribution in the field of Nematology, he was awarded and honoured by many societies including Nematological Society of India, New Delhi and Bioved Research Society, Allahabad.

I am thankful to all the contributors for writing authoritative and informative articles for this volume. The opinions and text contained herein are those of the authors and I have tried to honour their ideas in the original shape. I am sure, the information given in this volume will prove helpful to those interested in nematology and plant protection for many years to come and generate interest in an upcoming nematologist. I would very much appreciate receiving suggestions from readers so that shortcomings, if any, are corrected in future editions.

I wish to thank my wife Kusum, daughter Priyanka and son Rohit for their endurance during the compilation work of this volume. I appreciate the co-operation and support of Mr. Pawan Kumar of Scientific Publishers (India), Jodhpur for publishing the volume with patience, care and interest.

I am confident that the volume will be widely accepted by all the students, teachers and researchers in the field of plant nematology and plant pathology.

JAIPUR **P.C. Trivedi**

LIST OF CONTRIBUTORS

A. Kalra
Central Institute of Medicinal and Aromatic Plants. P.O. CIMAP. Lucknow - 226 015

A.K. Singh
Scientists (SS) Nematology. Crop Protection Programme. D.W.R. Karnal - 132 001

Akhtar Haseeb
Department of Plant Protection. Faculty of Agricultural Sciences. Aligarh Muslim University. Aligarh - 202 002

Archana Mittal
Indian Agricultural Research Institute. New Delhi 110 012

B.K. Goswami
Indian Agricultural Research Institute. New Delhi 110 012

B.N. Shukla
Department of Plant Pathology: Jawaharlal Nehru Krishi Vishwa Vidyalaya. Jabalpur

D.C. Gupta
Department of Nematology. CCS Haryana Agricultural University. Hisar - 125 004

H.B. Singh
National Botanical Research Institute, Lucknow

Hebsybai
College of Agriculture. Kerala Agricultural University. Vellayani - 695 522

I. Vadhera
Department of Plant Pathology. Jawaharlal Nehru Krishi Vishwa Vidyalaya. Jabalpur

I.J. Paruthi
Department of Nematology. CCS Haryana Agricultural University Hisar - 125 004

J. Bhatt
Department of Plant Pathology. Jawaharlal Nehru Krishi Vishwa Vidyalaya. Jabalpur

K. Sitaramaiah
Dept. of Plant Pathology. Acharya N.G. Ranga Agricultural University. S.V. Agricultural College. Tirupati - 517 502

M. Mashkoor Alam
Department. of Botany. Aligarh Muslim University. Aligarh 202 002

M. Nagesh
Division of Entomology. Indian Institute of Horticultural Research. Hessaraghatta Lake. Bangalore - 560 089

M.S. Rao
Division of Entomology, Indian Institute of Horticultural Research, Hessaraghatta Lake P.O., Bangalore - 560 089

M.S. Sheela
College of Agriculture, Kerala Agricultural University, Vellayani - 695 522

Mansoor A. Siddiqui
Department of Botany, Aligarh Muslim University, Aligarh - 202 002

N. Hasan
Indian Grassland and Fodder Research Institute, Jhansi - 284 003

P. Harinath Naidu
Dept. of Plant Pathology, Acharya N.G. Ranga Agricultural University, S.V. Agricultural College, Tirupati - 517 502

P. Parvatha Reddy
Division of Entomology, Indian Institute of Horticultural Research, Hessaraghatta Lake, Bangalore - 560 089

P. Sivaprasad
College of Agriculture, Kerala Agricultural University, Vellayani - 695 522

P.C. Trivedi
Department of Botany, University of Rajasthan, Jaipur - 302 004

P.P.K. Chahal
Department of Microbiology, Punjab Agricultural University, Ludhiana - 141 004

Prabhat Kumar Shukla
Department of Plant Protection, Faculty of Agricultural Sciences, Aligarh Muslim University, Aligarh - 202 002

R.A. Sikora
Institut Fur Pflanzenk rankheiten, der Universitat Bonn., Nessallee - 9, 53115, Bonn, Germanay

R.K. Jain
Department of Nematology, Haryana Agricultural University, Hissar - 125 004

Rakesh Pandey
Central Institute of Medicinal and Aromatic Plants, P.O. CIMAP, Lucknow - 226 015

S. Nageswari
Division of Nematology, Indian Agricultural Research Institute, New Delhi - 110 012

S. Pandey
Division of Plant Pathology, Indian Institute of Sugarcane Research, Lucknow

S.D. Mishra
Division of Nematology, Indian Agricultural Research Institute, New Delhi - 110 012

V.P.S. Chahal
Department of Microbiology, Punjab Agricultural University, Ludhiana - 141 004

CONTENTS

CHAPTER - 1

WHEAT SEED GALL NEMATODE (*Anguina tritici*) IN INDIA — A PROBLEM AND APPROACHES FOR ITS ERADICATION

I.J. Paruthi and D.C. Gupta

INTRODUCTION

Anguina tritici (Wheat seed gall nematode) is the oldest plant parasitic nematode reported in 1743 by Needham from England. In early part of 20th century, this nematode was reported from all wheat growing areas of the world and caused a threat to wheat cultivation. However, it was eradicated from the advanced European countries by ensuring sowing of cockle free clean seed persistently for several years (Kiryanova & Krall, 1963) but its trouble-some existence is still reported from several under developed and developing countries like India (Koshy and Swarup, 1971). The chapter deals with the problem of this nematode in India alongwith strategies and measures of its eradication.

1. GEOGRAPHICAL DISTRIBUTION

In India, though this nematode has been reported from all wheat growing areas of the country like Bihar, Delhi, Haryana, Himachal Pradesh, Punjab, Rajasthan and Uttar Pradesh (Koshy and Swarup, 1971), but recently, it has appeared as epidermic form in certain parts of Bihar and Madhya Pradesh (Nath and Pathak, 1993; Vadhera *et al.*, 1998).

2. HOST RANGE

Wheat is considered as its main host but some other crops like emmer, spelt, rye, barley, triticale and some weed plants like *Phalaris minor* and *Polypogon monspeliensis* have also been reported to be attacked (Suryanarayana and Mukhopadhyaya, 1971; Bhatti *et al.*, 1978; Dhawan and Swarup, 1979; Paruthi and Bhatti, 1981a, 1987a; Paruthi *et al.*, 1989).

3. ECONOMIC IMPORTANCE

The nematode causes quantitative as well as qualitative losses in wheat. Formation of cockles in place of healthy grains in earcockle disease and destruction of floral organs by the yellow bacterial slime in yellow ear rot (tundu) disease causes the direct loss in wheat yield. The average loss due to these diseases from India is about 2% but on certain farms it has been reported as high as 30-70% (Vasudeva and Hingorani, 1952; Singh *et al.*, 1953; Chakravarti and Singh, 1961; Paruthi and Bhatti, 1985; Paruthi and Gupta, 1987; Nath and Pathak, 1993).

In addition to the quantitative loss, contamination of cockles in the grains changes the colour of grains and flour and consequently reduces the market value of such products (Gupta, 1966; Paruthi and Bhatti, 1988). Unlike other species of *Anguina* particularly *A.agrostis*, this nematode does not produce any toxins (Midha and Swarup, 1974). However, Singh (1978) has mentioned the toxic effect of the cockles on man and animals, if consumed in large quantities. Although in yellow ear rot disease, no grains are produced but in some cases, the grains of poor quality are obtained from partially attacked earheads (Paruthi *et al.*, 1987).

4. ETIOLOGY

(I) Diseases

The nematode alone causes earcockle disease and is locally known as 'mamni' 'gegla' or 'sehun' but in association with a bacterium (*Clavibacter tritici*) causes another disease known as yellow ear rot or 'tundu' or 'tanan'.

(II) Disease Symptoms

(a) Earcockle : This disease is caused by *Anguina tritici*. The diseased plants show basal swelling of stem, crinkling, curling and twisting of leaves, stunted plant growth with profused tillering. These symptoms are conspicuous at the early stage of plant growth and normally disappear at later stages (Gupta and Swarup, 1968). Expression of such symptoms on leaves depends upon the level of contamination of the seed (Midha and Swarup, 1972).

Typical cockled earheads are smaller and broader than the healthy ones with or without awns on them. In such earheads, instead of formation of grains, cockles are formed which are initially green in colour but turn brown to black on maturity. However, there are certain symptomless diseased earheads (Paruthi and Bhatti, 1987) which are difficult to be diagnosed even by the experts. Under field conditions comparatively more number of earheads belong to this category. Under normal conditions, the galls are formed in place of the floral tissues. But in some cases larvae penetrate the non-floral organs like leaves, glumes, awns, lemma and palea etc. and form the galls there (Paruthi and Bhatti, 1986). The nematodes though complete their life cycle in such non-floral tissues but are not considered as important because of their rare occurrence (Christie, 1959). Besides these symptoms, Paruthi and Gupta, (1985) reported two abnormalities in earheads (sterility and elongation of distance between first and second spikelet) attacked by *A.tritici*.

(b) Yellow ear rot or 'tundu' : This disease is caused by *A. tritici* in association with a bacterium *Clavibacter tritici*. It was reported for the first time from India by Hutchinson 1917. Besides initial symptoms of *A.tritici* at vegetative phase of the plant, production of bright yellow gum or slime on the leaves, stem and earheads during humid and cloudy weather is common symptom of this disease (Gupta and Swarup, 1968). The floral primordia are destroyed and no grains are formed in this disease. However, some grains of poor quality are produced in partially diseased earheads (Paruthi *et al.*, 1987).

(III) Life cycle of nematode

Life history of *A.tritici* on wheat has been studied in details by Swarup and Gupta (1971) and Gokte and Swarup (1987). Whenever infested wheat seed is sown, on contact with moisture in soil, the larvae of *A. tritici* become active and are released in the soil in 10-15 days depending upon temperature and soil type (Paruthi *et al.*, 1990). These larvae make their ways to the growing point of young wheat plants and are carried mechanically upward to the inflorescence alongwith lengthening of the culm. At the time of initiation of floral primordia, these larvae invade them and stimulate the development of galls. At this stage, the nematode feeds endoparasitically on the floral initials and enclosed in the cavities formed by over growth of floral initials which eventually turn into galls. The larvae undergo three successive moults within 3-5 days after invasion of floral primordia and develop into adult males and females in the ratio of 1:2 (Midha and Swarup, 1974; Paruthi, 1982). The female starts laying eggs within 6-12 days. Each female lays about 2000 eggs which hatch into second stage larvae within the developing galls and become quiescent. This nematode completes one generation in a year (Swarup and Gupta, 1971).

(IV) Physiological strains/races

Existence of physiological strains or races in *A.tritici* has been speculated by Midha (1969) and Gokte and Swarup (1986). Gupta (1966), tested two populations of *A.tritici* from Delhi and Rajasthan and found that these were morphologically different but pathologically similar.

(V) Survival and spread

The larvae of *A.tritici* remain viable inside the galls in anhydrobiotic state for several years (Fielding, 1951; Limber, 1962 and 1973), but their viability in soil is less than six months under Indian conditions (Mukhopadhyaya *et al.*, 1970).

(VI) Influence of environmental factors on disease development

Epidemiological studies have revealed that the development of earcockle and yellow ear rot disease is dependent upon nematode inoculum level, bacterial concentrations, relative humidity, age of wheat seedlings and source of galls (Midha and Swarup, 1972). In general 'tundu' symptoms are more pronounced when the temperature is low and humidity is high (Midha, 1969). Vasudeva and Hingorani (1952a) reported that 'tundu' was more severe in early sown (October) crop and was further aggravated by flooding.

5. CONTROL MEASURES

(a) Use of gall free clean seed : These diseases can be eradicated by use of healthy (Cockle free) seeds obtained from national and state seed agencies (Paruthi and Gupta, 1986) due to their seed borne nature.

(b) Water flotation : Nematode galls contaminated wheat seed is immersed in water and stirred vigorously with a wodden stick, enables the cockles to float on the surface of water. These cockles can be decanted and sieved out and destroyed either by burning or burrying in soil (Paruthi and Bhatti, 1992). Plain water treatment is 90% efficient.

(c) Salt sedimentation : This method is similar to previously described water flotation method. Addition of 10-20% common salt (Vasudeva and Hingorani,

1952b; Singh *et al.*, 1953; Jain *et al.*, 1956; Paruthi and Bhatti, 1992) facilitates the removal of heavy and bold sized cockles. Seeds so treated should be washed in water before sowing. This method though removes all the cockles from the seed but is less preferred than water flotation method due to being more cumbersome and requires extra labour.

6. CONSTRAINTS

In spite of availability of very cheap and effective control operations (sieving, water flotation/salt sedimentation), these diseases could not be eradicated from our country due to the following reasons :

(a) Confusing disease symptoms : Although these diseases exhibit clear cut symptoms at the above ground parts of the plants. But in earcockle disease, some symptomless earheads (described earlier) are produced which remain unrecognised by the farmers. The farmer can recognise only 'tundu' attacked earheads at his field.

(b) Presence of small number cockles in the grains : Incidence of the cockles in the wheat grains is normally very low (0.01 to 1%) and is normally ignored or uncared by the farmers at sowing time due to their ignorance that such small number of cockles cause tremendous yield losses.

(c) Confusion of cockles with weed seeds : Due to their morphological resemblance, the presence of cockles in the grains is sometimes misconstrued due to confusion with the weed seeds commonly found in wheat seeds.

7. SOLUTION OR APPROACHES FOR ERADICATION

Like other European countries, these diseases can be eradicated from India by joint and simultaneous efforts at government level by the following methods:

(a) By educating the farmers to use clean (gall free) seed : A campaign for educating the farmers about the consequences of sowing cockle infested wheat seed at their farms and demonstration of easy and effective methods (sieving, water flotation/salt sedimentation) of removing the cockles from wheat seed is needed well before sowing time. This can very effectively be achieved through agricultural extension agencies of the states and centre; electronic media (radio, television, video etc.); print media (posters, hand outs, wall slogans, newspapers and farm magazines etc.); and other communicational channels (panchayats, social organisations, educational institutions etc.).

(b) By replacing old seed : To get the long lasting relief from these diseases, it is very essential to use gall free seed for sowing for a number of years. This can effectively be achieved by replacing the old seed of the farmers with clean (gall free) seed obtained from some national or state seed certification agencies by the state or centre governments either at free of cost or at subsidised rates or on loan basis. Though this approach is faught with practical difficulties but this is the only solution for eradiation/elimination of this dreaded pest from our country.

8. STEPS FOR EFFECTIVE IMPLEMENTATION OF THE PROGRAMME

The programme can be effectively implemented by the joint efforts of state agricultural universities and agricultural extension agencies of the state by the following methods :

(i) The farmers can be awakened about these disease by giving detailed symptomatology and economic importance of the diseases and demonstrating the control operations before sowing time. The efficacy of the programme can be enhanced by showing the documentary film of these diseases to the farmers before sowing.

(ii) Catchy slogans about these diseases, should be prepared in different local languages and can be given as advertisement in between the programmes of All India Radio and Doordarshan or in the leading newspapers at least one month before sowing. Also the stickers of these slogans can be prepared and pasted on the strategic public places in the village or on the back of agricultural equipments like tractors etc. These slogans can also be written on the walls of important buildings in the villages to catch the attention of the farmers.

(iii) At the time of maturity of the crop 'field days' should be organised at the farmers fields to educate them about these diseases and control operations.

9. CONCLUSION

Wheat seedgall nematode (*Anguina tritici*), though the oldest plant parasitic nematode and eradicated from the advanced European countries by following the clean gall free seed is still a troublesome pest in India. This nematode produces two diseases (earcockle and 'tundu' diseases) by self or in association with a bacterium (*Clavibacter tritici*) and can be very effectively be eradicated by using clean (gall free) seed at the farms for 2-3 years. In spite of availability of low cost and very effective control methods (sieving, water flotation/ salt sedimentation), the troublesome existence of the pest in our country reflects the backwardness of the farmers due to their lack of knowledge about these diseases. These diseases are mainly considered as poor man's problems where the farmers normally use their own old seed year after year is due to lack of their knowledge about these diseases. These diseases can be eradicated for ever by replacing their old seed with the new one obtained either through some seed certification agencies or by making their own old stock of seed free from cockles by the methods described above.

REFERENCES

Bhatti, D.S., Dahiya, R.S. and Dhawan, S.C., 1978. New record of tundu and earcockle in barley. *Nematologica.* 24 : 331-332.

Chakravarti, B.P. and Singh, D.K., 1961. 'Tundu' disease of wheat in *Bihar. Proc. Bihar. Acad. Sci.* 10 : 48-49.

Christie, J.R., 1959. Plant Nematodes, their bionomics and control. Florida Agricultural Experiment Station, 256 pp.

Dhawan, S.C. and Swarup, G., 1979. On *Anguina tritici* infestation in barley. *Indian J. Nematol.* 9 : 51-52.

Fielding, M.J., 1951. Observations on the length of dormancy in certain plant infection of nematodes. *Proc. Helminth. Sco. Wash.* 18 (2) : 110-112.

Gokte, N. and Swarup, G., 1986. Physiological specialisation in *Anguina tritici.* National Conference on Plant Parasitic Nematodes of India. Problems and Progress Dec. 17-20, 1986, IARI, New Delhi p.51.

Gokte, N. and Swarup, G., 1987. Studies on morphology and biology of *Anguina tritici. Indian J.Nematol.,* 17 : 306-307.

Gupta, P., 1966. Studies on earcockle and 'tundu' diseases of wheat. Ph.D..thesis, IARI, New Delhi India

Gupta, P. and Swarup, G., 1968. On the earcockle and yellow ear rot diseases of wheat. I. Symptoms and histopathology. *Indian Phytopath,* 21 : 318-323.

Gupta, P. and Swarup, G., 1972. Earcockle and yellow ear rot disease of wheat. II. Nematode bacteria association. *Nematologica.* 18 : 320-324.

Jain, J.S., Singh, R. and Mathur, R.S., 1956. Controlling earcockle of wheat by floating seed in salt solution. *Agric. Anim. Husb. India.* 5 : 3.

Kiryanova, E.S. and Krall, E.I., 1963. In : (Helminths of man, animals and plants and their control) *(K.I.Skrjabin 85th birthday vol.) Moscow, Izdat Azad Nauk SSSR*, pp. 477-479.

Koshy, P.K. and Swarup, G., 1971. Distribution of *Heterodera avenae, H.zeae, H. cajani* and *Anguina tritici* in India. *Indian J. Nematol.* 106-111.

Limber, D.P., 1962. Notes on the dormancy of *Anguina tritici* (Steinbuch, Filipjev, under constant humidity.) *Proc. Helminth. Soc. Wash.* 29 (1) : 91-92.

Limber, D.P., 1973. Notes on the longevity of *Anguina tritici* (Steinbuch, 1799) Filipjev, 1936 and its ability to invade wheat seedlings after 32 years of dormancy, *Proc. Helminth. Soc. Wash.*, 40 (2) : 272-274.

Midha, S.K., 1969. Studies on the earcockle nematode and the development of earcockle and tundu diseases of wheat. Ph.D. thesis IARI, New Delhi, India.

Midha, S.K., 1971. Investigations on the nature of association between *Anguina tritici and Corynebacterium tritici. Proc. 2nd Int. Symp. Pl. Path. IARI, New Delhi, India,* 8.

Midha, S.K. and Swarup, G., 1972. Factors affecting development of earcockle and tundu diseases of wheat. *Indian J. Nematol.*, 2 : 97-104.

Midha, S.K. and Swarup, G., 1974. Studies on the wheat seed gall nematode caused by *Anguina tritici. Indian J. Nematol.*, 4 : 129-130.

Mukhopadhyaya, M.C., Chand J.N. and Suryanarayana, D., 1970. Studies on the longevity of earcockles. *Punjab Agril. Univ. J. Res.* 7 : 625-627.

Nath, R.P. and Pathak, K.N., 1993. Widespread occurrence of earcockle disease of wheat in Bihar. *Indian J. Nematol.*, 23 : 129-130.

Needham, T., 1743. A letter concerning certain chalky tubitous concretions called malm; with some microscopical observations on the farina of red lily and of worms discovered in smutty corn. *Philos. Trans. Roy. Soc. Long.*, 42 : 173, 174, 634-641.

Paruthi, I.J., 1976. Studies on host-parasite relationship and chemical control of seed gall nematode *Anguina tritici* (Steinbuch) Filipjev, on wheat, M.Sc. thesis Haryana Agricultural University, Hisar, India.

Paruthi, I.J., 1982. Studies on *Anguina tritici* in wheat and barley, Ph.D. thesis, Haryana Agricultural University, Hisar, India.

Paruthi, I.J., 1998. Approaches for eradicating seed gall nematode from India. Pages 87-89 in Nematode pests in rice-wheat-legume cropping systems : Proceedings of a Regional Training Course, 1-5 September, 1997, CCS HAU, Hisar, Haryana, India (Sharma, S.B., Johansen, C., and Midha, S.K. eds.) Rice-Wheat Consortium Paper Series 4, New Delhi, India. Rice-Wheat Consortium for the Indo-Gangetic Plains.

Paruthi, I.J., Bajaj, H.K. and Bhatti, D.S., 1990. Studies on some factors affecting the emergence of *Anguina tritici* larvae from wheat seedgalls and their ascendence on to seedlings. *Indian J. Ecol.* 17 (1) : 48-50.

Paruthi, I.J., Singh, M. and Gupta, D.C., 1987. Quantitative and qualitative losses in wheat (*Triticum aestivum*) grains due to earcockle and 'tundu'. *Seed Res.*, 15 : 83-86.

Paruthi, I.J. and Bhatti, D.S., 1981a. Studies on *Anguina tritici* in barley. *Nematologica.* 27 : 463-466.

Paruthi, I.J. and Bhatti, D.S., 1981b. Estimation of losses in wheat yield due to *Anguina tritici* and seed gall infestation in market grains. *Indian J. Nematol.* 11 : 15-18.

Paruthi, I.J. and Bhatti, D.S., 1983. Studies on *Anguina tritici* in wheat and barley. *III All India Nematol. Symp. Solan India* (Abstract of paper p. 50.

Paruthi, I.J. and Bhatti, D.S., 1985. Estimation of loss in yield and incidence of *Anguina tritici* on wheat in Haryana. *Int. Nematol. Network. Newsl.* 2(3) : 13-16.

Paruthi, I.J. and Bhatti, D.S., 1986. Comparative morphology and contents of galls caused by *Anguina tritici. Nematoda* (*Anguinidae*) on various parts of wheat plants. *Nematologica*, 33 : 243-245.

Paruthi, I.J. and Bhatti, D.S., 1987a. Host differential responses of some cereals to *Anguina tritici. Int. Nematol. Network. Newsl.* 4 (3) : 8-10.

Paruthi, I.J. and Bhatti, D.S., 1987b. Variation in earcockle symptoms on wheat ears caused by *Anguina tritici. Nematologica.* 33 : 360-363.

Paruthi, I.J. and Bhatti, D.S., 1988. Effect on consumption and marketing of wheat contaminated with seed galls of *Anguina tritici. Haryana Agric. Univ. J. Res.* 18 : 173-176.

Paruthi, I.J. and Gupta, D.C., 1985. Two unrecorded abnormalities in wheat ears infested with *Anguina tritici. Nematologica* 32 : 357-360.

Paruthi, I.J. and Gupta, D.C., 1987. Incidence of tundu in barley and kanki in wheat field infested with *Anguina tritici. Haryana Agric. Univ. J. Res.* 17 : 78-79.

Paruthi, I.J. and Gupta, D.C. 1986. Control of earcockle and yellow ear rot diseases of wheat. *Indian Fmg.* 36 (6) : 32-35.

Paruthi, I.J.and Bhatti, D.S., 1992. Efficacy of different practices for the control of earcockle in wheat caused by *Anguina tritici. Current Nematology* 3 : 1-4

Singh, B., Singh, J. and Mathur, S.C., 1953. Earcockle or sehun disease of wheat. *Agric. Anim. Husb. India.* 3 (10): 7-9.

Singh, R.S., 1978. Plant Diseases (4th edn.) Oxford and IBH Publishing Co. New Delhi, India. 564 pp.

Suryanarayana, D. and Mukhopadhayaya, M.C., 1971. Earcockle and 'tundu' diseases of wheat. *Indian. J. Agri. Sci.,* 41 : 407-413.

Swarup, G. and Gupta, P., 1971. On the earcockle and 'tundu' diseases of wheat II. Studies on *Anguina tritici* (Steinbuch) Filipjev, and *Corynebacterium tritici* (Huntch), Burkh. *Indian Phytopath.* 24 : 359- 365.

Vadhera, I., Shukla, B.N. and Tiwari, S.P., 1998. Nematode pests in rice-wheat-Legume cropping systems in Madhya Pradesh, Pages 63-66. in N Nematode pests in rice-wheat-legume cropping systems : Proceedings of a Regional Training Course, 1-5 September, 1997, CCSHAU, Hisar, Haryana, India (Sharma, S.B., Johansen, C., and Midha, S.K. eds.) Rice-Wheat Consortium Paper Series 4, New Delhi, India. Rice-Wheat Consortium for the Indo-Gangetic Plains.

Vasudeva, R.S. and Hingorani, M.K., 1952a. 'Tanan' or 'tundu' disease of wheat. Indian Fmg. 2(3) : 14-19.

Vasudeva, R.S. and Hingorani, M.K., 1952b. Bacterial disease of wheat caused by *Corynebacterium tritici* (Huntchinson) Bergey *et al., Phytopathology.* 42 : 291-293.

CHAPTER – 2

NOVEL APPROACHES IN MANAGEMENT OF NEMATODE PROBLEMS IN WHEAT AND BARLEY

A.K. Singh

INTRODUCTION

Wheat, the foremost cereal crop occupies pre-empt position in Indian agriculture after rice as, it provides surplus food for our-never receding population, handsome amount even for export purposes and extends job opportunities for agricultural workers. In year (1999-2000) the production had touched a record level of 75.4 million tonnes hailing from approximately 27.2 million hectare area. Though, we are complacent but to any one's wonder, lot of produce is being waste due to several factors. There occur several biotic and abiotic factors, which if managed efficiently, our production level can go even quite up. Even among several biotic factors involved, nematodes have been playing substantial role in the various parts of India (like U.P., Bihar, M.P., Rajasthan and Haryana) in reducing the yields of wheat. The tragedy is that when we are noticing them damaging, they already have damaged the crop. Majority of farmers is not aware of their presence (nematodes), nor very satisfactory means are available to contain them. There are reports that more than 25 species of nematode associated with wheat rhizosphere and plants itself, six of them are endoparasites, but out of these, 2 species have got to be the major hurdles in realizing the full potential of wheat. These two are, *Heterodera avenae* (the molya disease causing agent) and *Anguina tritici* (the ear cockle disease causing agent/wheat seed gall nematode). Though the yield loss due to *Heterodera avenae* has been recorded to be around 50% or even total loss of the crop. but in case of *Anguina tritici*, in some cases and in some years, it goes up to 80% or more as it happened in 1997 in Bihar and in 1999 in several districts of M.P.

1. MOLYA DISEASE OR CEREAL CYST NEMATODE (CCN)

(a) Historical account, introduction and Symptoms

This nematode for the first time was reported from Neem Ka Thana, a village in Sikar district of Rajasthan by Vasudeva in 1958. Since then, it has been reported from Sikar, Jhunjhunu, Churu, Chomu, Bharatpur, Tonk, Ajmer, Dausa, Jaipur, Hanumangarh, Sawai Madhopur, and Udaipur. In Haryana, it is found in Ambala, Faridabad, Bhiwani, Gurgaon, Hissar, Mahendragarh, Rewari, Rohtak and Sirsa. The disease is characterized by typical patchy growth and stunted plants in the field. The patches tend to coalesce if the wheat growing is continued in the same field year after year. Infested plants exhibit yellowing, reduced leaf area, fewer leaf and roots, delayed emergence of ears, and reduced number of spikelets (resemble with nutrient deficiency symptoms). Also the affected plants have stiff, thin, and narrow leaf blade in comparison to healthy ones Under our condition, main root gets elongated with a bunch of rootlets at the extreme end. If the plants of wheat are uprooted after 75-80 days of sowing, white cyst bodies (females) are seen attached with the roots (Fig. 1) and roots show knotty appearance and also they just look like pearl.

The infested roots are short, with multiple branches giving appearance (Called coral roots) and often bearing small gall like formations. Since symptoms, as mentioned in preceding sentences, are non specific, presence of white cysts on roots are the only confirmatory indication. To confirm this (White cysts) in advanced stage harvesting stage of wheat, soils of suspected rhizosphere of wheat are washed on 60 mesh sieves in the month of late March or April. The collection is seen and if brown coloured lemon shaped cysts are

Fig. 1. White female bodies attached with root and roots showing knotty appearance

observed, it means plant is infested. The disease intensity mostly depends upon the environmental and ecological factors which influence both the host and the parasite. Damage to crop is governed by soil moisture, soil texture, inoculum load, and age of host plant. A positive correlation exists between degree of infestation of CCN and crop damage. This way it can be established that malady is due to *Heterodera avenae*. Further, there are two biotypes available in *Heterodera avenae*. Rajasthan and Haryana population constitutes one biotype and Punjab (Ludhiana & Hoshiarpur) other. This was proved by employing host differentials like Siri and Martin which were susceptible to Punjab population but resistant to Rajasthan and Haryana population whereas La-estanzuella was found to give reverse reaction with these population. Even in Haryana, there occur 2 biotypes representing Mahendragardh and Sirsa. If there are two populations in Haryana itself, then there is possibility of occurring 3 population in India. By resorting to differential, populations of these two biotypes have been diagnosed but inconsistency in reaction is observed year after year. For more authentic results on biotype variability, the population collected from different places may be resorted to molecular tools so that this pervasive confusion is resolved.

(b) Recent Approaches

Resistance to CCN has been identified in hexaploid wheat, *Triticum ventricosum*, Rye, Barley, Oats, and in *Triticum tauschii*. Two classes of resistance have been observed in *T. tauschii* and these confer either a high (CcnD1) or a moderate (CcnD2) level of resistance in allohexaploids. Several cultivars of wheat, Barley, Oats, and Rye have been produced incorporating resistance, but a significant limitation to breeding CCN resistant cultivars has been the time required, high labour input, and lack of consistency of the biological assays used to select resistant genotypes. An alternative strategy in breeding CCN resistant cultivars is to use molecular markers linked to the resistance genes. Such an assay can identify genetic loci independently, allowing two or more genes to be combined in a single genotype. Techniques frequently used in identifying molecular markers include DNA hybridization, and PCR, Restriction Fragment Length Polymorphism arising from sequence variation is revealed in DNA hybridization experiments as changes in the length of DNA fragments after cleavage by restriction enzymes take place. RFLPs are the most commonly used and informative molecular markers in cereal genome mapping. Again unlike RFLP analysis which is time consuming and expensive, molecular markers based on the polymerase chain Reaction (PCR) allow high throughout of material and have low start-up and running costs. The technique, PCR based allele-specific amplification assay from RFLP clone Tag 605 for use as a PCR based wheat chromosome 2B genetic marker has been employed. This region of 2B is of importance to wheat breeding programmes since it maps close to Cre1, a gene conferring resistance to cereal cyst nematode.

The conversion of the existing RFLP marker to a PCR based assay system represents a major goal in developing a cheap and rapid assay for screening for Cre-1 gene. In an experiment, the RFLP locus Xglk 605 identified by probe Tag 605 mapped to a proximal position on the long arm of wheat chromosome 2B about 7c M away from a gene conditioning resistance to cereal cyst nematode in the wheat line AUS 10894.

The development of CCN resistant cultivars may be accelerated through the use of molecular markers. A number of resistance genes against the pest are well known; one of them, the single dominant Ha 2 resistance gene, has been shown to be active against the Australian pathotype and maps to chromosome 2 of Barley. Segregation analysis identified two Restriction Fragment Length Polymorphism (RFLP) markers flanking the resistance gene in two double haploid populations of Barley. The Ha 2 gene confers resistance to *H. avenae*, arresting female development at about 15 days by causing the degradation of feeding sites in the root Stele. But the recent mechanization and computerization of the bioassay have reduced the cost but the test still takes 14 weeks and is confirmed to the winter month. The trend is going to identify molecular markers linked to the Ha2 gene for resistance to *H. avenae* that can be used for marker assisted selection in barley. The objective has been to determine if additional loci could be identified from alternative sources. Such loci would provide an opportunity to pyramid genes for resistance to CCN. RFLP markers can provide powerful tools for the indirect selection of agronomically important genes in breeding programme. The objective of the present study was to identify RFLP markers that could be used to select for CCN resistance in barley breeding programme.

Molecular markers can offer several important advantages over the bioassay for CCN resistance. RFLP markers described here can effectively replace the bioassay in all early generation screens. Of the particular significance will be availability of CCN resistant malting quality varieties permitting an increase in the cultivated area of malt barley and leading to higher growth returns (the area under Rajasthan, Haryana, eastern as well as southern part of UP can go up). Additionally, wheat and barley showing resistance to CCN will allow growers to effectively reduce the levels of nematodes in soil (carbofuran, a known nematicides, is costly and recently banned also). This will provide growers with more flexibility in rotations by being able to grow CCN susceptible, but otherwise, excellent, varieties of wheat and oats.

A high level of resistance was also transferred to wheat from *Aegilops triuncialis* (TR lines) using the cross (*T. turgidum and Ae triuncilialis*) and *T. aestivum*). Low fertility (3-5 viable kernels per plant) was observed during the process but the surviving hybrid plants were highly vigorous. To obtain stable resistant, further crosses to *T. aestivum* were performed. The resistance in TR lines seems to be transferred from the C genome of *Ae triuncilialis* (Genome CCUU). *Ae triuncilialis* was highly resistant to the two Spanish populations of *H. avenae* tested, as well as to four French races and two Swedish populations. The histological studies showed that hypersensitive reaction in the roots of a resistant TR line inoculated with the Ha 71 pathotype of *H. avenae*, whereas well formed syncytia were observed in the roots of the susceptible control. As a result, resistance to the *H. avenae* Ha 71 pathotype seemed to be inherited as determined by a single dominant factor in the crosses between resistant TR lines and susceptible cultivars. To target the *H. avenae* (cereal cyst nematode) resistance locus, a D genome amplified sequence (Dgas44), present in all chromosomes of the D genome, is being exploited to detect new sources of polymorphism. Another approach will apply Polymerase Chain Reaction analysis to bulk DNA pools from homozygous lines, resistant (CcnD1) and susceptible to CCN and generate RAPDs using 10 mer random and semi random primers.

Modest approach like cre gene of the wheat line AUS 10894 that confers resistance to the cereal cyst nematode can be applied. Using a pair of near isogenic lines that differed only in a small chromosome segment containing the cre locus, 58 group -2 probes were screened;

two (Tag 605 and CDO 588) were found that detected polymorphism between the NILs.

(c) Innovative Nematode Assay

Wheat, Barley, Oat and Rye seedlings are planted in 3 cms tubes filled in with sterile soil and inoculated five times at 3 day intervals @ 100 second stage juveniles per tube. After 12 weeks at 15°C the number of cysts formed on roots of each plant are counted. Then after cyst grading, the reaction is determined.

2. EAR COCKLE DISEASE

This disease is diagnosed once the symptoms have already appeared in the field, but at early stage, plants can be observed for having basal swelling (just after 20-25 days of sowing) of stem at the ground level with a whitish tinge. Twisting and crinkling of leaves and of stem is quite evident. Dwarfing of plants and less tillering are also observed. Affected plants show early emergence of earheads with spreading spikelets giving a broader look to the earheads in comparison to healthy ones. Again if cool weather prevails, tundu appears and ears get yellow slime mass. Few grains or whole grains in affected earheads get converted in to galls. Upon examining the seed after threshing or before sowing, if some small roundish, dark brown or black coloured cockles (Fig. 2) are found in grain or seed lot, it means disease may appear in ensuing crop. The appearance of the disease is highly dependent on weather factors. Thorough picking of cockled grains from the seed lot may help in checking the appearance of disease in the next crop. To manage this disease, the farmers are advised not to sow cockled grains, take healthy truthfully labelled or certified seeds from reliable sources in the country. The seed agencies must ensure that they are selling seeds only of pure nature, which do not have any contamination with ECN. The extension agencies must popularize the simple technique available for checking the

Fig. 2. Ear Cockles with healthy grain

disease. The techniques, available are salt sedimentation, water floatation, sieving, etc.

3. OTHER NEMATODES

There are certain other nematodes like *Helicotylenchus*, *Tylenchorynchus*, *Hoplolaimus*, *Meloidogyne graminicola*, *Merinius brevidens*, *Pratylenchus thornei* which are commonly found in the wheat fields. Though, there are no apparent damage caused by these nematodes but certainly they weaken the plants during the growth period. Sometimes, these nematodes predispose the plants to secondary infection by fungal, bacterial and viral pathogen. These nematodes have become more like root health issues in the wheat crop than the real inciter of the disease. If they are not managed efficiently, they may pose second generatin problems in wheat. Sometimes, at some point of time, they may act like vectors, for many viral, bacterial disease of wheat, which are not found in India. These nematodes are diagnosed by isolating them from soil and observing on root mass. Barring *Meloidogyne graminicola* and *Pratylenchus thornei*, all are seen in soil for which soil slurry is made in water and by using different sieves, the nematodes are isolated from soil and finally through tissue papers straining,. the suspension is collected. Thus collected suspension is seen under Stereobinocular Microscope, and identity of nematode is established. For *M. graminicola*, egg mass or root knots are observed on fresh root under Stereobinocular Microscope. Roots are observed in the same way for *P. thornei* also. The AICWIP (Wheat Nematology of All India Coordinated wheat Improvement Project) stationed at DWR, Karnal is closely monitoring since inception of Wheat Nematology programme at DWR the behaviour of these nematodes under wheat and wheat based cropping system in different wheat growing zones of the country. This can be termed as preparedness on the AICWIP part and also a base work for monitoring, evaluating and taking appropriate actions in tackling the damage caused by these micro entities.

4. EVALUATION OF NEMATODE PROBLEMS

(a) Heterodera avenae

The problem of molya disease is serious in Rajasthan and southern parts of Haryana. Though reports of occurrence are there from various parts of NWPZ, but seeing the soil type, weather and other related parameters, there is nothing alarming. Molya disease causes more damage to barley compared to wheat. Its ETL has been found to be as low as 5 juvenlies per cubic centimetres of soil to cause significant damage to wheat, Barley and Triticale. An unpublished report (1998-99) says that 20% increase in wheat yield due to the management of CCN in infested field by using carbofuran. The problem of CCN used to exist in Punjab earlier, but now due to change of cropping system in favour of rice-wheat, the problem is going down and down in the state. But in Rajasthan and southern Haryana, because of sandy nature of soil, same cropping system and also due to arid climate prevailing there, the molya problem is still glaring and deciding factor in output. To manage these diseases, a thorough management strategy basing on sound host resistant system is desired. Falling in the line AICWIP (DWR, Karnal) is thoroughly screening the entries against CCN every year at multilocations and under hot spot conditions. As a result, few materials have been generated and they are in the final stage of evaluation before they are released for Rajasthan condition by the Rajasthan government, until then, in the absence of such variety, the farmers in these states are not in position to grow wheat and barley crops efficiently rather they seem to be shifting for other crops (Mustard, pulses, Fenugreek etc.) which are not affected by this nematode and if they grow these crop for one or two years, nematode population reduces. Here comes the question of raising staple food crops. The wheat production has not found favour with the Farmers because of Molya disease in the state. Crores of rupees go waste as farmers remains unable to harvest more of harvesteable yield. The development and release of CCN resistant cultivars may act as a boon to resource starved, marginal farmers residing in harsh climatic locations in arid zone.

(b) *Anguina tritici*

In the year, 1997, Bihar farmers belonging to districts of Darbhanga and Madhubani could not harvest wheat because of ear cockle problem. Some fields were discarded because of misconception/ superstition that eating such cockled grain/bhusa (toori) might poison the eaters, be it human beings or cattle. In the 1999, ear cockle problem emerged in the state of MP also where it entrapped Panna, Satna, Chhatarpur, and Tikamgarh, districts. In few fields, the infection ranged up to 100% as a consequence there was hue and cry among the populace and at the start of the disease, no body could believe that it was due to nematode. The severity of the disease occurs as and when there is sowing of infested seed coupled with favourable weather occurs. The farmers remain unaware of the impending situation and take lightly the sowing the seed without having it properly cleaned. Now this aspect has to be thoroughly stressed so that no farmer could sow infested seed. The state government agencies, seed producing and distributing agencies, agricultural research institutes/stations, NGOs, DD, AIR, private TV channels, etc. located in the endemic areas should spread the message of clean cultivation to the farmers so that no yield is waste. For further details, these people can refer "Wheat Extension Bulletin" No. 10. Published from DWR, Karnal-132 001.

(c) Other nematodes

These are *Helicotylenchus, Tylenchorynchus, Meloidogyne graminicola, Hoplolaimus* and *Pratylenchus*. Though there are no apparent damage caused by these nematodes, but certainly they weaken the plant system during the growth period. Continuous monitoring of these nematodes is need of hour and DWR (AICWIP) is alive to it.

5. MANAGEMENT OF WHEAT NEMATODE

(a) Molya disease

(1) Deep summer ploughing (2-3 times) of the infested field during May-June to expose cysts to harsh sunlight that in turn kills the cysts thereby reduces the population to 40% in each season.

(2) Follow crop rotation if possible with Dhania, Fenugreek, Onion, Mustard, Gram, Carrot and Onion for bringing down the cyst population to 47-55% level. The cultivation of Wheat/Barley has increased the CCN population to the level of 160-180%. The thing is that whether these crops find favour with farmers or not.

(3) Host resistance-advanced material of wheat like CCNRV 1, CCNRV 3 and CCNRV 7 can be put to use. Cereal cyst nematode (CCN) resistant barley varieties like RD-387, RD-2052 and RD-2035 can be used in infested area.

(4) Apply, if required, carbofuran 3G @ 1.5 Kg ai/H or Sebufos 10G @ 2 kg a.i./H at the time of sowing in soil. Since carbofuran is costly, now banned, the alternative is limited.

(b) Ear cockle disease

(1) The only solution lies in sowing clean, healthy, uncontaminated seed in the field. For this, extension, developmental machinery and publicity wing of state/ central government required to be proactive.

(2) Once the symptoms have appeared, no need to go for any chemical application as this will be mere wastage and do not frightened as eating of cockled grain will not poison any more.

(3) If farmers are to sow their own saved seed and if, infestation is observed in the seed lot, they are advised to go for thorough sieving or brine solution treatment.

(c) Other nematode

(1) Insecticides applied for insect pest control (termite, white grub, cut worms etc) do take care of these entities.

(2) Continuous watch on dynamics of the population is desired to monitor the population.

6. INTEGRATED NEMATODE MANAGEMENT

(a) Molya (CCN)

Summer ploughing in May/June to expose cyst to weather factors. Apply Carbofuran or Sebuphos, if the field is having past history of CCN infestation. Farmer may opt non-graminaceous crops in the affected fields. The point to be remembered here is that if the farmer is going for crop rotation, he ought not to apply nematicide as it is a costly item. If possible, he may grow some genotypes that possess tolerance (like CCNRV-1, CCNRV3, CCNRV 7 of wheat, RD-387, RD-2052 and RD- 2035 of Barley).

(b) Ear cockle (ECN)

Farmers should use cockled free seed only. In case they are to use their own seed, they should ensure that it is free from galls. If galls are found in seed lot, they should resort to sieving or water floatation or brine solution treatment as the case demands. For brine solution (2-5% ordinary salt solutions) they can use at the rate of 200-250 g salt per 10 lit. of water. Once seed lot is put in the brine solution, galls will float on the surface and they can be easily skimmed off. The settled seed at bottom be washed 2-3 times in ordinary water and shade dried. Then they can be used as seed. In spite of dry or wet cleaning, some galls are bound to be with treated seed and they will certainly cause some symptoms to appear in the field, for that selective rouging is advised at early stages. Above all, the seed agencies are to play crucial role in checking this menace.

7. FUTURE STUDIES REQUIRED

(1) The riddle of existence of biotypes/ pathotypes needs to be thoroughly addressed through use of molecular tools.

(2) There are dire need of developing resistant varieties for which breeders have to put their efforts. The issue of pathotypes and resistant material development go hand in hand.

(3) Sound and effective publicity for preventing, containing ear cockle disease is required.

REFERENCES

All India Coordinated Wheat Improvement Projects reports (Nematology) 1985-86 to 2000-2001.

Barloy, D.J. Lemonie, F. Dredryver and J. Jahier 2000. Molecular markers linked to the *Aegilops variabilis* derived root knot nematode resistance gene Rkn-mn1 in wheat. *Plant Breeding* 118 : 169-172.

Kretschmer, J.M., K.J. Chaimers, S. manning, A. Karakousis, A.R. Barr, A.K.M.R. Islam, S.J. Logue, Y.W. Choe, S.J. Barker, R.C.M. lance, P. Langridge. 1997. RFLP mapping of the Ha 2 cereal cyst nematode resistance gene in Barley. *Theor, Appl. Genet.* 94 : 1060-1064.

Lagudah, E.S. and R. Apples 1993. A directed search for DNA sequences tightly linked to cereal cyst nematode resistance genes in *Triticum tauschii* Genome 37 : 311-319.

Singh, AK, AK Sharma and S. Nagarajan, 2000. "Ear Cockle Disease of Wheat - Management at Ease" Wheat Extension Bulletin No-10, DWR, Karnal - 132 001.

Twenty Years of Coordinated wheat Research 1961-1986. Editors, J.P. Tandon and A.P. Sethi, Wheat Project Directorate AICWIP, IARI, New Delhi, 12, pp- 189-206.

Williams, K.J. John, M. Fisher, and Peter Langridge 1996. Development of a PCR based allete-specific assay from an RFLP probe linked to resistance to cereal cyst nematode in wheat. Genome 39 : 798-801.

CHAPTER - 3

Aphelenchoides besseyi — A MAJOR THREAT TO RICE CULTIVATION

B.N. Shukla, I. Vadhera and J. Bhatt

INTRODUCTION

Rice is most important cereal food in the world occupying about 122 million hectares and producing about 150 million tones of grains. The significance of rice is that it is the staple food of the half world's population. Millions of people in Askia subsist almost entirely on rice and over 90 per cent of the world's rice is both grown and eaten in Asia. The association of nematodes with some pathological and physiological melodies of rice and threatening rice cultivation has been fully appreciated *Aphelenchoides besseyi*, an ectoparasite infects the rice plants from the emergence stage and has been found to cause stunting of plants and finally serious chaffyness of grains. The symptoms caused by this parasite were often confused with calcium and magnesium deficiency hence remained unattended with management point of view. Merge information, however is available on the extent of its distribution incidence, potentially to cause damage and effective management as a number of other parasitic nematodes are known to be associated with rice at the soil phase.

1. DISTRIBUTION

As early as 1936, Dastur described a nematode disease to have appeared in epidemic form in Chhattisgarh division of erstwhile state of C.P. and Berar (now part of M.P. and Maharashtra) India. A species of *Aphelenchoides* was found associated with the disease. The nematode *A. besseyi*, Christie, 1942, was later demonstrated to be the cause of 'white tip' symptoms in rice by Cralley (1949) in U.S. A. since then white tip nematode *A. besseyi* is known to occur in almost all major rice growing areas of the world viz. Egypt, Sierra, Leone, Senegal, Madagascar, Tchad, Toga, Ghana, Nigeria, Uppervolta, Mali, Malawi, Kenya, Uganda,

Tanzania, Comoro island (Africa) U.S.A., Cuba, Salvador and Mexico (North America), Japan, Indian, Bangladesh, Sri Lanka, Formosa, Thailand and Philippines (Asia), USSR, Hungary (Europe) (Fortuner 1975). Its prevalence from Brazil (South America) also, has now been reported (Huang *et al.*, 1977).

Todd and Atkins (1958) observed that Arkannas (USA) 1 to 2 nematodes per seed in a variety. Early prolific reduced the yield to the extent of 43.3 per cent while in resistant Bluebonnet losses were observed to the extent of 7.1 per cent. SImilarly, Fukano, cited by Huang *et al.*, (1979) reported sigtnificant damage to the rice crop at an infestation level of 132 nematodes and severe at 1241 nematodes per 100 seeds in Japan.

A. besseyi was widely prevalent in Tamil Nadu (Muniappan and Seshadri, (1969). It was recorded for the first time on rice from the state of Gujarat (*Thakar et al.*, 1987), from Thirupathisaram, Kanya-kumari district, Tamil Nadu (Sivakumar, 1987),at low frequency from Tripura (Nath *et al.*,); and 29% frequency of occurrence from Madhya Pradesh Shukla *et al.*, (1997).

2. IDENTIFICATION OF HOST RANGE

Amongst all species *A. besseyi* is the most widespread nematode throughout rice growing areas of the world. It has a very narrow host range mainly confined to rice *Setaria etalica* and strawberry etc.

A. besseyi is described and figured for *Dioscorea trifida* from Guadeloupe and its morphemeterics compared with those of specimens from the USA and Africa. This is new host record from foliage of tuber and Yam (*Dioscorea trifida*) in West Indies (Kermarrec and Anais, 1973, 1974). Light ear disease of foxtail millet (*Setaria italica*) harboured the population of *A. besseyi* (Gokte and Mathur, 1993; Shukla *et al.*, 1997).

Seeds of proso millet (*Panicum miliaceum*) received for Quarantine clearance, prior to export were found to be infested with *A. besseyi* and this is new host record from India (Gokte and Mathur, 1993; and Gokte *et al.*, 1990). Gokte and Mathur (1992) again reported that seed of *Stylosanthes hamato* from experimental station. Queensland, Australia, were found to harbour internal infection of *A. besseyi*. This seems to be the first report. *A. besseyi* occurred on *Dendrobuim* Lady Fay (Lin *et al.*, 1992), it also associated with root of black oat (*Avena strigosa*) in state of Parana (Curi *et al.*, 1991). A new disease of onion reported from Sri Lanka (Lamberti *et al.*, 1996), *A. besseyi* detected from roots of *marantha makoyana* imported from Netherlands (Tacconi, 1996). It also damaged the crop of chilli pepper on the farm in Sarawak. Malaysia (Hockland and Eng. 1997).

3. LOSSES

Loss is a function of the damage potential, and density of nematode in any area. Grain yield are compared of healthy plants with those with varying degree of infestation in fields or under presence of nematode and nematode free situation by cultural, chemical or varietal manipulations of plants and soil management.

Tikhonova (1966) has reported that white tip was first known in the USSR in 1939 in Krasnodarsk region, then the disease spread practically to the whole of rice growing zone in the USSR which suffered losses between 29 and 61% Nishizawa and Yamanoto (1951) observed reduction in Japan ranging from 14.5 to 46.7% according to varieties whereas at Taiwan, Hung (1959, 1962) found an average reduction between 20 to 25% in number of grains/panicle. At the same time Atkins and Todd (1959) in USA noticed a yield drop between 17-34% in susceptible rice varieties.

Losses are caused by reduction in chlorophyll, improper filling of kernel (21-46.3%) Nandakumar *et al.*, (1974) and sterility (20.4%) Muthukrishan *et al.*, (1974). In Tamil Nadu state rice cultivars tested for three years showed yield loss of up to 60% (Rajendran *et al.*, 1977). In, Astrakhan, USSR variety 'Dubovskii 129' and 'Kuban 3' reflects the yield loss 51.0 and 12% respectively and in Kuban region, USSR the yield loss was recorded to be 12-13% (Popova and Shesteperov 1978, Popov, 1984).

An epidemic in Andra Pradesh affected 60 per cent of rice varieties Jayaprakash and Joshi (1979) and discoloration of grain over 23 per cent severity in southern and eastern state causing losses upto 20 per cent (Prasad *et al.*, 1986; Prasad *et al.* 1987). The economic loss of rice was 7-17% in Huanggang country China (Zhang 1987). In Kanya Kumari, the yield loss was 6.1% in transplanted crop while directly seeded crops was 10.8% (Sivakumar, 1988). In Madhya Pradesh (Shukla *et al.*, 1997) highlighted 22% loss in yield due to this nematode. In Brazil nematode caused losses of upto 50% in upland paddy (Tsay, 1998).

4. SYMPTOMOLOGY

The disease syndrome in rice is the plants response to the histopathological, nutrition and physiological disorders induced by the plant parasitic nematode. Typical disease syndrome is manifested in inoculation with nematode species, the symptoms are diagnostic with the nematodes infesting leaves, stems and earheads etc.

The rice white tip nematode *A. besseyi* is ectoparasite feeding on the cells of the primordia leaves ear primordia and ears. The symptoms of attack have been described (Yoshi and Yomamoto, 1950; Todd and Atkins, 1958; Thorne, 1961; Hung, 1962). The most conspicuous symptoms occur at the tillering stage, crimping and chlorosis and, considerable changes on the configuration of lamina occurs (Artyukhova and Popva, 1981). Prasad *et al.*, (1987) noticed crinkling of boot leaves 50 days after transplanting of Jaya variety. Consequent upon feeding on young and unfurled leaves, withering from the tip downwards giving a whip like or whitish appearance to their tips during the wet season occurs. Chlorotic patches are also seen on the leaves. In dry season the leaves assume a bronze colour (Israel *et al.*, 1966). Before heading the leaf margins are often curled especially those of the flag leaves. Infected plants also become stunted and tillering is reduced. At the heading stage the flag leaf emerges twisted, earhead length is reduced and some glumes are rendered partly or entirely empty. Not all grains in one panicle are attacked, their may be misshapen or sterile grains along side the normal ones (Susan, 1974).

A. besseyi infected plant of strawberry showed dwarf or crimp symptoms on leaves and buds whereas in rice a typical crinkled leaflets with 1 or 2 cm leaf tip necrossi at 41 days of inoculation (Cheng and Fang, 1989; Lin *et al.*, 1992). On the contrary, Muthukrishan, 1974; Fomba, 1988 and Shukla *et al.*, 1997 reported that no specific symptoms were observed during growth phase of crop, however the malformed panicle or once the grain have ripened the typical distortion of spikelets provided a means of identification.

The nematode feed externally on lodicules of ovary leading to shrivelled and empty grain giving the appearance of chaffyness (Rajendran, *et al.*, 1977; Hung, *et al.*, 1979; Muthukrishan, *et al.*, 1974; Rao *et al.*, 1989). Rahman and Maiah (1989 and Found that disease incidence was highest with highest nematode population at booting or flowering stage of crop.

5. LIFE CYCLE

Amongst the plant parasitic nematodes *Aphelechoides* species are known to have shortest duration of life cycle. Under optimum condition these nematode complete their life cycle in one and a half weeks. The biology of parasite have been studied (Yoshi and Yamamoto, 1950; Goto and Fukatsu, 1952 in Japan; Cralley, 1949; Todd and Atkin 1958 in USA and Tikonova, 1966 in USSR). They reported that after germination, the parasite follows the young shoot as they grow, living as an ectoparasite it stays in the young leaves which are still coiled. The nematode then multiplies rapidly until heading. Then its larvae move on the spikelets where they first stay on the outside of the glumes then pass on to upper glumella after flowering. The number of nematodes on the stems and leaves decreases sharply. At harvest there are very few nematodes left in the straw. Most of them are coiled up on the surface of grains protected by glumellas and are anabiotic.

Nandakumar *et al.*, (1975) from India reported that white tip nematode usually remains coiled up in dormant condition on lodicules and inside the palea of the rice seed. Presoaking seed in water at 28°C for 12-15 hr was found essential to activate the dormant stages before processing for extraction. as the infected seed sprout the nematode reach the tender primordium for feeding and 12 days later migrate to the leaf as it uncurls. Typical white tip symptoms on leaves appear on 16th day after germination. As the plant grow up, the nematode move to the boot stage and panicles causing partial emergence of ear and ill filled or chaffy grains. A sizable population of adults and fourth stage junvenile of the nematode reach the ear under condition of humidity and dew in field. Panchangshang (1984) observed that *A. besseyi*, emerged from the rice seed three days after commencement of water soaking. The libereated larvae invaded the seedlings, penetrating the radicle and plumule. The larvae had 3 moults, they ascend the stem to spikelets which become sterile.

Male female of *A. besseyi* took 6-8 days to reach the adult stage from J2 stage at 29-32°C. The life cycle from egg to adult was calculated to be 10-12 days (Sivakumar, 1987) whereas, Gokte and Mathur (1989) found that *A. besseyi* reached the growing point of the rice seedling between 1 to 5 days depending on the time of inoculation. Post embryonic development was marked by multiplication of nuclei, both between and during the moults. Males developed faster and were recorded two days earlier than females. The life cycle from egg to egg was completed in 11 days. The nematode mainly located near the embryonic region of rice seed. The majority of nematodes recovered were 4th stage juveniles which constituted the main invasive stage (Gokte and Mathur, 1989, 1991). Tsay *et al.*, (1998) stated that *A. besseyi* feeds endoparasitically in the coleoptile for 7-10 days and ectoparasitically within the innermost leaf sheath during other plant growth stages. At late tillering nematode numbers increased rapidly and reduced to peak during the plant reproductive stage. Number of male of *A. besseyi* was most abundant at flower fertilization.

6. INTERACTION

Some species of *Aphelenchoides* involved with fungi or bacteria where they may aggravate the symptom appearance. *A. besseyi* destroyed the mycelium of *Fusarium roseum*, *Penicillium* sp., *Trichoderma* sp., *Gliocladium* sp.,

Rhizoctonia solani, Sclerotinia sclerotiorum, Sclerotium rolfsii and *Sclerotiorum cepivorum.* Production of spores and sclerotia was, however, increased when soil was inoculated with *A. besseyi* sp. and *Sclerotium rolfsii* germination in 3 successive bean (*Phaseolus vulgaris*) planting was higher than with the fungus alone (Sanudo, 1975). Nishizawa (1978) reported that discolation of rice kernel is due to association of *A. besseyi* with *Entobacter agglomeran.*

Green weight of Nova-76 plant was significantly reduced when inoculated with *A. besseyi* and *Sclerotium oryzae* but the reduction was comparatively lower when the plant were inoculated with either organisms alone or with distilled water. Weight of melrose plants was reduced significantly by treatment with *A. besseyi* alone and *A. besseyi* plus *S. oryzae* but not by *S. oryzae* alone. Plant weights were not reduced significantly by either organisms alone or by the two in combination McGawley, *et al.*, (1984). Rao, *et al.*, (1986) found that *Acrocylindricum oryzae* and *Corticium* sp. invaded the inter-veinal area of nematode *A. besseyi* affected leaf and caused enhanced injury to this crop.

Jagadiswari *et al.* (1994) found that presence of *Curvularia lunata* caused build up in the concentration of *A. besseyi* and increased the grain deformation of rice compared with infection by *A. besseyi* alone. A population level of 10 and 100 nematodes/seedling increased the grain discolouration and fungal infection in grains of rice. Higher population (1000/ seedling) inhibited the multiplication of *C. lunata* by vigorous feeding on the fungal mats.

7. SURVIVAL

The temperature and humidity are the two important factors that appeared to influence the activity of *A. besseyi.* This nematode can survive in the grains for several years, and Yoshi and Yomamoto (1950) found that 75% of nematode are remained alive in grains which have been kept dry for 3 years but almost all the nematode disappeared when grain are put in the soil for four months. Rao *et al.*, (1983) reported that under experimental conditions, preadults became quiescent in 4, 6 and 8 days at relative humidity of 30, 50-70 and 90-100 per cent respectively.

The quiescent preadults in rice grains can survive from 23 months (Atkins and Todd, 1958) to eight years (Todd and Atkins, 1959). Zem, *et al.*, (1977) found that the nematode remains active within the temperature range of 17.9-25.1°C. Attractiveness of rice seedlings to *A. besseyi* is 6 days and 30°C were most favorable (Gokte and Mathur, 1988). Sivakumar (1978) reported that the population declined by 27.1% during the first 15 months after harvest and was absent by 25th months.

Gokte and Mathur (1990) observed that the nematode can survive for 1 year inside the rice seed and 53 days in water under 10°C. The minimum temperature for nematode activity is 4°C and thermal death point is 49°C for 10 min. The nematode die totally at 40-44°C in water for 1 hour. The temperature and precipitation during 10-15 days after sowing can influence disease severity. Low temperature and more precipitation cause the disease to become more serious.

The number of nematodes detected increased significantly with storage period was higher at 18°C than 10°C. The temperature effects the number of nematodes and seed vigour but not the germination (Tenente, *et al.*, 1994). Choudhury and Choudhury (1996) studied the survival of *A. besseyi* in naturally infested with paddy seed stored at +10, +4, -4, and - 20°C. They found that the survival and recovery of nematode was greater in seed stored with low moisture content (8.1) as compared to the seed having higher moisture (15.9%) at most of the temperatures studied.

8. MANAGEMENT

Since the potentiality of the nematode as pest in rice culture has been well recognized steps to keep their population below economic threshold level has been necessary. Different methods viz. resistant varieties, physical and chemical have given varying degree of success against the nematodes.

(a) Source of resistance

Rice varieties differ greatly in susceptibility to *A. besseyi.* Variety Gurmatia and Chinoor has been reported resistant to *A. besseyi* (Dastur, 1936), variety Della has displayed resistant to *A. besseyi* (Joden and Sonnier, 1973), variety co-13, Rohini, T-292, RP 113-31, Ratna, Triveni, IET 228, Tkm 7, T-482, Asd. 14, Yamuna, Padma, IR 5 (2%), SR 26B, Cult. 7711, Vijay, Sona, Soorya, IR 30 (3%), Krishna, Bala (4%) has given resistant reaction (Rajendran *et al.*, 1977). Variety Tongil (Strain one), early Tongil, Tongil chal and Shirogane were highly resistant (Park *et al.*, 1976). Cultivar Blue belle, Labelle, EEA 404, EEA 406, and P 798 -B4-4-JT were moderately resistant (Oliveira and Rebeiro, 1980). Varieties viz. XS 603 and CMS XS 719 also showed resistant reaction (Silveira, *et al.*, 1982) whereas TPS2 was moderately resistant (Kalamani, *et al.*, 1987). Sivakumar (1988) reported that TPS-1, TPS-2, AS-19903, HKR-1, TTB-1, PTB-4 and Madhuri as moderately resistant. Oliveira and Oliveira (1989) found rice genotypes viz., Blue belle, BR-IRGA 409 and IRGA 172 F4 SS 39 showed resistant reaction in Brazil. The lines LS-10, 81-22, 82-279, 81-67, 8-55, 82-72, 82-76 and 82-8 of rice was considered tolerant, to *A. besseyi* (Silveira, *et al.*, 1990). Varieties Fortuna, Nira, Rexoro, Bluebonnet, Bella patha, Bluebonnet 50, bluebelle, Century patha and Vegold were identified as resistant to *A. besseyi* (Zelenskii and Popovo, 1990, 1991).

(b) Physical and Chemical control

The disease is primarily disseminated through infested grains, husk and plant remains. Eradication of the nematode from infested seed by soaking of seek in hot water and sun drying at different temperature and time interval has been demonstrated by many workers. Sun drying of seed for 6hr x 4 days and hot air (56-60°C) treatment of dry seed was found effective (Muniappan and Seshadri, 1964; Nandakumar *et al.* 1975). Presoaking of rice in hot water at 20 to 28°C for 16 to 18 hrs followed by 52°C for 10 min treatment of unsoaked seed in water at 57°C for 15 min. or hot air at 60°C for 15 min. effectively controlled *A. besseyi* (Anon, 1973). Similarly presoaking of seed in water at 28°C for 12-15 hrs followed by hot water treatment at 52°C for 10 min. has also been found effective without detrimental effect on embryo or quality of grain (Nandakumar, *et al.*, 1975). Soaking the seed in hot water at 55°C for 20 min. and immersing the seed in hot water at 52, 54, 57°C for 10, 15 min. was effective in control of *A. besseyi* (Merny, *et al.*, 1985; Tenente and Manso, 1994). Gokte and Mathur (1993) reported that complete eradication was achieved when infested seed was exposed to 50°C for 15 min.

Burning the stubble's prevented the carry over of the nematode to the next season (Sivakumar, 1985). While using chemicals (Anikeev, *et al.*, 1980, Gokte and

Mathur, 1993) demonstrated that presoaking of seed in Hydrogen peroxide and organic mercuric compound Granonzon halved *A. besseyi* infection. Similarly presoaking the seed of *Setaria italica* and *Panicum miliaceum* in 1% hydrogen peroxide for 3 hrs followed by hot water treatment at 48°C for 15 min. completely eradicated *A. besseyi.* Later, Sivakumar (1987) found that seed soaking in 1% KCl or 1% NaCl for 20 hrs. and then sun dried for 6 hrs. at almost original moisture level gave 95- 97% disinfestation of *A. besseyi.*

Seed soaking in 0.2% solution of mancozeb and monocrotophos followed by vacuum fumigation (Methyl bromide 32 g/m3) for 2 hrs at 30°C eliminated the nematode (Prasad and Varaprasad, 1992). Later, presoaking of seed with oxamyl or hot water treatment of seed reduced the infestation rate and increased the yield (Tsay *et al.,* 1998). Ishiy *et al.,* (1975) reported that seed disinifection with vydate 25.E (0.5%), methomyle (280g/100kg), thiabendazole (0.25 solution) were most effective in eliminating *A. besseyi* from rice, but Nagy (1987) found that 0.5-1% solution of vydate (oxamyl) 25 EC gave 94% mortality whereas Lannate (methomyl) 20 EC and Hostothion (triazophos) 40 EC was less effective.

Seed treatment with carbofuran 3G and UC 21865 (Oliveira and Ribeiro, 1980), carbofuran, thiram, carboxin and captafol (Silveira *et al.,* 1990), benomyl 2 or 4 oz (Templeton, *et al.,* 1977), MEP 50% EC, Fenthion, 50% EC and Dasuzin 34% EC and mixed treatment with fungicides, Benoram 40% WP Proraz 25% EC and TCMTB 30% EC (Cho, *et al.,* 1987); seed treatment combination using diazinon, prochloraz, benomyle, fention, phenthoate, and fenitrothion, carbofuran, thiobendazole or fenitrothion (Silva and Silva, 1992; Tenente and Manso, 1994), ethopophose 20 EC at 0.5% 100 lit (Tacconi, *et al.,* 1999) were highly effective against *A. besseyi* without any toxicity.

Spraying of chemicals viz., isofenphos and carbofuran gave best result (Prasad *et al.,* 1987). Monocrotophos sprayed at 1000 ml/ha at boot leaf stage reduced the white tip incidence and grain chaffyness. Phosphomidan, malathion, neem oil and copper oxychloride were effective to lesser degree (Kumar and Sivakumar, 1998). Romascu and Dragoescu (1979) reported that soil treatment with Temik 10 G at 10 to 15 kg/ha one week before planting was effective in the control of *Aphelenchoides.*

9. CONCLUSION

Aphelenchoides besseyi causing white tip of rice is potential parasite of rice and is widespread in all growing areas of the world. It is responsible for serious chaffyness of seed resulting grain yield loss to a maximum of 60%. The nematode has a narrow host range as compared to other plant parasitic nematodes. It has a very short life cycle of 12 days. It survives mainly on grains which serve as main source of dissemination. The nematode can be effectively controlled by using resistant varieties, field sanitation, physical and chemical methods of control. Therefore, an integrated nematode management package specific to an agroclimatic condition can manage this nematode effectively.

REFERENCES

Anikeev, A.S. and Shabel, nikov, Yu.G. 1980. The control of nematodes in rice. production. *Trudy Kubanskogo Selskokhozyaistvennogo Instituta.* 194. 244 : 63-67.

Anon, 1972, 1973. 301 pp: Plant nematology pp. 239-248.

Artyukhova, G.A. and Popova, M.B. 1981. Ultrastructural changes in rice leaves caused by *Aphelenchoides besseyi.* Byulleten Vsesoyuznogo Instituta Gel'mintologii im. K.I. Skryabina. Atkin 31 : 10-12.

Atkins, J.G. and Todd, E.H. 1958. White tip of rice I, Symptoms laboratory culture of nematode and pathogenicity test. *Phytopathology* 48 : 632-637.

Atkins J.G. and Todd, E.H. 1959. White tip disease of rice III yield tests and Varietal resistance. *Phytopathology* 49 : 189-191.

Cho, S.S.; Han, M.J. and Yang, J.S., 1987. Chemical control of rice white-tip nematode (*Aphelenchoides besseyi* C). by seed disinfectant and in the paddy field. *Korean Journal of Plant Protection* 26 (2) : 107-111.

Chaudhury, R. and Chaudhury, R. 1996. Effect of storage temperature and moisture on the survival of the white tip nematode, *Aphelenchoides besseyi* and infested paddy, *Oryza sativa* seeds. *Annals of plant Protection Sciences* 4 (1) : 16-21.

Christie, J.R. 1942. A discription of *A. besseyi* n sp. the summer dwarf nematodes of strawberry with comments on *A. substenuis* (Cobb) and A. hodsoni Goodey 1933. *Proc. Helminthol. Soc. Wash 9 : 82-84.*

Cralley, E.M. 1942. White tip of rice. *Phytopathology* 39 : 5.

Curi, S.M., Silveira, SGP, Da and Rodrigues, Netto, S.M. Da Silveira, SGP 1991. Nematodes associated with black oat (*Avena strigosa*), in the State of Parana. *Arquivos do Instituto Biologico Sao Paulo.* 58 (1-2) : 63-64.

Dastur, J.R. 1936. A nematode disease of rice on the central *province Proc. natn. Acad. Sci.*, 4 (2) : 108-121.

Fortuner, R. and Williams, K.J. 1975. Review of literature on *A. besseyi*, Christie, 1942, the nematode disease causing 'White tip disease in rice. *Helminth Abstr.* B. 44 (1) : 1- 31.

Fomba, S.N. 1988. Incidence of white tip nematode (*Aphelenchoides besseyi* Christie) in mangrove swamps at Rokupr, Sierra Leone, *WARDA Technical Newsletter.* 8 (1) : 12-13.

Gokte, N. and Mathur, V.K. 1988. On the attractiveness of paddy seedlings to *Aphelenchoides besseyi. Indian J. Nematol* 18 (2) : 239-243.

Gokte, N. and Mathur, V.K. 1989. Studies on embryonic and postermbryonic development of *Aphelenchoides besseyi. Nematol Medit* 17 (1) : 57-60.

Gokte, N. and Mathur, V.K. 1990. On the emergence of *Aphelenchoide besseyi* from infected rice seeds. *Indian J. Nematol* 20 (1) : 60-63.

Gokte, N. and Mathur, V.K. 1991. Infection of germinating paddy seed by *Aphelenchoides besseyi.* Nematol Medit. 19 (2) : 221.

Gokte, N. and Mathur, V.K. 1993. Treatment schedule for denematization of seeds of *Setaria italica* and *Panicum miliaceum* infested with *Aphelenchoides besseyi. Nematologica* 39 (2): 274-276.

Gokte, K. and Fukatsu, R. 1952. Studies on white tip disease of rice caused by *A. oryzae.* Yakooti number and distribution of the nematode on the affected plant. *Anm. Pytopath. Soc. Japan.* 16 (2) : 57-60.

Hockland, S. and Eng. L. 1997. *Capsicum annuum* v. *longuma* new host record for the rice white tip nematode, *Aphelenchoides besseyi* Intn J. Nematol 7 (2) : 229.

Huang, C.S.; Huang, S.P. and Chiang, Y.C. 1979. Mode of reproduction and sex ratio of rice. White tip nematode *A besseyi. Nematologica.* 25 : 255-260.

Hung, Y.P. 1962. The white tip nematode of rice. Kaohsiung district. Agric. Improvement. Stn. Pingtung, Taiwan, bull, series no. 6, 12 pp.

Ishiy, T.; Lordello, L.G.E. Lemes, L.A. and Francoso, S.A. 1975. Contribution to the study of the control of "white tip" of rice. *Revista de Agricultura, Piracicaba, Brazil.* 50 (3-4) : 141-145.

Israel, P. Rao, Y.S. and Rao, V.N. 1966. Rice parasitic nematodes (India). INT. RICE COMMN. Working party on Rice production and protection, 11th meeting Louisiana, FAO.

Jagadiswari, Rao; Kauraw, L.P. Anand P. and Rao, J. 1994. Deterioration of seed quality due to interaction between *Aphelenchoides besseyi* and *Curvularia lunata* in rice. *Indian Phytopathol.* 47 (1) : 92-93.

Kalaimani, S; Pillai, O.R. Manuel, W.W. and Subramaniam, M. 1987. TPS2 - a new variety for Kanyakumarki International Rice, *Research Newsletter* 12 (4) : 5.

Kermarrec, A. and Anais, A. 1973. Presence of *Aphelenchoides besseyi* Christie 1942 (Nematode, Aphelenchoididae) in foliage and tubers of yam (*Dioscorea trifida*) in Guadeloupe. *Turrialba* 23 (4) : 389-393.

Kermarrec, A and Anais, A. 1974. Presence of *Aphelenchoides besseyi* in foliage and tubers of yam (*Dioscorea trifida*) in the French West Indies. Nematropica. 4 (1) : 2-3.

Kumar, S. and Sivakumar, C.V. 1998. Management of rice white tip nematode, *Aphelenchoides besseyi. Indian J. Nematol* 28 (1) : 85-87.

Lal, A and Mathur, V.K. 1988. Record of *Aphelenchoides besseyi* on Setaria italica. *Indian J. Nematol* 18 (1) : 131.

Lin, Y.Y., Wang, K.M. and Tsay, T.T. 1992. The occurrence of *Aphelenchoides besseyi* Christie on Dendrobium Lady Fay. *Plant Protection Bulletin Taipei,* 34 (3) : 202-215.

Marlatt, R.B. 1972. Relationship of disease method of inoculation in *Ficus elastica* with *A. besseyi. Plant. Dis. Reptr.* 56 : 471-472.

Gokte, N; Mathur, V.K. and Rajanlal, A. 1990. *Panicum miliaceum - a new host record for Aphelenchoides besseyi Indian. J. Nematol* 20 (1) : 111-112.

McGawley, E.C., Rush, M and CandoHollis, J.P. 1984. Occurrence of *Aphelenchoides besseyi* in Louisiana rice seed and its interaction with *Sclerotium oryzae* in selected cultivars. *J. Nematol.* 16 (1) : 65-68.

Merny, G.; Billard G. and Pelletier, R. 1985. Technique for eradication of *Aphelenchoides besseyi* (Nematoda Aphelenchina) from seeds of *Panicum maximum. Revue de Nematologie* 8(2) : 155-160.

Muniappan, R. and Seshadri, A.R. 1964. On occurrence of white tip nematode of rice, A. besseyi Chirstie 1942 in Madras state. *Madras Agric. J.* 51: 510-511.

Muniappan, R. and Seshadri, A.R. 1969. The white tip nematode of paddy. *A. besseyi* Christie 1942 and its control under South Indian condition. All India Symp ICAR 21st, 22nd, Aug. pp 72.

Muthukrishnan, T.S.: Rajendran, G. and Chandrasekaran, J. 1974. White tip nematode of rice, *Aphelenchoides besseyi* in Tamil Nadu. *Indian J.Nematol.* 4 (2) : 188-193.

Nagy, J. 1987. Increased efficacy of seed-dressing agents by ultrasound exposure of rice nematodes. *International Agrophysics* 3 (4) : 291-300.

Nandakumar, C. Prasad, J.S. Rao, Y.S. and Rao, J. 1975. Investigation on white tip nematode (*A. besseyi* Christie 1942) of rice. *Indian J. Nematol* 5 : 62-69.

Nath, R.C., Mukherjee, B. Dasgupta, M.K. and Siddiqui M.R. 1994. Prevalence and distribution of plant parasitic nematodes in rice fields of Tripura : *Afro Asian J. Nematol* 4 (2) : 147-150.

Nishizawa, T. 1976. Occurrence of abnormal rice kernels associated with the infestation of white tip nematode. *A. besseyi. J. Nematol.* 6 : 73-79.

Nishizawa, T. and Yamamoto, S. 1951. Studies on varietal resistance of rice plant to the rice nematode disease. *Kyushu agric Res.* 8 : 91-92.

Oliveira, J.V. and Ribeiro, A.S. 1980. Study on the nematode *Aphelenchoides besseyi* Christie in irrigated rice. *Lavoura-Arrozeira* 33 (322): 40-43.

Oliveira, J.V. de and De. Oliveira J.V. 1989. Evaluation of the resistance of four irrigated rice genotypes to the nematode *Aphelenchoides besseyi*, Christie, 1942. *Agronomia Sulriograndense* 25 (1) : 11-18.

Pan, Changshang, and Pan, Cangsang. 1984. Studies on plant parasitic nematodes on economically important crops in Fujian. II. Studies on the biology of *Aphelenchoides besseyi* in rice plants. *Acta Zoologica Sinica* 30 (4) : 368-374.

Park, J.S.; Lee, J.O. Bak, J.S. and I.J.O. 1976. Varietal resistance of rice to the white-tip nematode *Aphelenchoides besseyi. International Rice Research Newsletter.* 1 (1) :15.

Popova, M.B. 1984. Principal factors determining the intensity and pathogenicity of *Aphelenchoides besseyi* infestation on rice. *Byulleten'Vsesoyuznogo Instituta Gel'mintologii im. K.I. Skryabina.* 36 : 39-41.

Popova, M.B. and Shesteperov, A.A. 1978 : Comparative assessment of the susceptibility of local varieties of rice to *Aphelenchoides besseyi* in the Astrakhan region. *Fitogel'mintologicheskie issledovaniya.* 107-110.

Prasad, J.S. Panwar, M.S. and Rao, Y.S. 1986. Effect of seed soaking with chemicals on the parasitic nematodes of rice. *Indian J. Nematol* 16 (1) : 119-121.

Prasad, J.S.; Panwar, M.S. and Rao, Y.S. 1987. Chemical control of white tip nematode, *Aphelenchoides besseyi* Oryza. 24 (4) : 391-392.

Prasad, J. and Varaprasad, K.S. 1992. Elimination of white-tip nematode, *Aphelenchoides besseyi* from rice seed. *Fundamental and Applied Nematology.* 15 (4) : 305-308.

Rahman, M.L. and Miah, S.A. 1989. Occurrence and distribution of white tip disease in deepwater rice areas in Bangladesh. *Revue de Nematologie* 12 (4) : 351-355.

Rajendran, G.; Muthukrishnan, T.S. and Balasubramanian, M. 1977. Rice varieties and the white tip nematode. *International Rice Research Newsletter* 2(2) : 3.

Rao, Y.S. Prasad, J.S. and Panwar, M.S. 1986. Nematode Problem in rice, crop losses, symptomology and management. In plant parasitic nematode : Problem and progress. pp. 279-290.

Romascu, E. and Dragoescu, M. 1979. Degree of resistance of rice varieties to *Aphelenchoides besseyi* Christie. 1942. The efficacy of the preparation Temik 10G on the control of nematodes on tobacco plantations. *Analele Institului Cercetari Pentru Protectia Plantelor.* 15 : 219-226.

Sanudo, S.B. 1975. Some aspects of the nematode *Aphelenchoides* sp. and its effect on the development of various soil fungi. *Noticias Fitopatologicas* 4 (2) : 148- 159.

Shukla, B.N. Vadhera, I. and Tiwari, S.P., 1997. Two decade of nematology in Madhya Pradesh. Research bulletin DRS I JNKVV. Jabalpur. pp. 52.

Silveira, SGP, Da; Curi, S.M. Tisseli, O. and Da Silveira, SGP. 1990. Reaction of rice breeding lines to seed nematode *Aphelenchoides besseyi* Christie *Nematologia Brasileira* 14 : 54-60.

Silva, G.S. da. and Da, Silva, G.S. 1992. White tip and national rice production. *Informe Agropecuario Belo Horizonte.* 16 (172) : 57-59, Special issue; *Nematodes : o inimigo oculto da agricultura.*

Sivakumar, C.V. 1987. The rice white tip nematode in Kanyakumari district, Tamil Nadu, India. *Indian J. Nematol* 17 (1) : 72-75.

Sivakumar, C.V. 1987. Disinfestation of white tip nematode in rice seeds. *Indian J. Nematol.* 17 (1) : 148-149.

Susan, D.F. 1974. Nematode. *In pest control in rice.* Ad. Susan, DF Pans Mannual 3. Centre for querseas pest Research, Foreigen and Commonwealth office querseas Development. Administration, London pp. 99-100.

Tacconi, R.; Santi R. and Gironi R. 1999. Population dynamics of *Aphelenchoides besseyi* on rice and effect of seed treatments on seed for propagation. *Nematologia Mediterranea.* 27 (2): 291-293.

Templeton, G.E., Johnston, T.H. and Daniel, J.T. 1971. Benomyl controls rice white tip disease. *Phytopathology* 61 (12) : 1522-1523.

Tenente, R.C.V. and Manso, ESBGC. 1994. Chemical and heat treatments of rice seeds infested with *Aphelenchoides besseyi Nematologia Brasileira.* 18 (1-2) : 28-34; Paper presented at the 9th meeting of the Brazilian Congress of Nematology, in Piracicaba, SP, 1985.

Tenente, R.C.V.; Wetzel, MMVS, Manso, ESBGC, S. Marques, A.S. Dos; and Dos S. Marques A.S. 1994. Survival of *Aphelenchoides besseyi* in infested rice seeds stored under controlled conditions. *Nematologia Brasileira* 18 (1-2) : 85-92. Paper presented at the Brazilian Congress of seeds. Brasilia, DF, Brazil, October 1985.

Thakar, N.A.; Patel, B.K. Patel, H.R. and Patel, C.C. 1987. Occurrence of white tip disease of rice in Gujarat Gujarat Agricultural University *Research Journal.* 13 (1) : 65.

Thorne, G. 1961. Principles of nematology. McGraw Hill Book Company, Inc. N.Y. Toronto, Lond. 553 pp.

Tikhnova, L.V. 1966. *A. besseyi* (Christie 1942) of rice and its control. *Zool ZL* 45 (12) : 1759 - 1766.

Todd, E.H. and Atkins, J.G. 1958. White tip disease of rice, 1 Symptoms, Laboratory culture of nematode and pathogenicity test. *Phytopathology* 48 : 632-637.

Todd, E.H. and Atkins, J.G. 1959. White tip disease of rice II. Seed treatment studies. *Phytopathology* : 49 : 184-188.

Tsay, T.T.; Cheng, Y.H. Teng, Y.C. Lee, M.D. Wu, W.S. and Lin, Y.Y. 1998. Bionomics and control of rice white tip disease nematode, *Aphelenchoides besseyi.* Plant protection Bulletin Taipei. 40 (3) : 277-286.

Yamamoto, S. and Yoshii, H. 1950. Ear blight of Italian millet and its causal nematodes. *Ann. Phytopath. Soc. Japan.* 14 : 81-83.

Yoshi, H. and Yamamoto, S. 1950. A rice nematode disease. I Symptiom and pathogenic nematode. J. Fac. Kyushu Univ. 9 (3) :209- 222.

Zelenskii, G.L. and Popova, M.B. 1990. Source material for breeding nematode-resistant rice varieties. *Problemy teoreticheskoi I prikladnoi genetiki v Kazakhstane : Materialy Respublikanskoi konferentsii, Alma Ata, 18 22 Noyabrya* 1990, 1990, 41.

Zelenskii, G.L. and Popova, M.B. 1991. Breeding rice for resistance to the rice leaf nematode in the USA *Selektsiya I Semenovodstvo Moskva* (5) :59-60.

Zem, A.C. Momtelra, A.R. and Bahia, J. 1977. The white tip nematode also occur in rice seed. *Revista de Agriculture. Piracicabu Brazil* 52 : 81-82.

Zhang, Y.L. 1987. The occurrence and control of *Aphelenchoides besseyi. Hubei Agricultural Science.* (1) : 15-16.

CHAPTER – 4

PLANT PARASITIC NEMATODES OF *MENTHA* AND THEIR MANAGEMENT

Akhtar Haseeb and Prabhat Kumar Shukla

INTRODUCTION

Mints, *Mentha* spp. (Family - Lamiaceae) has been grown and utilized since ancient times. The plant is believed to have originated in the Mediterranean basin and spread to rest of the world. Distillation of mint oil was mentioned even in 410 AD in an account written by Synesius of Alexendria, the Bishop of Ptolemais. Spearmint (*M. spicata* L.) was first to be cultivated in convent gardens about the 9th century. Peppermint (*M. piperita* L.) was first described by John Ray in 1696 in England. The commercial cultivation of mint moved from England to several European countries and U.S.A. by 1800 AD. Japanese mint (*M. arvensis* L) was introduced in Japan from China in 1700 A.D. By 1892, mint oil was considered as an outstanding stemachie, stimulant antispasmodic and carminative. These mints were introduced in India during 1954 at Regional Research Laboratory, Jammu (Husain, *et al.*, 1978). SInce then, as a result of the efforts made by the Central Institute of Medicinal and Aromatic Plants (CIMAP), Lucknow and Essential Oil Association of India, India now rank first in the export of mint oil (Anon. 1999).

Cultivation of mint has been done on a large scale in many tropical and sub-tropical countries of the world including India, China, Brazil, Japan and U.S.A. etc. In India, the highest area under commercial cultivation of mint has been in Uttar Pradesh but during past few years it has spread through out the country especially in north and northwestern states including Madhya Pradesh. Among different oil yielding species of mint, *M. arvensis* occupy the highest area under mint cultivation followed by peppermint (*M. piperita*), spearmint (*M. spicata*), Bergamot mint (*M. citrata*) and Scotch spearmint (*M. cardiaca*) respectively. *M. viridis* L. (Podina) is grown for vegetable purpose.

Japanese mint yielding essential oil on hydro distillation from fresh herb is used as a source of menthol, a chemical which finds wide use in medicine, food and cosmetics. The oil of peppermint, which contains less methol but has a sweeter aroma and taste, is used in toothpaste, chewing gums, candies, high grade liquors, medicines and other pharmaceutical preparations.

Spearmint oil having carvone and limonene is mainly used for flavoring whereas, Bergamot mint oil is mainly used in perfumery and flavouring due to higher content of linaly acetate and linalool.

The first report regarding the association of plant parasitic nematodes with mint was appeared in 1938 (Buhrer, 1938). Since then a lot of research have been done on various aspects of nematode problems of this crop. Work regarding the nematode diseases of this crop has mainly been done at CIMAP, Lucknow (since 1983) and recently (since 1997) started at A.M.U., Aligarh. Among the recognized species of plant parasitic nematodes, root knot and root lesion nematodes has been considered the most important. However, several other species have also shown potential to cause significant damage to these crops (Table 2).

1. MENTHOL (JAPANESE) MINT, *Mentha arvensis*

Mentha arvensis is the best, among the high oil yielding species of mint. Several recently bred cultivars (Gomti, Shivalik, Himalaya, Kosi, etc.) have revolutionized its cultivation by way of increasing herb and oil yield to a great extent. These cultivars have been recognized world-over and Himalaya was patented (Kumar *et al.*, 1998; Anon. 1999). It is well established that root-knot nematodes, *Meloidogyne incognita* and *M. javanica* are the most severely damaging nematodes of *M. arvensis*, followed by rot lesion nematode, *Pratylenchus thornei, Rotylenchulus reniformis, Helicotylenchus indicus, Tylenchorhynchus vulgaris, Hoplolaimus indicus, Paratylenchus* sp., etc. (Haseeb, 1992, 1994; Shukla, 1997, 1998).

(i) Root knot nematodes, *Meloidogyne* spp.

Root knot nematodes are considered the most potential nematode pests of many crops (Sasser, 1979) due to their worldwide distribution, extensive host range and highly advanced life long parasitism. They produce specific symptoms in the form of knots of different size on different crops. The main symptoms of the disease on mints under field conditions are occasional yellowing of leaves, stunting and wilting of plants. Initially symptoms of the disease on mints under field conditions appears in patches as the reduced plant growth with smaller leaf size and temporary wilting under slightest stress of water during hot sun. As the crop grow, especially after first harvest, symptoms become more severe

Table 1. Average yield and composition of oil of different species of *Mentha*

Mentha species	Oil content of fresh herb (%)	Major components in oil (%)
M. arvensis	0.75-0.90	Menthol (80-90), Menthyl acetate (7-15), Menthone (5-10)
M cardiaca	0.40-0.55	Carvone (65-80)
M. citrata	0.45-0.65	Linalool (35-53), Linalyl acetate (45-60)
M. piperita	0.40-0.55	Menthol (40-50), Menthone (20-30), Iso-menthone (25-30)
M. spicata	0.50-0.60	Carvone (60-75), Limonene (25-35)

(Basis : Fresh herb from pot experiments was distilled in Clevenger apparatus and analyzed)

Table 2. Plant parasitic nematodes associated with different species of *Mentha*

	Mentha/Nematode species	Reported by	From
1	***Mentha arvensis***		
	Meloidogyne sp.	Buhrer, 1938	USA
	M. hapla	Horner & Jensen, 1954	USA
	M. incognita	Haseeb & Pandey, 1989	India
	M. javanica	Haseeb & Pandey, 1989	India
	Pratylenchus thornei	Anon., 1984, 1985; Haseeb, 1992	India
	Ditylenchus destructor	Henderson, 1951	USA
	Tylenchus sp.	Anon., 1984, 1985; Haseeb, 1992	India
	Tylenchorhynchus vulgaris	Anon. 1984, 1985; Haseeb, 1992	India
	Hoplolaimus sp.	Anon., 1984, 1985, Haseeb, 1992	India
	Rotylenchulus reniformis	Anon., 1984, 1985; Haseeb, 1992	India
	Helicotylenchus sp.	Anon., 1984, 1985; Haseeb, 1992	India
	H. indicus	Shukla, 1998	India
	Paratylenchus sp.	Shukla, 1998	India
	Criconemoides sp.	Shukla, 1998	India
	Hirschmanniella sp.	Shukla, 1998	India
	Xiphinema sp.	Anon., 1984, 1985; Haseeb, 1992	India
	Longidorus pisi	Anon., 1984, 1985; Haseeb, 1992	India
2	***Mentha cardiaca***		
	Meloidogyne chitwoodi	O'Bannon *et al.*, 1982	USA
	M. hapla	Horner & Jensen, 1954	USA
	M. incognita	Haseeb & Pandey, 1989	India
	Pratylenchus thornei	Anon., 1984, 1985; Haseeb, 1992	India
		Skotlant & Menzies, 1957	USA
	P. minyus	Skotland & Menzies, 1957	USA
	P. penetrans	Skotland & menzies, 1957	USA
	Paratylenchus hamatus	Skotland & Menzies, 1957	USA
	Tylenchus sp.	Anon., 1984, 1985; Haseeb, 1992	India
	Tylenchorhynchus vulgaris	Anon., 1984, 1985; Haseeb, 1992	India
	Hoplolaimus sp.	Anon., 1984, 1985; Haseeb, 1992	India
	Rotylenchulus reniformis	Anon., 1984, 1985; Haseeb, 1992	India
	Helicotylenchus sp.	Anon., 1984, 1985; Haseeb, 1992	India
	Longidorus pisi	Anon., 1984, 1985; Haseeb, 1992	India
	Xiphinema sp.	Anon., 1984, 1985; Haseeb, 1992	India
3	***Mentha citrata***		
	Pratylenchus thornei	Anon., 1984,1985; Haseeb, 1992	India
	Tylenchus sp.	Anon., 1984, 1985; Haseeb, 1992	India
	Tylenchorhynchus vulgaris	Anon., 1984, 1985; Haseeb, 1992	India
	Hoplolaimus sp.	Anon., 1984, 1985; Haseeb 1992	India
	Rotylenchulus reniformis	Anon., 1984, 1985; Haseeb, 1992	India

Table 2 contd...

Table 2. contd...

	Mentha/Nematode species	Reported by	From
	Helicotylenchus sp.	Anon., 1984, 1985; Haseeb, 1992	India
	Longidorus pisi	Anon., 1984, 1985; Haseeb, 1992	India
	Xiphinema sp.	Anon., 1984, 1985; Haseeb, 1992	India
4.	***Mentha piperita***		
	Meloidogyne sp.	Buhrer, 1938	USA
	M. hapla	Horner & Jensen, 1954	USA
		Maqbool *et al.*, 1985	Pakistan
	M. Chitwoodi	O'Bannon *et al.*, 1982	USA
	Pratylenchus thornei	Anon., 1984, 1985; Haseeb, 1992	India
	P. minyus	Skotland & Menzies, 1957	USA
	P. penetrans	Skotland & Menzies, 1957	USA
		Esmenjaud *et al.*, 1990	France
	Pratylenchoides laticauda	Esmenjaud *et al.*, 1990	France
	Paratylenchus hamantus	Skotland & Menzies, 1957	USA
	P. macrophallus	Horner & Jensen, 1954	USA
	Paratrichodorus sp.	Marrifield & Inghum, 1992	USA
	Longidorus pisi	Anon., 1984, 1985; Haseeb, 1992	India
	L. elongatus	Jatala & Jensen, 1974	USA
	L. sliphus	Jensen & Horner, 1956	USA
	Criconemella xenoplax	Marrified & Inghum, 1992	USA
	Aphelenchoides plesistus	Goodey, 1940	USA
	A pareitinus	Horner & Jensen, 1954	USA
	Tylenchus	Anon., 1984, 1985; Haseeb, 1992	India
	Hoploilamus sp.	Anon., 1984, 1985; Haseeb, 1992	India
	Rotylenchulus reniformis	Anon., 1984, 1985; Haseeb, 1992	India
	Helicotylenchus sp.	Anon., 1984, 1985; Haseeb, 1992	India
	H. indicus	Shukla, 1998	India
	Paratylenchus sp.	Shukla, 1998	India
	Criconemoides sp.	Shukla, 1998	India
	Hirschmanniella sp.	Shukla, 1998	India
	Xiphinema sp.	Anon., 1984, 1985; Haseeb, 1992	India
	Tylenchorhynchus vulgaris	Anon., 1984, 1985; Haseeb, 1992	India
5	***Mentha spicata***		
	M. hapla	Horner & Jensen, 1954	USA
	M. chitwoodi	O'Bannon *et al.* 1982	USA
	Pratylenchus thornei	Anon., 1984, 1985; Haseeb, 1992	India
		Skotland & Menzies, 1957	USA
	P. minyus	Skotland & Menzies, 1957	USA
	P. penetrans	Skotland & Menzies, 1957	USA
	P. scribneri	Rhoades, 1983	USA

Table 2 contd...

Table 2 contd...

Mentha/Nematode species	Reported by	From
Paratylenchus macrophallus	Horner & Jensen, 1954	USA
P. hamatus	Skotland & Jensen, 1957	USA
Paratrichodorus christiei	Rhoades, 1983	USA
Belonolaimus longicaudatus	Rhoades, 1983	USA
Dolichodorus heterocephallus	Rhoades, 1983	USA
Aphelenchoides ritzemabosi	Goodey, 1940	USA
Longidorus pisi	Anon., 1984, 1985; Haseeb, 1992	India
Tylenchus sp.	Anon., 1984, 1985; Haseeb, 1992	India
Tylenchorhynchus vulgaris	Anon., 1984, 1985; Haseeb, 1992	India
Hoplolaimus sp.	Anon., 1984, 1985; Haseeb, 1992	India
Rotylenchulus reniformis	Anon., 1984, 1985; Haseeb, 1992	India
Helicotylenchus sp.	Anon., 1984, 1985; Haseeb, 1992	India
H. indicus	Shukla, 1998	India
Paratylenchus sp.	Shukla, 1998	India
Criconemoides sp.	Shukla, 1998	India
Hirschmanniella sp.	Shukla, 1998	India
Xiphinema sp.	Anon., 1984, 1985; Haseeb, 1992	India

and appears as stunted growth with yellowing of leaves, while veins remain green. The symptoms of its attack on roots and suckers appear as small sized galls with large egg-masses. Under severe infection initiation of lateral roots and root-lets on suckers is checked, as a result, uptake of nutrients is inhibited, which intern produces the deficiency symptoms on the aerial portion of the plants (Haseeb and Pandey, 1989).

Buhrer (1938) reported for the first time, *M. arvensis* cv. *piperascens* as the host of *Meloidogyne* sp. In 1954, horner and Jensen reported very high populations of *M. hapla* in the rhizosphere of *M. arvensis* in western Oregon. In greenhouse tests they reported *M. arvensis* cv. *Canadensis* as good host of *M. hapla*.

Root knot disease of *M. arvensis* caused by *M. incognita* and *M. javanica* was identified for the first time in the world in the experimental beds of CIMAP, Lucknow and the in the Tarai region of Uttar Pradesh (Anon. 1985-87; Haseeb and Pandey, 1989). They reported that in the farmer's fields both the species of *Meloidogyne* attack together the root system of *M. arvensis*. However, *M. incognita* dominated over *M. javanica*.

Pathogenic potential of *M. incognita* on different cultivars of *Mentha arvensis* was determined. Results indicated that the increase in initial number of nematodes per pot, resulted increased reduction in plant fresh and dry weights, leaf chlorophyll content, photosynthetic efficiency of leaves and essential oil content in fresh herb (Anon., 1986, 1987; Pandey *et al.*, 1992; Haseeb and Shukla, 2000a, b, Pers. Comm.). Among some improved cultivars and lines of Japanese mint, MAS-1 and HY-77 were observed to be highly susceptible, white Shivalik, Gomti, Kosi and Himalaya showed slight tolerance to the infection by *M. incognita*. Reproduction of *M. incognita* and extent of galling on roots and suckers were observed directly

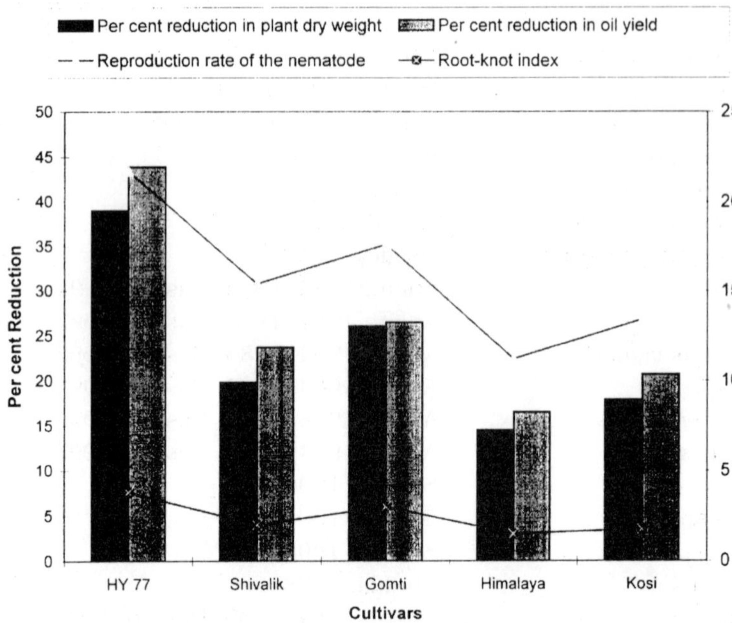

Fig. 1. Effect of *Meloidogyne incognita* (5000 J2 larvae/pot) on plant dry weight and oil yield of different cultivars of *Mentha arvensis* and its reproduction

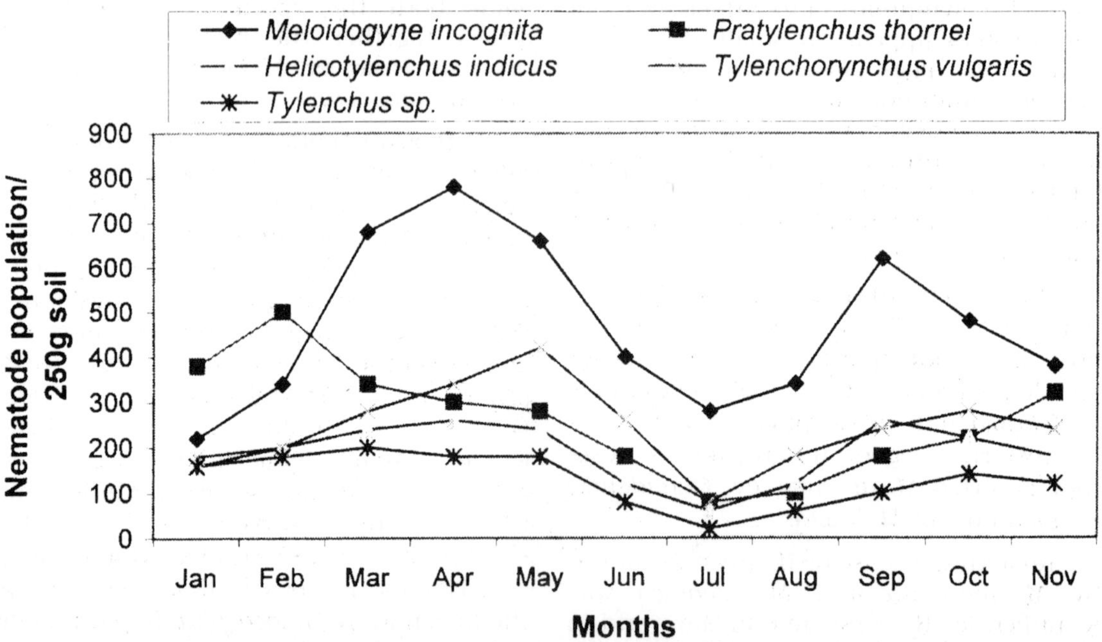

Fig. 2. Vertical distribution of plant parasitic nematodes in relation to seasonal fluctuations on *Mentha arvensis* (0-10 cm)

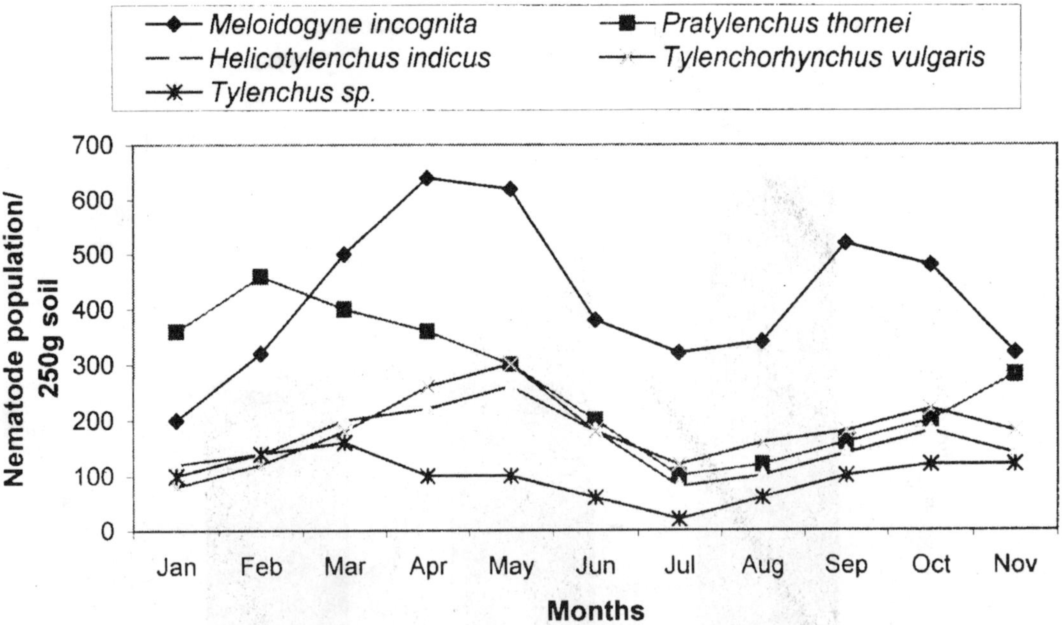

Fig. 3. Vertical distribution of plant parasitic nematodes in relation to seasonal fluctuations on *Mentha arvensis* (11-20 cm)

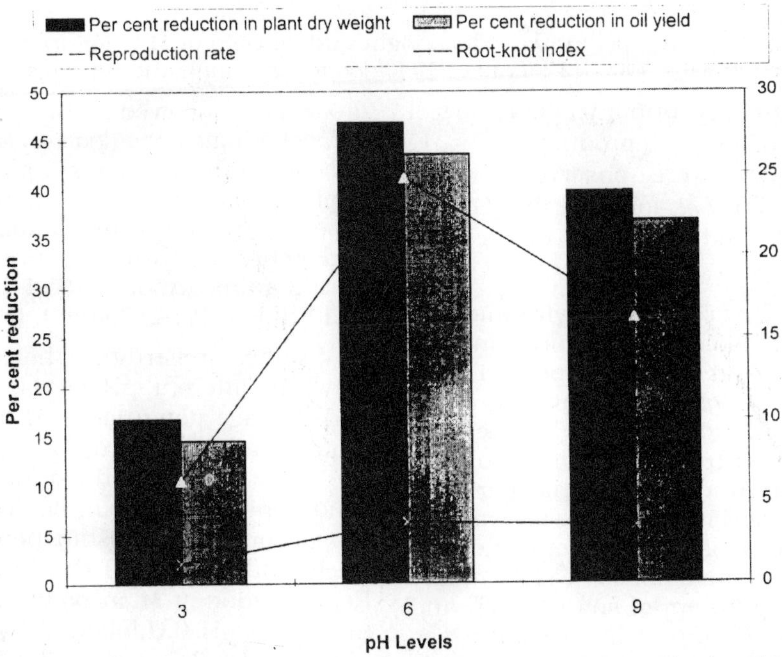

Fig. 4. Effect of pH levels on the reduction in plant dry weight and oil yield of *Mentha arvensis* plants inoculated with 5000 J2 larvae of *Meloidogyne incognita* and its reproduction

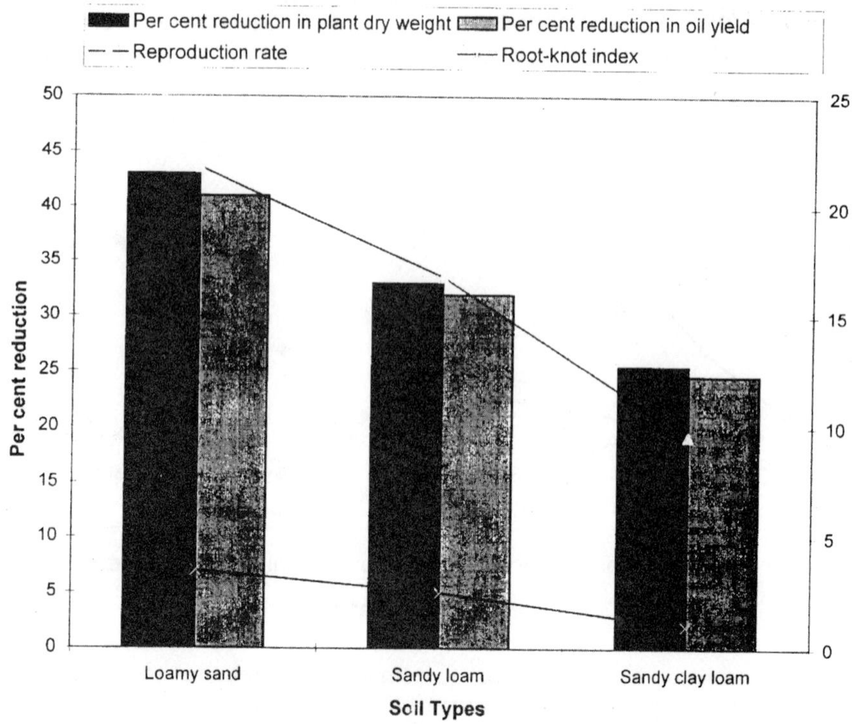

Fig.5. Effect of soil type on plant dry weight and oil yield of *Mentha arvensis* plants inoculated with 5000 J2 larvae of *Meloidogyne incognita* and its reproduction

proportional to susceptibility of cultivars of *Mentha*. Highest reproduction and root-knot index were observed on *M. arvensis* cv. HY-77, followed by Gomti, Shivalik, Kosi and Himalaya respectively (Fig. 1).

Effect of several soil, environmental, cultural and biological factors on the reproduction and damage potential of *Meloidogyne incognita* on various crops have received considerable attention (Norton, 1978). Studies were made on the vertical distribution of plant parasitic nematodes associated with *M. arvensis* in relation to seasonal fluctuations. Results indicated that nematodes were mainly confined in top 20 cm of soil (Haseeb and Shukla, pers. Comm.). *M. incognita* was observed to dominant nematode populations in the rhizosphere of *M. arvensis* followed by *P. thornei*, *Tylenchorhynchus vulgaris*, *Helicotylenchus indicus* and *Tylenchus* sp. respectively. Reproduction of these nematodes was observed high during October-November and February through April. and lowest during rains (Anon., 1984-1986; Haseeb and Shukla, Pers. Comm.), (Fig. 2 and 3).

Studies regarding the relationships between different pH levels, reproduction and damage potential of *M. incognita* and growth and oil yield of *M. arvensis* cv. HY 77 were conducted under controlled conditions. Results indicated a directly proportional relationship between the pH levels and severity of the disease (Fig. 4). Reproduction of *M. incognita* was observed highest at pH 6.0 followed by 9.0 and 3.0 respectively (Haseeb and Shukla, Pers. Comm.).

Studies were also done on the effect of soil types/soil texture and its organic matter content on the reproduction and damage potential of *M. incognita* to *M. arvensis* cv. HY 77. Results indicated that high clay and organic matter content suppressed the nematodes and enhanced the growth and oil yield (Fig. 5), (Haseeb and Shukla, Pers. Comm.).

Effect of various factos on the severity of root knot disease of Japanese mint was also studied through the surveys conducted in the mint growing areas of Tarai region of U.P., Results revealed that with the increase in clay content of soil, severity of root- knot disease was reduced to a considerable extent. Cultivation of some non host crops of root knot nematode in rotation with Japanese mint, water logging, paddy cultivation during rainy season caused the reduction in the existing populations of root knot nematodes (Shukla, 1997, 1998; Shukla *et al.*, 1997a).

The severity of root knot disease of Japanese mint was increased under field conditions in the presence of several soil borne fungi, particularly the species of *Fusarium* and *Rhizoctonia*. Studies conducted under controlled conditions indicated that *Meloidogyne* - *Rhizoctonia* disease complex caused severe reduction in growth and oil yield at lower initial inoculum densities, and at higher inoculum levels, young plants could not survive (Haseeb and Shukla, Pers. Comm.).

Information regarding the management of root knot nematodes infesting *M. arvensis* is very meager. Studies conducted at CIMAP, Lucknow, under glasshouse and field conditions revealed that application of different nematicides and oil cakes improved the growth of the plant coupled with oil yield and reduced the root knot development and nematode population. However, the best results were achieved when soil was treated with neem cake and carbofuran. Results also indicated that the application of hydro-distillation wastes of *M. arvensis*, *M. piperita*, *Cymbopogon martinit (palmarosa)*, *C. winterianus* (Citronella), etc. in the soil (one month before transplantation) was although found effective to reduce nematode populations but enhanced termite attack under field conditions. Gokte and Mathur (1990) suggested hot water treatment (40°C for 30 minutes) of *M. arvensis* planting material for complete eradication of *M. incognita* without any adverse effect to suckers and roots. Overnight soaking of suckers into 0.03% a.i. carbofuran solution has also been found effective in eradicating the nematodes from infected suckers (Haseeb and Shukla, Pers. Comm.).

Therefore, an integrated approach has been suggested for the management of root-knot nematodes infesting *M. arvensis*. Tolerant cultivars should be grown in nematode-free/less infested nursery beds, if necessary, suckers should be treated before transplanting into field, and pre-transplant nematode population in soil should be managed to a minimum level.

(ii) Root lesion nematode, *Pratylenchus* spp.

Root lesion nematodes are the next potential pests after the root knot nematodes, which produce the characteristic symptoms on roots and aerial portion of *M. arvensis*. The symptoms on the aerial portion caused by *Pratylenchus* species are almost similar to those of root knot nematode. However, this nematode, cause the specific symptoms of its infection in the form of light brown to dark coloured lesions on the suckers and roots. In severe cases whole underground portion becomes black and rotting of cortical portion takes place, first due to the nematode infection and later because of the invasion of other pathogen or saprophytic attack on necrotic cells (Anon. 1984-1986; Haseeb 1992-1994). The necrosis of cells takes place either by physical injury due to nematode feeding and movement through the roots or by the action of B glucosidase enzyme released by the nematodes during feeding.

Pathogenicity of *P. thornei* on *M. arvensis* cv. HY- 77 was established under controlled conditions. Results indicated significant reduction in plant fresh and dry weight, chlorophyll, photosynthetic rate, total sugar, phenol and oil content in leaves at very low (250 nematodes/7.5 kg soil) initial inoculum levels, Which increased with the increase in inoculum densities (Haseeb and Shukla, 2000b). Highest rate of reproduction of the nematodes was observed at lowest Pi, Which gradually decreased with the increase in Pi (Fig. 6).

Recently, studies were made to determine the interaction between *M. incognita* and *P. thornei* on *M. arvensis* (Haseeb and Shukla, 2000b). Results that reduction in growth and oil yield of plants inoculated with *M. incognita* was higher than that from *P. thornei* at equal initial inoculum densities (Pi). At Pi of 5000 *P. thornei* or *M. incognita* or 2500 *P. thornei* + 2500 *M. incognita* reduction in root shoot fresh and dry weights, oil yield, leaf chlorophyll, sugar and phenol content was highest in plants inoculated with *M. incognita* alone followed by *M. incognita* prior to *P. thornei, M. incognita* + *P. thornei* simultaneously, *P. thornei* alone and *P. thornei* prior to *M. incognita* respectively. Reproduction rate (Rf) of *M incognita* was observed higher than that of *P. thornei* when inoculated separately, and in combined inoculations, Rf of both the nematodes was suppressed as compared to single inoculations (Fig. 7).

Effect of several soil, environmental, cultural and biological factors on the reproduction and damage potential of *Pratylenchus* spp. on various crops have received considerable attention (Norton, 1978). However, still not much information is available on *M. arvensis* Results of vertical distribution studies, shown in Fig. 2 & 3, indicate that *P. thornei* reproduced at higher rate during February and March, and was suppressed during rains. Studies also indicate that soil texture do not play a major role in its damaging potential on *M. arvensis* unless the soil is compact in nature (Haseeb and Shukla, Pers. Comm.).

Studies regarding the management of *P. thornei* infesting *M. arvensis* indicated that similar measures should be adopted as described for root-knot nematodes.

(iii) Other Nematodes

In general, plant parasitic nematodes, other than *M. incognita* and *P. thornei*, associated with *M. arvensis* do not produce characteristic symptom(s). However, the potentiality of the particular plant parasitic nematode was correlated with the presence of high population of individual nematode species in the rhizosphere soil under field conditions and subsequent establishment of pathogenicity of particular nematode under controlled conditions. Thus, *Rotylenchulus reniformis, Helicotylenchus indicus, Tylenchorhynchus vulgaris, Hoplolaimus indicus* and *Paratylenchus* sp. were found to reproduce well on *M. arvensis* and to cause significant reduction in herb and oil yield (Haseeb and Shukla, Pers. Comm.).

2. PEPPERMINT, *Mentha piperita*

Peppermint has been considered the most useful species of mint since about a century back, and been grown all over the tropical and subtropical parts of the world. It is only recent past, particularly in India, when *M. arvensis* dominated mint cultivation. This is why, that maximum research work on nematode diseases of mints has been done on this crop.

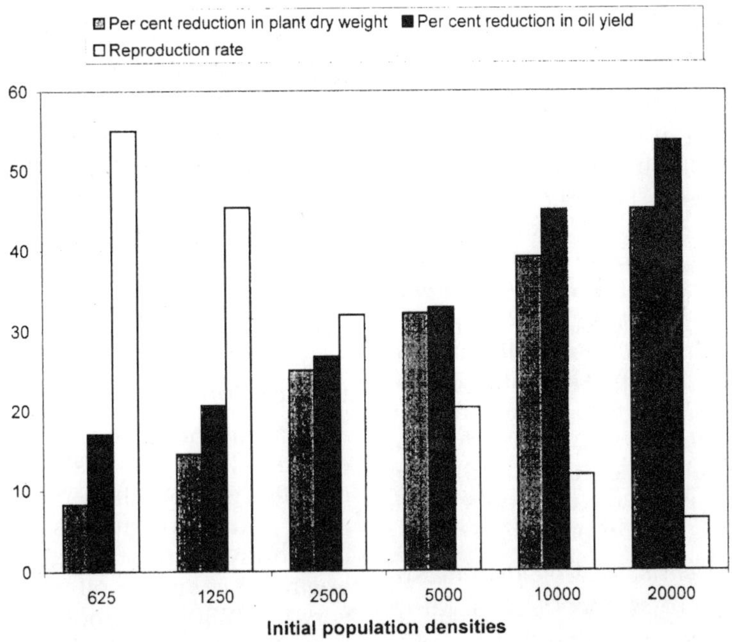

Fig. 6. Effect of different initial population densities of *Pratylenchus thornei* on growth and oil yield of *Mentha arvensis* and its reproduction

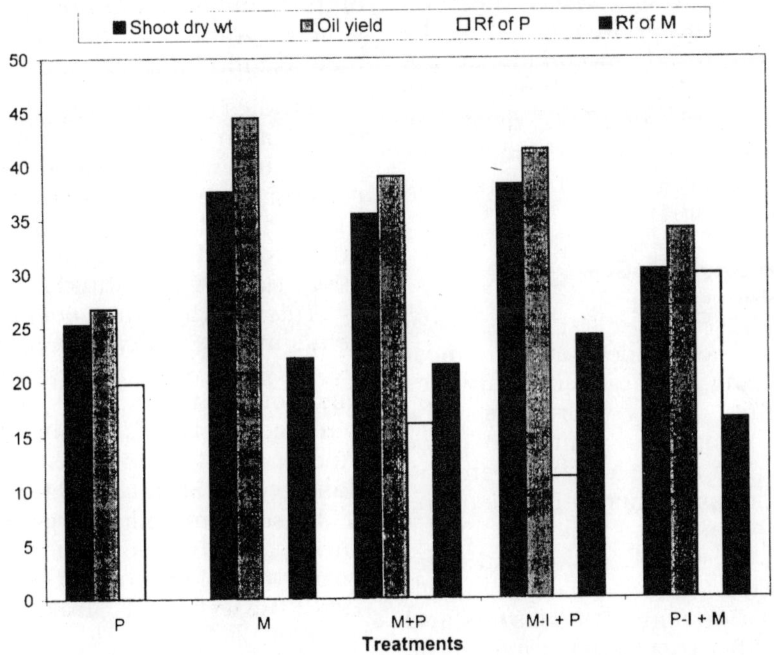

Fig. 7. Interactive effect of *Pratylenchus thornei* (P) and *Meloidogyne incognita* (M) on the per cent reduction in shoot dry weight and oil yield of *Mentha arvensis* (Pi = 5000)

(i) Root knot nematodes, *Meloidogyne* spp.

In 1938, peppermint was first time included in the list of hosts of *Meloidogyne* sp. (Buhrer, 1938). In green house tests, Horner and Jensen (1954) reported *M. piperita* cvs. Mitcham and American as good hosts of *M. hapla*. Skotland and Menzies (1957) reported that *M. hapla* was the most prevalent nematode species associated with *M. piperita* cv. Mitcham in the Yakima valley and the Columbia basin. O'Bannon *et al.* (1982) reported the association of *M. chitwoodi* with *M. piperita*. In Pakistan, Maqbool *et al.* (1985) also reported severe galling on the roots of *M. piperita* due to infestation of *M. hapla*.

Consistent presence of *Meloidogyne* sp. in the rhizosphere of *M. piperita* cvs. MP-1 and MPS-1 has been reported in the Tarai region of Uttar Pradesh and experimental beds of CIMAP, Lucknow (Haseeb, 1992, 1993, 1994; Shukla *et al.*, 1997a). However, *M. piperita* cv. Kukrail was found to be highly resistant to *M. incognita*. In pot experiments, 1,00,000 J2 larvae were inoculated/pot and after 90 days, no larvae/female was recovered from soil or root of the above mentioned cultivars of *M. piperita* (Shukla, 1997).

(ii) Root lesion nematodes, *Pratylenchus* spp.

Root lesion nematodes are the most severely damaging migratory endoparasites of peppermint. They produce the characteristic symptoms, similar to those on *M. arvensis*, on above and underground plant parts. The association of *Pratylenchus* spp. with peppermint was reported for the first time from U.S.A. (Skotland and Menzies, 1957). They observed high populations of *P. minyus*, *P. penetrans* and *P. thornei* in the rhizosphere soil of *M. piperita* cv. Mitcham in the Yakima valley of Washington. Bergeson (1963) conducted pathogenicity experiment of *P. penetrans* on *M. piperita* and reported the reduction in foliage and root/stolon growth up to 34 and 66% respectively. Bergeson and Green (1979) reported that *P. penetrans* was mainly responsible for reduction in herb weight and root growth of peppermint in Indiana. They further reported that all the cultivars of peppermint grown in Indiana were equally susceptible to *P. penetrans*. Pinkerton (1984) reported severe losses to *M. piperita* cv. Murry Mitcham and Todd Mitcham while cv. Black Mitcham was tolerant to *P. penetrans*.

Pathogenicity of *P. thornei* on *M. piperita* cv. MPS-1 was established under controlled conditions. Results indicated significant reduction in plant fresh and dry weight, chlorophyll, photosynthetic rate, total sugar, phenol and oil content in leaves at very low (250 nematodes/7.5 kg soil) initial inoculum levels, which increased with the increase in inoculum densities (Haseeb and Shukla, 1996). Highest rate of reproduction of the nematodes was observed at lowest Pi, which gradually decreased with the increase in Pi (Fig. 8).

Studies regarding the vertical distribution of plant parasitic nematodes associated with *M. piperita* cv. MPS-1 in relation to seasonal fluctuations indicated that nematodes were mainly confined in top 20 cm of soil (Anon., 1984-1986; Haseeb, 1993; Haseeb and Shukla, 1994a) followed by *Tylenchorhynchus vulgaris*, *Helicotylenchus indicus*, *Tylenchus* sp. and larvae of *Meloidogyne* spp. respectively. *P. thornei* was found to be the most dominant nematode in the rhizosphere of *M. piperita* (Anon., 1984-1986; Haseeb, 1992-1994; Haseeb and Shukla 1994a). Reproduction of these nematodes was observed high during October-November and February through April (Anon., 1984- 1986; Haseeb, 1993; Haseeb and Shukla, 1994a), (Fig. 9-10).

In general, nematodes are highly specific to host and soil environmental

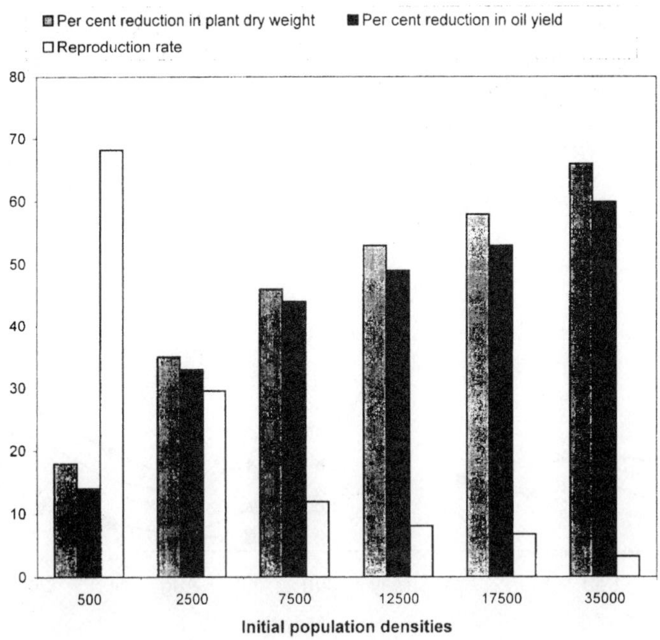

Fig. 8. Effect of different initial inoculum densities of *Pratylenchus thornei* on the growth and oil yield of *Mentha piperita* and its reproduction

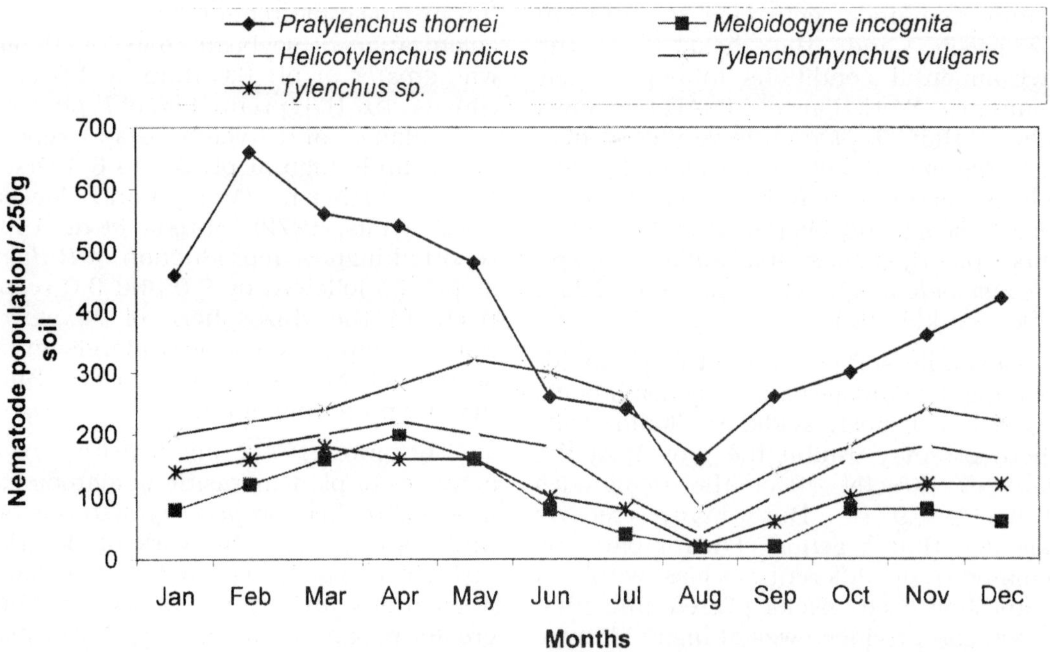

Fig. 9.Vertical distribution of plant parasitic nematodes in relation to seasonal fluctuations on *Mentha piperita* (0-10 cm)

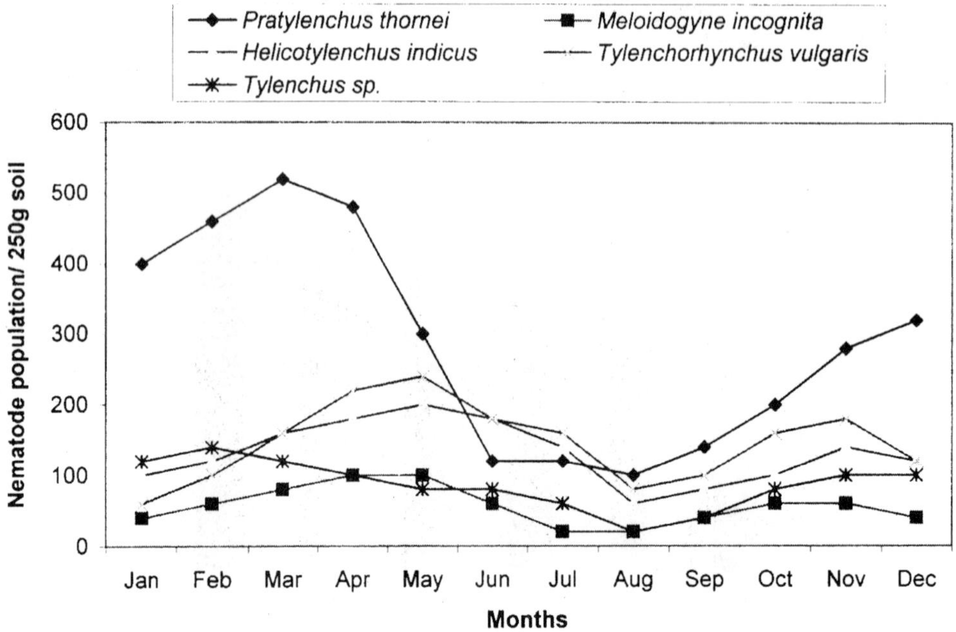

Fig. 10. Vertical distribution of plant parasitic nematodes in relation to seasonal fluctuations on *Mentha piperita* (11-20 cm)

factors. Nematode occurrence in nature in polyspecific and the host and the environmental conditions influence their dominance. Work done at CIMAP, Lucknow, indicate that *M. piperita* is very good host of *P. thornei, T. vulgaris, Helicotylenchus indicus*, good host of *R. reniformis, Hoplolaimus indicus*; and poor host of *Paratylenchus* sp., *Tylenchus* sp., *Xiphinema* sp., *Criconemoides* sp. etc. (Haseeb, 1993, 1994, Shukla, 1997).

Generally soil pH is not a major factor affecting the damage caused by nematodes, except in extremely acidic or alkaline soils where pH may inhibit the growth of the plant or directly affect the nematodes (Wallace, 1971). But many reports indicated that a genus of plant parasitic nematodes or different species within a genus show their specific pH requirements. *P. thornei* reproduced well at high pH range (7.1-8.5) (Grandison and Wallance, 1974), while *P. brachyurus* multiplied well at pH 3.8 in pineapple roots (Sarah *et al.* 1991), colonization of soybean roots by *P. alleni* was greater at pH 6.0 than at 4.0 or 8.0 (Burns, 1971), reproduction of *P. penetrans* on alfalfa and vetch was reported significantly high at pH 5.2 to 6.4 than at lower or higher pH (Morgan and MacLean, 1968; Willis, 1972). Shukla *et al.* 1998c) reported highest reproduction of *P. thornei* at pH 6.0 followed by 9.0 and 3.0 respectively in the rhizosphere of *M piperita*. Relationship between the pH levels and the severity of the disease was observed to be directly proportional (Fig. 11).

The impact of soil texture on the activities of plant parasitic nematodes and associated affect on plant growth has been under study since the work of Bessey in 1911. It is widely accepted that nematode infestations and crop damage are much greater in sandy soils than in clay or heavy soils (Barker *et al.*, 1981; Shane and Barker, 1984; Jordan, *et al.*, 1989;

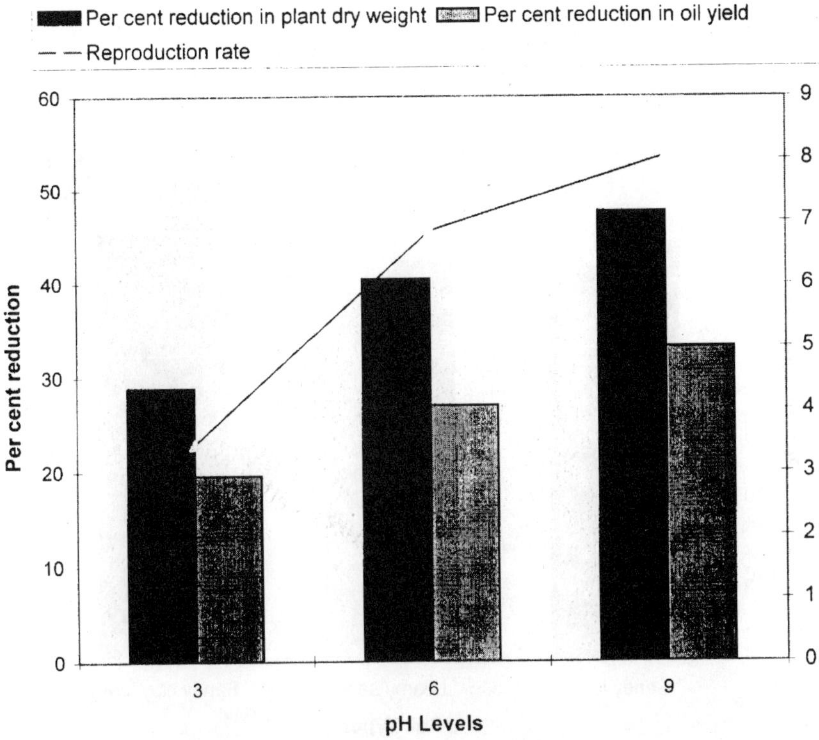

Fig. 11. Effect of pH levels on the average reduction in plant dry weight and oil yield of *Mentha piperita* plants inoculated with 10000 *Pratylenchus thornei*

Wallance, 1989; Nakasono *et al.*, 1990). Shukla *et al.*, (1998a) reported that soil texture and its organic matter content may also play a very important role on the avtivities of *P. thornei* in the rhizosphere of *M. piperita* (Fig. 12). Seasonal changes such as temperature and rainfall (Haseeb and Shukla, 1994a), crop rotation, fallowing and water logging (Shukla, 1997, 1998) have been considered important to influence the losses to pepermint due to lesion nematode. Thus a soil with high clay and organic matter content with lower pH may be most suitable for the cultivation of mints with least damage due to lesion nematode.

Very meager information is available regarding the interrelationships between root lesion nematodes and other soil borne pathogens on the development of disease(s) in peppermint. However, the available reports indicated that the severity of several diseases caused by fungi and viruses was increased in the presence of nematodes. Bergeson (1963) was the first to test the Pathogenicity of *Pratylenchus penetrans* alone and in combination with *Verticillium albo-atrum* on *Mentha piperita*. He reported that under sterile conditions, *P. penetrans* reduced foliage and root growth, but when *P. penetrans* infected plants were inoculated with *V. albo-atrum* diagnostic symptoms of Verticillium wilt appeared approximately two weeks earlier than the plants inoculated with fungus in the absence of nematode. This did not occur with simultaneous inoculations of fungus and nematodes. Faulkner and Skotland (1963) reported that the *Pratylenchus minyus* has been found through out the

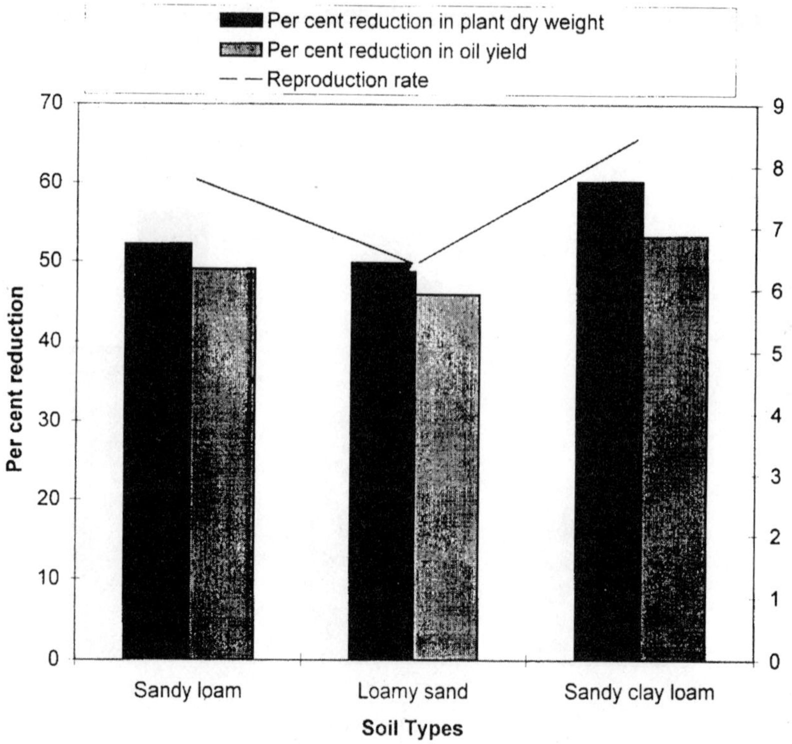

Fig. 12. Effect of different soil types on the reduction in plant dry weight and oil yield of *Mentha piperita* plants inoculated with 12500 *Pratylenchus thornei*

Yakima valley of Washington in association with the incitant of peppermint wilt fungus, *V. dahliae* (Kleb.) f.sp. *menthae* Nelson. A synergistic interaction was observed between *V. dahliae* (Kleb) f.sp. *Menthae* and *P. minyus* in the causation of Verticillium wilt of *M. piperita* cv. Mitcham was reported by Faulkner and Skotland (1965). Presence of the nematode increased both incidence and severity of the disease. The period for the development of Verticillium wilt was reduced for 2 to 3 weeks by the presence of the nematode. Reproduction of nematode was increased in the presence of fungus. Dry weights of plants reduced only up to 22% when either of the pathogen was present, which increased up to 68% when inoculum contained both the pathogen in combinations.

Faulkner and Bolander (1969) studied the disease severity on peppermint at 18, 21, 24, 27 and 30°C soil temperature when plants were inoculated with the *V. dahliae f. sp. menthae* and *P. minyus*. Symptoms expression was highest at 27°C in plants inoculated with both the pathogens simultaneously. Whereas, the optimal expression of wilt symptom occurred at a soil temperature of 24°C. Final population of the nematodes was highest at 24°C, when both the pathogens were present simultaneously, whereas lowest at 30°C when plants were inoculated with nematodes alone.

Faulkner *et al.*, (1970) demonstrated that *P minyus* does not only help *V. dahliae* infection by providing physical injuries on roots of *M piperita* but also altered the physiology/biochemistry, which increased the susceptibility of plant to *V. dahliae* infection.

Information regarding the control of *Pratylenchus* spp. infesting peppermint is mainly confined to the chemical measures. Pinkerton *et al.*, (1988) reported that decrease in population of *P. penetrans* and increase in herb yield of peppermint was observed in treatments of oxamyl, aldoxycarb and carbofuran @ 5.5 kg a.i./ha, when these were applied as broadcast and spray during spring. Single treatment of in early April was as effective as multiple applications during April, May or April, May and June. Fall application of oxamyl and carbofuran neither reduced the multiplication of nematode nor enhanced the spring growth.

In field micro-plot trials Marrifield and Ingham (1992) applied ethoprop @ 6.6 kg a.i./ha) for the control of *P. penetrans* infesting peppermint in oregon. While Ingham *et al.*, (1992) applied the same nematicide @ 6.6, 13.2 or 26.4 kg a.i./ha in peppermint fields. They reported that the application of 6.6 kg a.i./ha ethoprop was highly effective to reduce the populations of *P. penetrans*.

Studies regarding the management of *P. thornei* infesting peppermint by the application of neem, mustard and linseed cakes @ 1500 kg/ha, and aldicarb, carbofuran and ethoprop @ 4 kg a.i./ha were conducted under glass house conditions. Results indicated that neem cake was the most effective followed by mustard cake, aldicarb, ethoprop, carbofuran and linseed cake respectively. However, all the treatments significantly improved the growth and oil yield of *M. piperita* as compared to untreated inoculated plants, and reduced final nematode populations in root and soil (Shukla and Haseeb, 1996).

(iii) Other plant parasitic nematodes

Goodey (1940) reported the association of *Aphelenchoides olesistus* with *M. piperita*. Horner and Jensen (1954) reported that single shoot tip of the runners of *M. piperita* possessed about 400 *Aphelenchoides* spp. and *A. parietinus*. They further reported that in western Orgeon, *Paratylenchus macrophallus* parasitized commercial peppermint and the infected plants become stunted and chlorotic. About 8000 nematodes were isolated per quart of soil.

Skotland and Menzies (1957) found that in addition to other plant parasitic nematodes, *Paratylenchus hamatus* was also associated with *M. piperita* cv. Mitcham in the Yakima valley and in the Columbia basin. Horner and Jensen (1954) observed extensive stunting and dying of *M. piperita* due to infestation of *Longidorus* sp. throughout many of the older mint growing areas of western Oregon during 1953. In experimental tests, they found that *Longidorus* caused stunting and a root die back condition. Latter, Jensen and Horner (1956) established the pathogenicity of *L. silphus* and controlled the disease by soil fumigation. Jatala and Jensen (1974) controlled the *L. elongatus* infestation in peppermint by application of oxamyl as foliar spray. Such treatments were effective only when oxamyl was applied before, during and after the exposure to nematode. Pinkerton and Jensen (1983) reported that the treatment of oxamyl as spray and combined with aldicarb and oxamyl in soil as drench resulted into significant increase in oil yield and hay in first after application.

Esmenjaud *et al.*, (1990) reported reduction in the yield of *M. piperita* due to the infestation of *Pratylenchoides laticauda* in France. They suggested quarantine regulations to limit the introduction of this nematode to new areas through planting materials. Merrifield and Ingham (1992) reported that the application of ethoprop (6.6 kg/ha) had no effect on the Populations of *Paratylenchus* sp., *Criconemella xenoplax*, *Longidorus elongatus* and *Paratrichodorus* sp. infesting peppermint.

Studies regarding the vertical distribution of plant parasitic nematodes associated with peppermint in relation to seasonal fluctuations indicated that nematodes were mainly confined in top 20 cm of soil (Haseeb, 1993; Haseeb and Shukla, 1994a) (Fig. 9- 10). Several species of plant parasitic nematodes such as : *Tylenchus* sp., *Tylenchorhynchus vulgaris*, *Hoplolaimus indicus*, *Helicotylenchus indicus* were present throughout the year, and *Rotylenchulus reniformis*, *Paratylenchus* sp., *Criconemoides* sp., *Hirschmanniella* sp., *Xiphinema* sp., *Longidorus pisi* etc. were found occasionally (Haseeb, 1992, 1993, 1994; Haseeb and Shukla, 1994a). Reproduction of these nematodes was observed high during October - November and February through April (Haseeb, 1993; Haseeb and Shukla, 1994a).

In general, nematodes are highly specific to host and soil envrionmental factors. Nematode occurrence in nature is polyspecific and the host and the environmental conditions influence their dominance. Work done at CIMAP, Lucknow, indicate that *M. piperita* is very good host of *P. thornei*; good host of *R. reniformis*, *T. vulgaris*, *Helicotylenchus indicus*; and poor host of *Hoplolaimus indicus*, *Paratylenchus* sp., *Tylenchus* sp., *Xiphinema* sp., *Criconemoides* sp. etc. (Haseeb, 1993, 1994; Shukla, 1997, 1998; Shukla and Haseeb, Pers. Comm.).

The impact of soil texture and soil pH, and water holding capacity of soil on the activities of plant parasitic nematodes associated with peppermint was determined under controlled conditions and through survey of farmer's fields (Haseeb 1993; Shukla *et al.*, 1997a; Shukla, 1997, 1998, 1999). It was observed that high clay content of soil, highest and lowest pH levels and WHC were suppressive to the majority of plant parasitic nematodes.

So far no published report is available on the interrelationships of different species of plant parasitic nematodes excluding root knot and root lesion nematodes and soil borne fungi and viruses on the growth and oil yield of peppermint grown under field conditions. This area needs attention for future research.

Management of different plant parasitic nematodes associated with different species of mint has been successfully achieved by the application of chemical means (Horner and Jensen, 1954; Faulkner and Skotland, 1963; Jatala & Jensen, 1974; Pinkerton & Jensen 1983; Inserra and Rhoades, 1989; Marrified and Inghum, 1992; Haseeb and Shukla, pers. Comm.). Recently in field micro plot trials, ethoprop and carbofuran @ 4kg a.i/ha, and neem cake @ 1000 kg/ha were found highly effective in managing the plant parasitic nematodes associated with *M. piperita* cv. MPS 1 (Haseeb and Shukla, Pers. Comm.).

3. SPEARMINT, *Mentha spicata*

Spearmint is the next important (after *M. arvensis* and *M. piperita*) species of mint, which has been cultivated on large scale for its oil in India as well as in other countries. Nematological aspects of *M. spicata* are very much similar to those of *M. piperita*.

(i) Root knot nematodes, *Meloidogyne* spp.

Horner and Jensen (1954) conducted glasshouse experiments and reported spearmint as a good host of *M. hapla*. Later, O'Bannon *et al.*, (1982) reported the association of *M. chitwoodi* with spearmint. In india although larvae of *Meloidogyne* sp. were found present in the rhizosphere soil of *M. spicata* (Haseeb, 1992-1994; Haseeb and Shukla, 2000a), but in pathogenecity experiments it was found immune to *M. incognita* (Shukla, 1997).

(ii) Root lesion nematodes, *Pratylenchus* spp.

Root lesion nematodes are the most damaging nematode pests of spearmint. They produce the characteristic symptoms, similar to those on *M. arvensis* on above and underground parts of the plants.

The association of *Pratylenchus* spp. with spearmint was reported for the first time from U.S.A. (Skotland and Menzies, 1957). They observed high populations of *P. minyus*, *P. penetrans* and *P.thornei* in the rhizosphere of *M. spicata* cv. Native in the Yakima valley of Washington. Bergeson and Green (1979) reported that *P. penetrans* was mainly responsible for reduction in herb weight and root growth of spearmint in Indiana. Rhoades (1983) reported that the growth of spearmint was reduced and severe stunting of plants was observed due to the infestation of *P. scribneri* in central Florida. He established the pathogenicity under glasshouse conditions and reported significant reduction in plant fresh weight and chlorosis of leaves after 4-5 months of inoculation.

Pathogenicity of *P. thornei* on *M. spicata* cv. MSS - 5 was established under controlled conditions. Results indicated significant reduction in plant fresh and dry weight, chlorophyll, photosynthetic rate, total sugar, phenol and oil content in leaves at very low (250 nematodes/7.5 kg soil) initial inoculum level, which increased with the increase in inoculum densities (Haseeb and Shukla, 1994b). Highest rate of reproduction of the nematodes was observed at lowest Pi, which gradually decreased with the increase in Pi (Fig. 13).

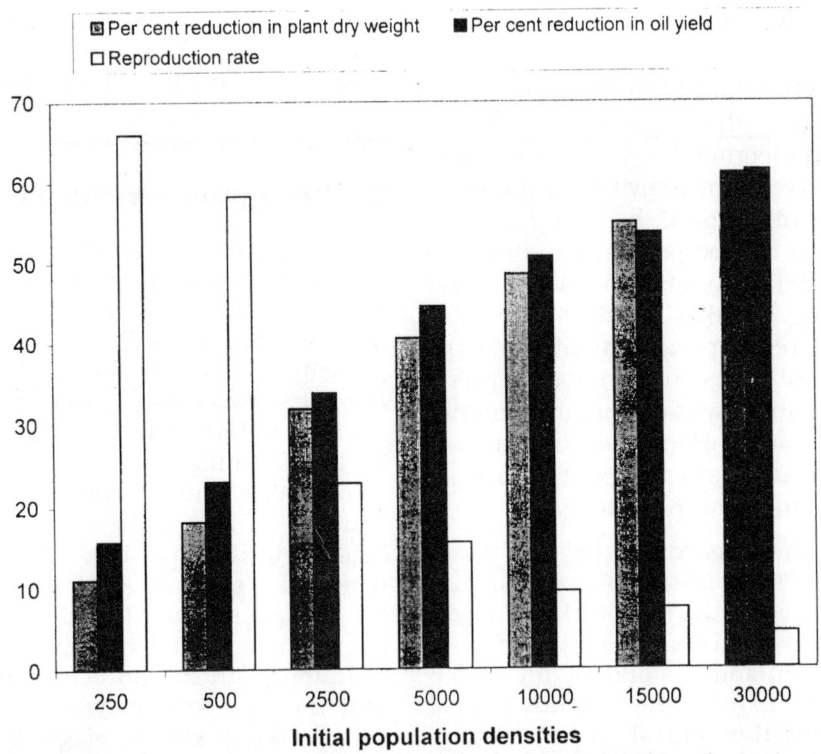

Fig. 13. Effect of different initial inoculum densities of *Pratylenchus thornei* on the growtl and oil yield of *Mentha spicata* and its reproduction

Studies regarding the vertical distribution of plant parasitic nematodes associated with *M. spicata* cv. MSS-5 in relation to seasonal fluctuations indicated that nematodes were mainly confied in top 20 cm of soil (Anon., 1984-1986; Haseeb, 1993; Haseeb and Shukla, Pers Comm.). *P. thornei* was found to be the most dominant nematode in the rhizosphere of *M. spicata* (Anon., 1984-1986; Haseeb 1992-1994; Haseeb and Shukla, pers. Comm.). Reproduction of these nematodes was observed high during October-November and February through April (Anon. 1984-1986; Haseeb, 1993; Haseeb and Shukla, Pers. Comm.).

Shukla *et al*, (1997b) reported highest reproduction of *P. thornei* at pH 6.0 followed by 9.0 and 3.0 respectively in the rhizosphere of *M. spicata*. Relationship between the pH levels and the severity of the disease was observed to be directly proportional.

Shukla (1998) and Shukla and Haseeb (2000) reported that soil texture and its organic matter content may also play a very important role on the activities of *P. thornei* in the rhizosphere of *M. spicata*. Seasonal changes such as temperature and rainfall (Haseeb, 1993), crop rotation, fallowing and water logging (Shukla, 1997, 1998) have been considered important to influence the losses caused by lesion nematode. Thus a soil with high clay and organic matter content with lower pH may be most suitable for the cultivation of spearmint with least damage due to lesion nematode.

Information regarding the control of *Pratylenchus* spp. infesting spearmint is mainly confined to the chemical measures. Rhoades (1984) have studied the effect of carbofuran, fenamiphos and oxamyl @ 5.6 and 11.2 kg a.i./ha and terbuphos @ 11.2 kg a.i./ha for the control of *P. scribneri*/ infesting *M. spicata*. The population of nematode was significantly reduced by all these nematicides, while fenamiphos being the most effective. Both the rates of carbofuran and oxamyl were less effective in increasing the herb yield than the fenamiphos and terbufos treatments. Ingham *et al.*, (1992) applied ethoprop @ 6.6, 13.2 or 26.4 kg a.i./ha in spearmint fields. They reported that the application of 6.6 kg a.i./ha ethoprop was highly effective to reduce the populations of *P. penetrans*.

Studies regarding the management of *P. thornei* infesting spearmint by the application of neem, mustard and linseed cakes @ 1500 kg/ha, and aldicarb, carbofuran and ethoprop @ 4 kg a.i./ha were conducted under glass house conditions. Results indicated that neem cake was the most effective followed by mustard cake, aldicarb, ethoprop, carbofuran and linseed cake respectively. However, all the treatments significantly improved the growth and oil yield of *M. spicata* as compared to untreated inoculated plants, and reduced final nematode populations in root and soil (Shukla and Haseeb, 1996).

(iii) Other plant parasitic nematodes

Goodey (1940) reported the association of *Aphelenchoides ritzemabosi* with *M. spicata*. Skotland and Menzies (1957) found that in addition to other plant parasitic nematodes, *Paratylenchus hamatus* was also associated with *M. spicata* cv. Native in the Yakima valley and in the Columbia basin. Faulkner (1962) also reported infestation of *P. hamatus* in Scotch and Native spearmint in Washington. Generally, 3000 to 10000 nematodes per mint soil were isolated from the rhizosphere of infected plants. Severely infected plants did not show any specific symptoms other than occasional dwarfing and the root system was much reduced with varying stages of decay. Faulkner (1964) further reported that 10-20% and 20-36% reduction in fresh and dry weight of the plants and delayed flowering was noticed in inoculated plants.

Rhoades (1983) reported that the severely stunted beds of *M. spicata* in central Florida were found to be heavily infested with sting (*Belonolaimus longicaudatus*), awl (*Dolichodorus heterocephalus*) and stubby root (*Paratrichodorus christiei*) nematodes. In glasshouse experiments. All the species of nematodes were observed to reproduce well on *M. spicata* and caused significant reduction in shoot and root growth. Inserra and Rhoades (1989) described the symptoms and the damage caused by *B. longicaudatus*, *D. heterocephalus* and *P. christiei* separetely on *M. spicata*. They have also suggested the use of non volatile nematicides for managing the nematode population and increasing the yield.

Studies regarding the vertical distribution of plant parasitic nematodes associated with spearmint in relation to seasonal fluctuations indicated that nematodes were mainly confined in top 20 cm of soil (Haseeb, 1993). Several species of plant parasitic nematodes such as : *Tylenchus* sp., *Tylenchorhynchus vulgaris*, *Hoplolaimus indicus*, *Helicotylenchus indicus* were found present throughout the year, and *Rotylenchulus reniformis*, *Paratylenchus* sp., *Criconemoides* sp., *Hirschmanniella* sp., *Xiphinema* sp., *Longidorus pisi*, etc. were found occasionally (Haseeb, 1992, 1993, 1994; Haseeb and Shukla, pers. Comm.). Reproduction of these nematodes was observed high during October-November and February through April (Anon., 1984-1986; Haseeb, 1993).

Studies conducted under controlled conditions indicate that *M. spicata*, is very good host of *P. thornei*; good host of *T. vulgaris*, *H. indicus*, *R. reniformis*, and poor host of *Hoplolaimus* spp.; *Paratylenchus* sp., *Tylenchus* sp., *Xiphinema* sp., *Criconemoides* sp. etc. (Haseeb, 1993, 1994; Shukla, 1997, 1998).

The impact of soil texture and soil pH, and water holding capacity of soil on the activities of plant parasitic nematodes associated with spearmint was determined under controlled conditions and through survey of farmer's fields (Haseeb 1993; Shukla *et al.*, 1997a, Shukla, 1997, 1998, 1999). It was observed that high clay content of soil, highest and lowest pH levels and WHC were suppressive to the majority of plant parasitic nematodes.

So far no published report is available on the interrelationships of different species of plant parasitic nematodes excluding root knot and root lesion nematodes and soil borne fungi and viruses on the growth and oil yield of spearmint grown under field conditions. This area needs attention for future research.

Management of different plant parasitic nematodes associated with different species of mint has been successfully achieved by the application of chemical means (Horner and Jensen, 1954; Faulkner and Skotland, 1963; Jatala & Jensen, 1974; Pinkerton & Jensen, 1983; Inserra and Rhoades, 1989; Marrifield and Inghum, 1992; Haseeb and Shukla, Pers. Comm.). Recently in field micro plot trials, ethoprop and carbofuran @ 4 kg a.i./ha, and neem cake @ 1000 kg/ha were found highly effective in managing the plant parasitic nematodes associated with *M. spicata* cv. MSS 5 (Haseeb and Shukla, Pers. Comm.).

4. SCOTCH SPEARMINT, *Mentha cardiaca*

Association of a plant parasitic nematode with Scotch spearmint was first by Horner and Jensen (1954). Since then a lot of research have been done on nematode problems of this crop. It is very good host of *M. incognita*, *M. Javanica* and *P. thornei*;

good host of *R. reniformis, T. vulgaris, Helicotylenchus indicus* and *Paratylenchus* sp.; poor host of *Hoplolaimus indicus, Longidorus pisi, Xiphinema* sp., *Criconemoides* sp. etc. (Haseeb, 1992, 1994, Haseeb and Shukla, 2000a).

(i) Root knot nematodes, *Meloidogyne* spp.

Horner and Jensen (1954) reported very high population of *Meloidogyne hapla* in the rhizosphere of *M. cardiaca*. They conducted pathogenicity tests under green house conditions and reported *M. cardiaca* as good host of *M. hapla*. Skotland and Menzies (1957) also reported that *M. hapla* was most prevalent nematode species associated with *M. cardiaca* in Yakima Valley and Colombia basin. O'Bannon *et al.*, (1982) reported that association of *M. chitwoodi* with *M. cardiaca*.

Pathogenicity of *M. incognita* on *M. cardiaca* also showed the high degree of susceptibility and severe reduction in plant growth, oil content, leaf chlorophyll content, photosynthetic efficiency of leaves and sugar and phenol content in fresh leaves (Haseeb and Shukla, 2000a, 2001). Results of other studies, particularly effect of soil factors on disease development and the management of this nematode on *M. cardiaca*, were very much similar to *M. arvensis*

(ii) Root lesion nematodes, *Pratylenchus* spp.

Association of root lesion nematodes (*Pratylenchus thornei, P. penetrans* and *P. minyus*) with *M cardiaca* was first reported by Skotland and Menzies (1957). Later *P. thornei* was found to be associated with this species of mint in Lucknow and the tarai region of Uttar Pradesh (Anonymous, 1985; Haseeb, 1992, 1994). Recently experiments were conducted to determine its damage potential alone and combination with *Meloidogyne incognita* (Haseeb and Shukla, 2001), effect of soil factors on disease development and management. Results indicated that it cause severe infection of roots and suckers, and reduction in growth and oil yield. When *P. thornei* and *M. incognita* were inoculated on same plant, severity of disease and damage were further increased. Effect of various soil factors and management of this nematode was observed similar to that of *Mentha arvensis*.

(iii) Other plant parasitic nematodes

Among other plant parasitic nematodes, several potential nematodes e.g. *Rotylenchulus reniformis, Helicotylenchus indicus, Hoplolaimus indicus, Tylenchorhynchus vulgaris*, etc. have been found associated with this species of mint (Anonymous, 1984, 1985; Haseeb, 1992, 1994) but no systematic studies have been conducted so far.

5. BERGAMOT MINT, *Mentha citrata*

Work regarding the association of plant parasitic nematodes with *M. citrata* has only been done at CIMAP, Lucknow, India. This plant has been reported as the most resistant species of mint against *Meloidogyne* spp. (Haseeb, 1992, 1994). Later, Shukla (1997) reported *M. citrata* cv. Kiran immune to *M. incognita*. On the other hand, root lesion nematode, *P. thornei* has been considered the most damaging nematode of this plant (Haseeb, 1992, 1993, 1994; Haseeb and Shukla, 1995). Studies regarding the vertical distribution (Shukla and Haseeb 1991), pathogenicity (Haseeb and Shukla, 1995), factors affecting disease development (Shukla *et al.*, 1997c, 1998b) and Management (Shukla and Haseeb, 1996) of *P. thornei* on *M. citrata* cv. Kiran were conducted and very much similar results were observed as described in detail on *M. piperita*. Among other plant parasitic nematodes, *T. vuglaris* and *Helicotylenchus indicus* were found to cause significant damage to *M. citrata*.

(i) Other species of *Mentha*

Several more species of Mentha such as *M. crispa*, *M. diemenica*, *M. gentilis*, *M. longifolia*, *M. mirennae*, *M. pulegium*, *M. suaveolens*, *M. sylvestris*, *M. verticillata*, *M. viridis*, etc. are being cultivated at small scale in various parts of world. However, no systematic work has so far been done on their nematode problems.

REFERENCES

Anonymous, 1984. Annual Report 1983-84, Central Institute of Medicinal and Aromatic Plants, Lucknow, India.

Anonymous, 1985. Annual Report 1984-85, Central Institute of Medicinal and Aromatic Plants, Lucknow, India.

Anonymous 1986. Annual Report 1985-86, Central Institute of Medicinal and Aromatic Plants, Lucknow, India.

Anonymous, 1987. Annual Report 1986-87, Central Institure of Medicinal and Aromatic Plants, Lucknow, India.

Anonymous, 1999. Menthol mint research brings more recognition. CIMAP Newsl. 26 (3) : 50-51.

Barker, K.R., F.A. Todd, W.W. Shane, and L.A. Nelson, 1981. Interrelationships of *Meloidogyne* species with flue cured tobacco. *J. Nematol.* 13 : 67-79.

Bergeson, G.B. 1963. Influence of *Pratylenchus penetrans* alone and in combination with *Verticillium albo-atrum* on growth of peppermint. *Phytopathology* 53 : 1164-1166.

Bergeson, G.B. and R.J. Green Jr. 1979. Damage to cultivars of peppermint, *Mentha piperita*, by the lesion nematode, *Pratylenchus penetrans* in Indiana. *J. Nematol.* 7 : 154.

Bessey, E.A. 1911. Root knot nematode and its control. Bureau of Plant Industry, U.S. Deptt. of Agric., Washington D.C. Bull., pp. 217.

Buhrer, E.M. 1938. Additions to the list of plants attacked by root knot nematode (*Heterodera marioni*). *Plant Dis. Reptr.* 22 : 216-234.

Burns, N.C. 1971. Soil pH effects on nematode population associated with soybean. *J. Nematol.* 3 : 238-247.

Esmenjaud, D.R., Voisin, J.C. Minot and G.De Guiran. 1990. *Pratylenchoides laticauda* on peppermint (*Mentha piperita* Huds.) in Southern Alps : Host range, population densities in different crop rotations and host suitability of three mint sub species. *Med. Fac. Landbouww. Rijksuniv. Gent.* 55 : 779-786.

Faulkner, L.R. 1962. Pathogenicity and population dynamics of *Paratylenchus hamatus* Thorne and Allen on *Mentha* spp. *Phytopathology* 52 : 731.

Faulkner, L.R. 1964. Pathogenicity and population dynamics of *Paratylenchus hamatus* on *Mentha* spp. *Phytopathology* 54 : 344-348.

Faulkner, L.R. and W.J. Bolander. 1969. Interaction of *Verticillium dehliae* and *Pratylenchus minyus* in Verticillium wilt of peppermint : Effect of soil temperature. *Phytopathology* 59 : 868-870.

Faulkner, L.R., W.J. Bolander and C.B. Skotland, 1970. Interaction of *Verticillium dehliae* and *Pratylenchus minyus* on Verticillium wilt of peppermint. Influence of the nematodes determined by a double root technique. *Phytopathology* 60 : 100-103.

Faulkner, L.R. and C.B. Skotland, 1963. Control of Verticillium wilt of mint and plant parasitic nematodes associated with mint by soil fumigation. *Plant Dis. Reptr.* 47 : 662-665.

Faulkner, L.R. and C.B. Skotland, 1965. Interaction of *Verticillium dahliae* and *Paratylenchus minyus* in Verticillium wilt of peppermint. *Phytopathology* 55 : 583-586.

Gokte, N. and V.K. Mathur, 1990. Thermal therapy for the eradication of *M. spicata* roots infected with *Meloidogyne incognita*. FAO Plant Prot. Bul. 8 : 213-215.

Goodey, T. 1940. The nematode parasites of plants catalogued under their hosts. *Imp. Bur. Agric. Parasit. (Helminthol.)*, St. Albans, England, 80 pp.

Grandison, G.S. and H.R. Wallace, 1974. The distribution and abundance of *Pratylenchus thornei* in fields of strawberry clover (*Trifolium fragiferum*). *Nematologica* 20 : 283-290.

Haseeb, A. 1992. Nematode pests of medicinal and aromatic plants. In : Bhatti, D.S. & Walia, R.K. (Eds.) *Nematode Pests of Crops*, pp .239-249. CBS Publication, New Delhi, India.

Haseeb, A. 1993. Studies on certain aspects of pathogenicity of *Pratylenchus thornei* on mint and its control. Final Technical Report - An Indian Council of Agricultural Research (ICAR), New Delhi, India Sponsored Project (F.No. 2-5/88 PP dt. 29.03.1989).

Haseeb, A. 1994. Plant parasitic nematodes of medicinal and aromatic plants. In : Singh, T. & Trivedi P.C. (eds.) Vistas in Seed Biology., pp. 98-119. Printwell publication, Jaipur, India.

Haseeb, A. and R. Pandey, 1989. Observations on *Meloidogyne incognita* affecting Japanese mint – new disease record. *Nematropica* 19 : 93-98.

Haseeb, A. and P.K. Shukla, 1994a. Studies on vertical distribution of plant parasitic nematodes associated with *Mentha piperita* L. Proc. II Afro Asian Nemato. Symp., Menofiya, Cairo, Egypt, pp. 121-124.

Haseeb, A. and P.K. Shukla, 1994b. Effect of *Pratylenchus thornei* on growth, physiology and oil yield of *Mentha spicata. Afro-Asian J. Nematol.* 4 : 51-53.

Haseeb, A. and P.K. Shukla, 1995. Influence of *Pratylenchus thornei* on growth, photosynthetic rate, chlorophyll, total sugar and oil yield of *Mentha citrata. Nematol, Medit.*, 23 : 89-91.

Haseeb, A. and P.K. Shukla, 1996. Suppressive effect of *Pratylenchus thornei* on the growth, oil yield, physiological and biochemical parameters of *Mentha piperita. Nematropica* 26 : 81-85.

Haseeb, A. and P.K. Shukla, 2000a. Nematode disease of mints and their control. In : *Advances plant Disease management*." Eds. Udit Narain, Kumud Kumar and M. Srivastava. Advance Publishing Concept, New Delhi pp. 229-248.

Haseeb, A. and P.K. Shukla, 2000b. Effects of *Pratylenchus thornei* and *Meloidogyne incognita*, alone and in combinations, on the growth and oil yield of *Mentha arvensis. Nematol. medit.*

Haseeb,. A. and P.K. Shukla, 2001a. Influence of soil types on the reproduction and damage potential of *Pratylenchus thornei* on *Mentha spicata*. Current Nomato.

Haseeb, A. and P.K. Shukla, 2001b. Reproduction and damage potential of *Pratylenchus thornei*, alone and in combination with *Meloidogyne incognita* on growth and oil yield of *Mentha cardiaca. J. Plant Prot. Sci.*

Henderson, V.E. 1951. Some host relationships of the potato rot nematode, *Ditylenchus destructor* Thorne, 1945. *Nature* 167- 952.

Horner, C.E. and H.J. Jensen, 1954. Nematodes associated with mints in Oregon. *Plant Dis. Reptr.* 38 : 39-41.

Husain, A.P. Singh, O.P. Virmani, B. Khanna, and G.N. Srivastava. 1978. Mint farming in India. *Farm Bull. No. 11*, Central Institute of Medicinal and Aromatic Plants, Lucknow, India.

Ingham, R.E., K.J. Merrifield and M.A. Morris. 1992. Effect of ethoprop on *Pratylenchus penetrans* and yield of mint in Oregon. *J. Nematol.* 24 : 599.

Inserra, R.N. and H.L. Rhoades, 1989. Some nematode problems of spearmint in Florida. *Fla. Dept. Agric. & Consumer Serv., Div. Plant Indust. - Nematol. Circ.* 167 : 4.

Jatala, P. and H.J. Jensen, 1974. Oxamyl controls *Longidorus elongatus* on peppermint in glasshouse experiments. *Plant Dis. Reptr.* 59 : 591-593.

Jensen, H.J. and C.E. Horner, 1956. A decline of peppermint caused by an ectoparasitic nematode, *Longidorus sylphus. Phytopathology* 46 : 637.

Jordan, E.M., D. De Waele and P.J. Rooyen. 1989. Endoparasitic nematodes in maize roots in the western Transwall as related to soil texture and rainfall. *J. Nematol* 21 : 356-360.

Kumar, S., B.R. Tyagi, J.R. Bahal, S.P.S. Khanuja, A.K. Shasany, R.S. Shukla, A. Sattar, D. Singh, A. Haseeb, V.P. Singh, P. Ram, K. Singh, S. Singh, S.P. Singh, N.K. Patra, M. Alam, A.A. Naqvi, M. Ram, K.K. Agarwal, and K. Singh (1997). Himalaya - a high menthol yielding hybrid clone of *Mentha arvensis. J. Med. & Aromatic Plant Sci.* 19 : 729-731.

Maqbool, M.A., S. Hashmi and A. Ghaffar, 1985. New host record of root knot nematodes (*Meloidogyne* spp.) in Pakistan and identification of physiological races. Pak. J. Nematol. 3 : 49- 52.

Marrifield, K.J. and R.E. Ingham, 1992. Effect of ethoprop on nematode population dynamics and root growth in Oregon peppermint. *J. Nematol.* 24 : 607.

Morgan, G.T. and A.A. MacLean, 1968. Influence of soil pH on an introduced population of *Pratylenchus penetrans. Nematologica* 14 : 311-312.

Nakasono, K., Z. Sano, M. Araki and Y. Itoh. 1990. Effects of *Meloidogyne incognita* on the growth of tomato plants in different soil types (2). *Japanese J. Nematol.* 20 : 37-43.

Norton, D.C. 1978. *Ecology of Plant parasitic Nematodes.* John Wiley & Sons. New York. 268 pp.

O'Bannon, J.H., G.S. Santo and A.P. Nyczepir. 1982. Host range of the Columbia root-knot nematode. *Plant Disease* 66 : 1045- 1048.

Pandey, R., A. Haseeb and A. Husain. 1992. Distribution, pathogenecity and management of *Meloidogyne incognita* on *Mentha arvensis* cv. MAS-1. *Afro Asian J. Nematol* : 2 : 27-34.

Pinkerton, J.N. 1984. Relationship of *Pratylenchus penetrans* (Cobb, 1917) population density and yield of peppermint, *Mentha piperita* L. *Dissertation Abst. Internat. B. (Sci. & Eng.)* 44 : 3593B.

Pinketon, J.N. and H.J. Jensen. 1983. Chemical Control of *Longidorus elongatus* on peppermint with non-volatile nematicides. *Plant Disease.* 67 : 201-203.

Pinkerton, J.N., H.J. Jensen, G.B. Newcomb and R.E. Ingham, 1988. Management of *Pratylenchus penetrans* damage to peppermint with selected nematicides. *Plant Disease* 72 : 167-170.

Rhoades, H.L. 1983. Effect of *Belonolaimus longicaudatus, Dolichodorus heterocephalus,* and *Pratylenchus scribneri* on the growth of spearmint, *M. spicata* in Florida. *Nematropica* 13 : 145-151.

Rhoades, H.L. 1984. Control of *pratylenchus scribneri* on spearmint, *Mentha spicata,* with non fumigant nematicides. Nematropica 14 : 85-89.

Sarah, J.L., B. Osseni and R. Hugan, 1991. Effect of soil pH on development of *Pratylenchus brachyurus* population in pineapple roots. *Nematropica* 21 : 211-216.

Sasser, J.N. 1979. Pathogenicity, host range and variability of *Meliodogyne.* In : F. Lamberti and C.E. Taylor (eds.) *Root knot Nematodes (Meloidogyne species) systemics, Biology and Control. pp. 256-268. Academic Press, New York.*

Sasser, J.N. and D.W. Freckman, 1987. A world perspective on Nematology : The role of the society. In : J.A. Veech and D.W. Dichson (eds.) *Vistas on Nematology,* pp. 7-14. Society of Nematologists, Inc. Hyattsville, Maryland.

Shane, W.W. and K.R. Barker, 1984. Effect of temperature, plant age, soil texture and *Meloidogyne incognita* on early growth of soybean. *J. Nematol.* 18 : 320-327.

Shukla, P.K. 1997. Studies on the nematode diseases of Mentha citrata, M. Piperita and M. spicata, and their management. Annual Report - A Council of Scientific and Industrial Research (CSIR), New Delhi Research Associateship Project (No. 9/112 (299)/97-EMR-I).

Shukla, P.K. 1998. Studies on the nematode diseases of *Mentha citrata, M. piperita* and *M. spicata,* and their management. Annual Report - A Council of Scientific and Industrial Research (CSIR), New Delhi Research Associateship Project (No. 9/112(299)/97-EMR-I).

Shukla, P.K. 1999. Studies on the nematode disease of *Mentha citrata, M. Piperita* and *M. spicata,* and their management. Annual Report - A Council of Scientific and Industrial Research (CSIR), New Delhi Research Associateship Project (No. 9/112(299)/97-EMR-I).

Shukla, P.K. and A. Haseeb, 1991. Studies on the vertical distribution of plant parasitic nematodes under *Mentha citrata* Ehrh. *Indian J. Plant Pathol.* 9 : 19-23.

Shukla, P.K. and A. Haseeb 1996. Effectiveness of some nematicides and oil cakes in the management of *Pratylenchus thornei* on *Mentha citrata, M. piperita* and *M. spicata.* BITE 57 : 307-309.

CHAPTER 5

PLANT PARASITIC NEMATODES — MANAGEMENT OPPORTUNITIES AND CHALLENGES

K. Sitaramaiah and P. Harinath Naidu

ECONOMIC IMPORTANCE

Plant parasitic nematodes infect almost all cultivated crop plants. They can directly cause plant disease or may act as predisposing agents to several fungal and bacterial diseases. They also act as vectors of many plant viruses. The degree of damage to a particular crop is influenced by the crop itself, the nematode species involved, the level of infestation and several environmental factors. The importance of plant parasitic nematodes as one of the limiting factors in crop production started receiving proper attention in Europe, North America and South East Asia including India, only after the volatile fumigants nematicides were introduced in 1940's and improved yields could be demonstrated by soil fumigation.

Almost all agricultural and horticultural crops are infected by one or more kinds of plant parasitic nematodes. The direct and indirect damage due to nematodes in crop production is more than indicated by present records. Several estimates have been made on crop losses caused by these pests. However, many of these estimates are often questioned for want of precise techniques and adequate survey information. A committee of the society of Nematologists (USA) estimated the crop losses due to nematode damage. The parameters considered in making these estimates were reduced yield, reduced market quality due to nematode damage, increased yield or improved quality or both after chemical treatment and cost of chemical control measures. In the United States alone it has been estimated that plant parasitic nematodes cause annual damage amounting to about 1.6 million dollars i.e., about 72 per cent of the total annual crop value based on 1967-68 price. In field crops losses amounting to 110 million dollars or 6 per cent, in fruits and nuts 12 per cent or 225 million dollars per year were estimated. The potato cyst nematode alone causes losses in potato yield to about 20 million dollars annually in the United Kingdom. In cotton, the United States losses about 53 million dollars annually. The burrowing nematode, *Radopholus similis* causes the spreading decline disease of citrus which is a destructive disease. In Florida (USA) it causes pepper yellows or slow wilt in Indonesia. Coconut, arecanut, pepper and banana are infected by this nematode in India also. Citrus nematode, *Tylenchulus semipenetrans* is of world wide occurrence and contributed to decline of citrus trees. Sugarbeet cyst nematode, *Heterodera schachti* is a major problem in sugarbeet growing regions of Europe and USA.

In India some attempts have been made to estimate the losses due to nematodes in important commercial crops. Annual losses to the extent of Rs. 40 million in wheat and Rs. 30 million in barley due to 'molya disease' caused by *Heterodera avenae* are considered common in Rajasthan alone. The ear cockle of wheat, *Anguina tritici* causes an annual loss of over Rs. 75 million even at low infestation levels of one per cent. Around Rs. 20 millions worth of loss in coffee due to root lesion nematode (*Pratylenchus coffeeae*) occurs annually. Potato seed imported from scotland and Netherlands brought the potato cyst nematode, *Globodera rostochiensis*. Nilgiri hills of Tamil Nadu where more than 7,000 acres out of the 25,000 acres under potato cultivation are heavily infected by this nematode. Perennial crops like citrus, coconut, arecanut, tea, coffee, and banana and annual crops like vegetables, tobacco, cotton, rice and sugarcane grown under intensive cultivation in irrigated areas are particularly liable to damage by parasitic nematodes. Tomato, brinjal, okra, potato, papaya, jute and groundnut suffer most. Infestation of root knot nematodes to the extent of 85 per cent has been observed in some areas under these crops. The losses due to root knot nematodes have been estimated from 25 to 50 cent depending upon the crops cultivated. Complete failure of vegetable crops is also not uncommon. The reniform nematode, *Rotylenchulus reniformis* has also been reported to cause heavy losses in vegetable crops, cotton and also in pulse crops. It has been estimated by Indian Agricultural Research Institute, New Delhi that about 10 per cent of crop loss is due to attack of plant parasitic nematodes. About 100 nematode diseases are considered for major economic significance in the world. According to Albert L Taylor (1967) in many developing countries nematodes are more serious and complex problem than in developed countries because most of the developing countries lie in the tropical regions where the climate is suitable for continued activity and reproduction of nematodes throughout the year. Due to socioeconomic conditions in these countries, neither there is money to undertake costly disinfestation measures nor follow suitable cropping practices. Most of the crops are sown in the same field year after year. The problem is increased due to lack of quarantine regulation to prevent the internal spread of the nematode infested planting material. Taylor (1967) further cautioned that developing countries must increase the availability of their agricultural products as rapidly as possible if the present decline in per capita agricultural production is to be reversed. Prevention or reduction of crop losses caused by nematodes is one way of increasing crop yields.

1. NEMATODE MANAGEMENT - OPPORTUNITIES

There are several methods of nematode management. Some of the most important methods are discussed here.

(a) Regulatory methods :
(Legislative measures)

Exclusion : This implies excluding the possibility of entry of a nematode pest in a region where it is not known at the time of regulatory operation but it is likely to establish in the absence of the same.

Quarantine : Aims at restricting human interference leading to spread of pests and diseases. In hundred years the golden nematode of potato has already spread to 40 countries. Plant quarantine legislative methods traditionally used to restrict the movement of live specimens/cultures infested/infected plant, seed, soil and their containers from one country to another (foreign or international quarantine) and from one region to another within a country (domestic quarantine) quarantines may also be exclusive (embargo against all entries of a plant/seed from a given area) or regulatory (entry permitted after inspection and certification at the points of orgin or destination). For more details of quarantine regulation the reader is referred to consult Das Gupta, 1998 and Singh and Sitaramaiah, 1993.

(b) International cooperation

The principle of exclusion is also the basis of many non- legislative measures against important nematode pests based on international cooperation or collaboration, intrastate cooperation or through voluntary participation and the community activity. International collaborative research is a pre-requisite to quarantine in respect of important plant parasitic nematodes. FAO sponsored an international plant protection convention in 1951. Under article 3 of this convention the plant protection agreement for South East Asia and Pacific was sponsored by FAO in 1956 and India became a party to it (Australia, Sri Lanka, U.K., Netherlands, Laos, Indonesia, Portugal, Vietnam and India are members). The agreement India signed made it obligatory on her part to do a number of things.

(a) official inspection of growing plant and nurseries.

(b) Inspection of consignment in international traffic

(c) Disinfestation and destruction

(d) Issue of phytosanitary certificate

(e) Establishment of a National quarantine organisation.

The main objective of this agreement is to take steps to prevent the introduction of pest and disease not present in the region but which may be introduced into it from areas outside it. The short comings and solutions of quarantine regulations have been given by ·Mathur in Singh and Sitaramaiah (1993).

2. PHYSICAL METHODS

Plant parasitic nematodes, can be killed by heat, desiccation, irradiation, high osmotic pressure etc., but it is more different to employ physical methods in killing nematodes in soil and planting materials. It is virtually impossible when they are in growing plants. Further, in such conditions, the nematodes are rather protected and less vulnerable. Field scale treatments as of soil are also difficult. Heat treatment is one of the oldest methods of nematode control. Since the enzymes are inactivated at short time exposure to about 50°C. Nematodes can be killed at 44°-48°C *in vitro*. Autoclaving or steam chamber treatment is not suitable for field use unless organizing for clone health certification. Bare root hot water treatment has to be specifically determined for different nematode species. Solarisation is a modern technique to be developed for each.

(a) Soil solarization

Soil solarization and solar pasturization results in the accumulation of heat by transmission of shortwave solar radiation but prevention of long wave radiation from soil, resulting in thermal inactivation at high temperature and production of heat shock proteins and irreversible heat injury. High soil moisture by intermittent irrigation improves thermal conductivity and the water vapours of high latent heat. Greater thermal sensitivity of hydrated organisms and antagonistic microorganisms have lower thermal sensitivity and faster colonization capacity enabling shift in favour of biocontrol. Due to physico-chemical and microbiological processes, some toxic gases and minerals are released and through increased mineralisation, some others provide nutrients to crops to induce greater tolerance to them. Prolonged exposure to high temperatures weakens the pathogens. Induced suppressiveness is of long term nature. Solarization results in 92-98% reduction in *M. incognita* populations at 0-30 cm depth. Soil solarization may be combined with other control methods and form a part of integrated nematode management. Both chemical and biocontrol methods are compatible with soil solarisation. Better control of *G. rostochiensis* and *Pasteuria penetrans* in pasturized soil was possible with soil solarization combined with application of farm yard manure or *Pongania glabrea* cake which resulted in greater kill of *M. incognita* and *R. reniformis* although no synergitic effects were there. Better survival of biocontrol agents was observed in pasturised soil. Most spectacular effects were obtained on the control of weeds including *Cyperus rotundus*, *Orbanche aegyptica*. Solarization is a modern technique to be developed for each crop nematode system in a given location.

(b) Summer ploughing and fallowing

Summer ploughing may help to reduce plant parasitic nematodes in the tropical countries. Ploughing exposes nematodes to solar heat and desiccation. Accentates effects of fallowing kills atleast a good part of the population of soil borne insect pests, pathogens and weeds, and fits well in agricultural practices adopted by farmers. Summer ploughing reduces nematode population to a great extent and including upto 42.4 per cent reduction in the population of *Heterodera avenae* and consequent yield increase by upto 97.5% of wheat in Rajasthan (Mathur *et al.*, 1987). Summer ploughing is also effective against *Heterodera cajani*.

Fallowing is the common practice of keeping the land free of all vegetation for a certain period by occasional ploughing of soil during hot and dry weather. This method ensures that the plant parasitic nematodes will have no host plants to feed and there will be no reproduction. For the absence of host plants the nematodes will die after reserved food in the nematode body is depleted (Sitaramaiah, 1993). For details the reader is to refer sitaramaiah (1993).

3. USE OF BIOLOGICAL AGENTS

Biological control of plant parasitic nematodes mainly deals with the application of biological agents for the reduction of parasitic nematode populations, either by introduction or by stimulation of natural enemies of the nematodes in the soil environment. Elaborating the definition given by Garrett (1965), Cook and Baker (1983) have defined biological control as "the reduction of inoculum or disease producing activity of a pathogen accomplished by or through one or more organisms other than man". This definition excludes man as a biological agent suppressing nematodes. Thus, biological control agents include not only parasites and predators of plant parasitic nematodes but also the host (resistant varieties) and other plants (trap crops, cover crops, and enemy or antagonistic plants). Antagonistic microorganisms mainly the parasites and predators of nematodes are discussed in detail (Singh and Sitaramaiah, 1993).

An antagonist, broadly defined is an opponent or adversary. In Biology, the term refers to a member of an interaction that interferes with another member. In biological control of plant pathogens, antagonists are biological agents with the potential to interfere in the life processe of plant pathogens (Cook and Baker, 1983). In entomology, such antagonists have been called "natural enemies" and the control achieved through them as "natural control". Antagonists include virutually all classes of organism and viruses, fungi, bacteria, nematodes, protozoa, seed plants etc. While antagonism has three facets, viz., antibiosis (production of antibiotics to halt the other organisms), competition (for food and other resources) and exploitation (parasitism and predation). In nematode control only the last has been considered of importance.

This aspect of nematode control has been reviewed by many workers from time to time, Boosalis and Mankau, (1965), Duddington (1960), Mankau (1980), Sayre (1971), Singh and Sitaramaiah (1970), Singh and Sitaramaiah (1973) and Webster (1972). Till the 1960's most of the attempts of biological control were only observational records on parasites and predators attacking nematodes. Detailed studies on the nature and behavioural aspects were intensified only after the 1960's. More recently attention has been diverted to the study of the influence of ecological factors on the activity of parasites and predators and also the mechanism(s) of activating them in the soil environment. The details of nematode destroying fungi : including nematode trapping fungi, endoparasitic (endozoic fungi), egg parasites were discussed in detail by Singh and Sitaramaiah (1993).

(a) Nematode control

Nematode destroying fungi have long been considered promising biological agents for control of nematodes. The predacious behaviour on agar plates was mainly responsible for detailed investigations. The predacious fungi are extremely common in all types of soils. However, technical problem in observing and assessing the efficacy of these organisms and also the complexities of soil habit made it difficult to study these fungi. Specialised techniques are required to isolate them from soil and some fungi are obligate parasites and others donot sporulate easily on media.

The destruction of nematodes by fungal parasites was recorded by several workers from different environments. Drechsler (1933), (1937) and (1941) added greatly to the list of nematode trapping fungi and to an understanding of their trapping mechanisms as biological control agents. Linford and Yap (1938) and Linford *et al.*, (1938) made exhaustive studies on predacious fungi in Hawaii against root-knot nematodes on pineapple. They used *Arthrobotrys oligospora, A. musiformis, Dactylaria thaumasia, Dactylella* spp. and *Dactylella ellipospora.* When they added chopped pineapple green tops to the pots to stimulate fungi they observed an increased population of free living nematodes as well as predacious fungi. Simultaneously root-knot nematode population fell down significantly. Intensive studies were conducted in France during Second World War. The use of several nematophagous fungi (viz., *A. oligospora, Dactylella ellipospora* and *D. bembiocodes*) was successfully demonstrated in reducing population of nematodes. During 1950-1960 a series of field experiments were conducted in England against cereal cyst nematode, *Heterodera avenae* and the golden nematode of potato *Globodera rostochiensis.* Several methods of inoculation with predacious fungi were tried with or without the addition of organic soil amendments. Duddington and Duthiot (1960) obtained statistically significant reduction in the mean number of larvae of *H. avenae* invading oat seedling by green manure with chopped cabbage leaves and showed that this was due to stimulation of the predacious fungi. The fact that green manure without the addition of fungi to the plots failed to decrease nematode invasion lead them to conclude that it was the fungus rather than the manure which was effective. Cooke (1968) has reviewed the ecological relationships of these fungi. He postulated that before predation can occur, mycelial growth and trap formation must occur. Both these process require energy which can be supplied by a readily available carbon source (Organic amendments). The factors that initiate trap formation of predacious fungi have been investigated. She has isolated a few valyl peptides which could effectively trigger trap formation. However, such compounds donot appear to elicit as strong and rapid response as do living nematodes. Many predacious fungi form traps spontaneously and consistently even in pure culture, while others produce traps very weak even in the presence of nematodes. Nordbring Hertz and Mattiasson (1979) found evidence for the presence of a lectin on the traps of *A. oligospora* which binds to a carbohydrate on the nematode cuticle and initiates an enzymatic alteration of the nematode surface leading to the penetration of the cuticle by hypha. According to Mankau (1980) an average fertile field soil contains approximately 10^9 bacteria, 10^5 to 10^6 actinomycetes, 10^5 to 10^6 fungi and 10^4 to 10^5 protozoa reproductive units per gram, plus a large number of fungivorus nematodes, tardigrades, collemboles, mites and other assorted micro and macrofauna. It is difficult to believe that adding one additional organism will have an immediately notable or even measurable effect. Therefore, it is apparent that predacious fungi are successful in many instances, but unsuccessful in other for control of plant parasitic nematodes. Since the fungi are susceptible to antagonism and fungistasis, their activity depends on certain specific conditions where in they become independent of these conditions and have a predominant position to exert their predacious role. Such conditions may not usually be available even if organic matter has been used as amendment because the decomposition process of the latter may be directed by many other edaphic factors (Singh and Sitaramaiah, 1993).

(b) Predacious nematodes

Predacious nematodes are one of the most important and effective natural enemies of plant parasitic nematodes. *Mononchids* and *Dorylaimids* and small *Seinura* spp. are effective nematode predators. The *Dorylaimids* such as *Discolaimus*, and *Actinoclaimus* pierce the body of nematodes and their eggs with their stylet and suck-out the body contents. The Dorylaimids are largely omnivorus and feed on fungi, algae, nematodes and also on soil microfauna particularly Enchytraeids. The Mononchids feed on nematodes rotifers, protozoans, algae and probably on some fungal spores. Diplogasteroids can be both microbial feeders and predators. Based on the feeding habits the predatory nematodes can be arranged into three major groups. (i) The nematode prey is swallowed as a whole and ingested in toto Ex. *Tripyla monhystera*. The stoma is usually unarmed and often very small. (ii) Some nematodes have open stoma and feed by puncturing or rasping a hole in the nematode prey Ex. *Mononchus papillatus* (iii) Nematodes like *Seinura christie* and *S. tenuicaudata* are equipped with stylet with which they feed on nematodes similar to feeding of plant parasitic nematodes. The nematodes insert their stylet and then withdraw the body contents. Prey preferences of predacious nematodes in water agar culture was also observed. The lesion nematodes *Pratylenchus penetrans*, *P. vulnus*, *Paratylenchus curvitatus* and *Meloidogyne* spp. were susceptible to predation by several Dorylaimids. Whereas *Hoplolaimus tylenchiformis*, *Belonolaimus longicaudatus* and *Criconemoides* spp. were fairly resistant to predation. Thorne (1927) observed that *Mononchus papillatus* fed most readily on the adult and juveniles of sugarbeet cyst nematode *Heterodera schachtii*. Despite the apparent success of predacious nematodes, they probably do not occur in the soil in sufficiently large numbers in the crucial early phase of build up to plant parasitic nematode population to be efficient natural regulators of nematodes. More research is needed on the factors affecting the population dynamics of predatory nematodes (Singh and Sitaramaiah, 1993).

(c) Viruses and other soil organisms

There is suspicion that plant parasitic nematodes to suffer from viral infections but conclusive proof of pathogenicity and presence of virus particles is lacking. One example is that *Meloidogyne incognita* in which the juveniles showing virus infection as sluggish in motion were able to transmit the disease to healthy ones when brought in contact with diseased specimens. After eliminating bacteria and fungi as possible causes it was concluded that filterates of suspension containing the sluggish nematodes contained as infective agent. In another case the phenomenon of swarming (congregation of nematodes in masses) was observed in *Tylenchorhynchus martini*. Electron microscopic studies of the cuticle of *T. martini* revealed abnormal and disruptive morphological changes in the cuticle layers that suggested a diseased condition. Virus like particles were observed in the hypodermis and muscle layers, digestive and reproductive systems and also on the surface of cuticle. Normal healthy nematodes which were not exposed to the virus exhibited intact tissue without any suspected virus particles. However, in these studies the virus has not been isolated or transmitted to healthy nematodes.

(d) Bacteria infection to nematodes

Bacteria have been detected within the body cavity, gut and gonads of plant parasites and in free living nematodes. There is doubt whether they were present as saprophytes or parasites. The nematode parasite *Bacillus penetrans* was first described as a protozoan *Duboscquia penetrans* from *Pratylenchus pratensis*. In addition to *P. pratensis* other nematodes *P. crenatus*, *P. penetrans*, *Rotylenchus robustus*, *Tylenchorynchus dubius* and *Meloidogyne arenaria* were also infected with this parasite. A big forward was taken by Prasad and Mankau (1969). They observed *D. penetrans* in a glass house populations of *M. javanica*. They recognised that it was not a microsporidian as evidenced by absence of a polar filament. Now it is recognised as *Bacillus penetrans* (Thorne), an obligate parasite of nematodes.

Spores adhere to second stage juveniles in soil and germinate after nematodes enter roots and initiate feeding. Germ tubes penetrate the cuticle, colonies develop in the pseudocoelom and ultimately some of the vegetative cells differentiate into spores which fill the body cavity. Infected nematodes do not reproduce. All stages of nematode development (*M. javanica*) were favoured by temperature. At 30°C the parasite proliferates extensively through females before they reach maturity. *B. penetrans* infects the larvae in the soil and also within the root galls. The infected females do not produce any eggs. The tomato root system containing large number of *Melodogyne* females parasitized by *B. penetrans* were obtained by inoculation. The roots were air dried and powdered. This preparation has potential as microbial nematicide. Since *B. penetrans* is an obligate parasite, the spores parasitize root- knot nematode entering roots. In green house tests, air dried soil infected with spores of the bacterium significantly reduced and demonstrated the use of *B. penetrans* in controlling *Melodogyne* spp. The major problem in utilising *B. penetrans* as a biological control agent is the inability to culture the bacterium *in vitro* on any of the standard bacteriological media. Neither vegetative cells nor spores of the organisms can be harvested in sufficient number to test extensively in the laboratory or to infest soils for large scale fields tests to determine its influence. Further research is needed relating to the survival and dispersal of *B. penetrans* in field soil. Intracellular bacterium like micro- organisms were also found in *Heterodera gottingiana*. *H. glycines* and *Globodera rostochiensis* (Singh and Sitaramaiah, 1993).

(e) Tardigrades

Tardigrades commonly called water bears are known to feed on several nematodes and other soil organimsm (Doncaster *et al.*, 1962). The tardigrade (*Hypsibius myrops* was cultured on *Panagrellus redivivus* in laboratory for several months and significantly decreased population of *M. incognita* and *Ditylenchus dipsaci* (Sayre, 1971).

(f) Protozoa

A large amoeboid proteomyzano organisms (*Theratromyxa weberi* frequently observed ingesting nematodes is considered to be of practical importance in the control of plant parasitic nematodes (Zwillemberg, 1953). This protozoa was seen engulfing larval stages of *G. rostochiensis*, *Melodogyne* spp. and *Pratylenchus* spp. *Urostyla* spp. another common predator has also been found on nematodes. Other protozoa probably have an incidental predatory relationship to nematodes (Cannings, 1973).

(g) Turbellarians

Turbellarians have shown better promise as nematode controlling agents. *Turbellarian, Adenoplea* spp. consume significantly high numbers of nematodes in certain soil types. Under natural conditions they are unlikely to be present in sufficient numbers to regulate the populations of plant parasitic nematodes (Syne and Powers, 1966).

(h) Collembola

Most collembola feed on decaying plant material. *Onychiurus armatus, Isotoma viridis, Hypogastrura* spp. are active predators of nematodes. In one experiment with 1:5 ratio of *O. ornatus* and *Heterodera crucifera* cysts about 6.9 per cent of the cysts were damaged by the collembola.

(i) Mites

Linford and Oliveira (1938) recorded six mites, five *parasitinae* and one *Rhizoglyphus* as possible enemies of nematodes. An orbatid mite *Pergalumna* spp. was reported to eat *Tylenchorhynchus martini* or *T. dubius* in large numbers. A predacious mite *Lasioscies scapulatus* was isolated from soil around citrus roots. The mite voraciously feed on a variety of nematodes, had a rapid life cycle of 8-10 days and reproduce asexually (Singh and Sitaramaiah, 1970).

(j) Enchytraeids

Enchytraeids worms *Fridericia* and *Enchytraeus* were found to feed on sugarbeet cyst nematode *Heterodera schachtii* and decreased its population in pot experiment (Schaeffenberg and Tendl, 1951).

Remarks

The above account of the activity of natural enemies does prove the effective suppression of nematode populations. Nevertheless, it is also evident that so far no promising biological agent has been developed for field application. Eventhough predators and parasites of nematodes are potential means of destroying plant parasitic nematodes in soil, the feasibility of their use on field scale seems remote. The complex nature of soil environment, intense antagonism among soil organisms and lack of knowledge of ecological behaviour of these agents are the main limitations. Even if under laboratory conditions some of these agents can be shown to have profound effect on nematode activity, it will be very costly method to prepare large quantities of their inoculum for use on a field scale.

In biological control the most desirable characters of natural enemies are : (a) mobility and the ability to searchout its prey, (b) adaptability to the environment (c) host specificity, (d) synchronization with the host and (e) ability to survive host free periods. The studies of nematode parasites and predators are inconclusive and inadequate with regard to some of these basic characters. Due to lack of information on these aspects the biological agents have not been used properly on a large scale. The use of biological agent require a thorough knowledge of their biology and ecology. The search for effective biological agents should continue for their practical use. An integration of natural control, already operating under certain conditions, when supplemented with a biological control agents may yield acceptable results (Singh and Sitaramaiah, 1993).

4. PESTICIDES AND THEIR USE

Nematicides have been in large scale use since 1945 in Europe and North America and their use has been increasing every year. The use of fumigant or non fumigant nematicides is always restricted to cash crops with high economic returns. The chemicals are also used in nursery beds to obtain nematode free plants and to demonstrate the nematode damage and extent of losses by comparing the yields of treated and untreated plants. Based on large free supply of shell DD, the Indian Council of Agricultural Research (ICAR) approved a scheme which involved treatments of number of crops suspected to be suffering from unexplained causes.

The fumigant nematicides are highly injurious to seeds or plants if they are planted immediately after soil fumigation. Discussed theoretical consideration of soil fumigations. Contact and systemic nematicides are relatively less toxic and they can be applied at any time of planting. They are effective at much lower dosages. The systemic nematicides are also effective against insects, mites as well but not against soil borne fungi. The systemic granular pesticides introduced in 1970's provided very effective chemicals for nematode control and had replaced the fumigant nematicides in most of the countries (Sitaramaiah, 1993).

There is plenty of information in developed countries on the impact of soil fumigation on food production through control of nematodes. Improved plant growth in fumigated soil is the general effect observed in most crops. The increase in yields is reported to be as high as 87 per cent. Although it can be expected that nematode control is the major factor, involved in this increase, there are many other associated factors that account for the yield increase. Fumigation reduces the losses from diseases in which nematodes and other plant pathogens are jointly involved in disease complex (Sitaramaiah and Pathrak, 1992). Biologically bound nutrients are released after the effect of fumigation on soil microflora and fauna and are utilised by the plants. Plants in pesticide treated plots show distinctly improved and uniform growth with extensive root system and feeder roots. The increased root system enhances uptake of nutrients and water resulting in better yields. The healthy plants growing in treated plots are also less susceptible to other plant diseases and can tolerate adverse growing conditions better (Sitaramaiah, 1993).

(a) Side effects of pesticides

All the soil fumigant nematicides are general biocides. Halogenated compounds can temporarily upset normal soil nitrification process which results in increased ammonia accumulation with the result that plants may suffer from ammonia toxicity in nitrate starvation. Brominated compounds like methyl bromide or Ethylene Dibromide have residue problem. Bromine residues have been found in a number of crops like citrus, carrots, potatoes, groundnut and tomato following fumigation with these nematicides. Increased uptake of bromide or chlorine by some plants has been reported. Increased halogen content in tobacco leaves reduces the leaf quality and adversely affects the burning quality of tobacco grown in fumigated fields. Nematicides with fungicidal properties can adversely affect nematophagous fungi and other fungal parasites and predators of nematodes. Normal infection and development of mycorrhizal fungi are suppressed by nematicides having fungicidal properties. Citrus seedlings grown in fumigated soil are commonly stunted and chlorotic. Heavy fertilization partially overcome stunting. For many years, the problem was attributed to soil toxicity developed by fumigation later showed that stunting was caused by inadequate nutrition due to destruction of vesicular arbuscular mycorrhizal fungi. Inoculation of fumigated soil with VA mycorrhizae partially solved stunting problem and reduced need for frequent fertilization.

Little information is available on the long term ecological effects of frequent and repeated use of soil fumigation in cultivation fields. Investigations at the University of California in the USA indicated that Ethylene Dibromide (EDB) Dibromochloropropane (DBCP) and Dichloropropane and Dichloropropene (DD) mixtures are not persistent in soil and have no adverse effects on the physical and biological properties of the soil, on the environment and on the nutritional value of the crop grown in fumigated fields. The fumigants are quickly degraded by normal biological activities in the soil (Singh and Sitaramaiah, 1994).

Plant growth regulators, phenolics, aromatic acids, aminoacids and fatty acids have been tested either as foliage treatment, soil drench or as bare root dip treatmen to reduce nematode infestation of plant. These chemicals though effective are still in experimental stage and no commercial formulation have been proposed/developed for practical use (Sitaramaiah and Pathrak, 1979, 1981, Sitaramaiah and Prasadji, 1989). (Sitaramaiah and Singh, 1978) and (Prasadjee and Sitaramaiah, 1992).

5. APPLICATION OF ORGANIC AMENDMENTS AND FERTILIZERS

Organic matter is an important component of all soils. The value of decomposition of organic amendments in soil in the reduction of nematode damage was first demonstrated by Linford and his associates, since then a large number of reports have been published showing that addition to a variety of organic materials to nematode infested soils results in definite reduction of several plant parasitic nematode populations. The organic amendments used include manures, green crop residue and green leaf manures, cellulosic soil amendments, mature crop residues, oil cakes, oils and sugars (Singh and Sitaramaiah, 1970, 1973), Sitaramaiah (1987), Sitaramaiah (1990) and Webster (1972).

(a) Organic manures

A brief review of the effects of manuring on population of nematodes in soil was given. Addition of farm yard manure and compost to nematode infested soil reduced the infestation level of *Globodera rostochiensis* in potato roots. These manures hampered the development of nematodes in plant roots, where they developed more slowly than in plants grown in unmanured plots. Obviously there was some type of biochemical resistance imparted to the plants by nutrients from these manures. Deep ploughing during hot summer months followed by application of farm yard manure at planting spots increased the yield of tomato in root- knot infested soil. Reduction in citrus nematode populations by amendments with sheep manure and chicken manure has also been reported. Organic nitrogen in the form of sewage sludge was more effective than ammonium nitrate for reducing population of *Belanolaimus longicaudatus* and *Hypsoperine graminis* in turf grass.

(b) Green manuring

Green manuring is a conventional practice in which green plants are ploughed into the soil to rot and provide nutrients for the succeeding crop. The organic matter, thus applied during the decomposition brings about important physical, chemical and biological changes in the microsites where microflora and microfauna are active. Linford and his associates (1938) had as early as 1938 demonstrated that the incidence of root-knot could be significantly reduced by incorporating into soil chopped pineapple leaves. In India, Patel and Desai (1964) also found that green manuring in root knot incidence. Chopped leaves of Karanji (*Pongamia glabra*) and *Crotalaria juncea* when applied to root - knot infested soil reduced the disease severity. The decomposition of green leaves of *Azadirachta indica, Melia azadirachta, Sesbania aculeata* to infested soil was also very effective (Singh and Sitaramaiah, 1967, 1970).

(c) Cellulosic soil amendments

Excellent success in the control of plant parasitic nematodes has been obtained with the use of cellulosic soil amendments, Singh and Sitaramaiah (1973) have reported field reduction of root knot nematodes of okra and tomata by amending the nematode infested soil with saw dust supplemented with nitrogenous fertilizers. Supplemental nitrogen in saw dust amend soil caused several fold increase in yield. Reduction of citrus nematode, *Tylenchulus semipenetrans* in cellulose amended soil was attributed to increase in the number of microbivorus and fungivorus nematodes.

(d) Mature crop residues

Application of mature crop residues such as dried straw is not a conventional manuring practice. However, a substantial quantity of the dry crop debris is ploughed into the soil at the end of every cropping season. According to Johnson (1963) every type of dried crop residue introduced into the nematode infested soil reduced root-knot severity and the greatest control was obtained when the crop residues were allowed to decompose for atleast 30 days. The dried crop residues of alfalfa, oats, flax, rye straw, timothy hay, buckwheat bull, cocoa, bean hull, lucerne hay, peanut hull were all effective in reducing several kinds of plants parasitic nematodes (Sitaramaiah, 1970).

(e) Oil cakes

Fertilization of soil with oil cakes is a common and routine practice in many parts of the world. Lear (1959) reported a significant reduction in population of plant parasitic nematodes when large quantities of castor been pomace were mixed with soil. Populations of lesion nematodes are reduced by amendments with soybean and corn meal, However, soybean meal had no effect in sterilized soil or without microorganisms indicating that microbial degradation of the meal was essential for bringing out nematode suppression. The microbial decomposition of nitrogenous substances especially the process and decrease in nematode population. Oil cakes of margosa, groundnut, castor linseed, mustard, mahuva are effective on a field scale in controlling vegetable root knot *M. javanica* and *M. incognita*. From a study conducted over several years it was found that the soils amended with different oil cakes reduced root knot galling on a number of vegetable crops and also the soil population of *Meloidogyne* spp. *Tylenchorhynchus* spp. *Hoplolaimus* spp. *Helicotylenchus* spp. *Rotylenchulus reniformis*. Many of the oil cakes used in these studies are common animal feeds and therefore are not available as organic fertilizers and only a few can be used for reducing the losses due to nematodes. Oil cakes of Karanji and Margosa have these potentialities and these oil cakes also proved to be economical.

(f) Oil and Sugars

Use of oils and carbohydrates such as sugar, dextrose etc., although unusal soil amendments, have been shown to be effective against several plant parasitic nematodes. Adverse effect of black mustard oil containing allyl isothiocyanate on potato cyst nematode was demonstrated by Ellen by (1945). In field experiments application of 0.1 cwt oil/acre increased potato yields by 100 per cent and decreased the emergence of larvae from cysts. Spraying of 1.5 per cent water soluble hydrogenated fish oil on surface of soil reduced the population of *Radopholus similis*. in citrus roots. Several vegetable oils such as corn oil, cotton seed oil, groundnut oil and soybean oil when applied at one per cent (w/w) reduced the population of lesion nematodes corn oil being most effective. Osmotic destruction of nematodes in the presence of high concentrations of sugar in soil and solution has been reported by several workers. As little as 1000 ppm of dextrose was nematicidal. Use of sugar containing substances such as sugarcane bagasse and molases with 0.2 per cent sucrose concentration was also effective giving 50 per cent larval mortality of *M. javanica* in about nine days. Most of these investigators utilised large quantities of different organic materials and only few materials can be used for the control of plant parasitic nematode. Different explanations of the possible mechanisms involved in the control of nematodes are found in these reports (Sitaramaiah, 1990).

(g) Mechanism of control

The value of organic amendments of soil in reducing the damage caused by plant parasitic nematodes has been proved in temperate subtropical and tropical areas. Due to intensive microbial activity during decomposition of organic matter, the amendments will prove useful in tropical regions. The effect of these organic amendments will vary from soil to soil and from region to region. There is a need for better understanding of the mechanisms involved in this type of control to guarantee its success. The available literature suggests the following mechanisms.

Parasitism and predation of plant parasitic nematodes by nematophagous fungi, bacteria and soil microfauna :

The role of nematode trapping fungi was considered as one of the mechanisms of reduction of plant parasitic nematodes in organic manures amended soils. The approach has usually been to inoculate the nematode infested soil with a predacious fungus and then supplementing the soil with different organic materials in order to stimulate the fungal growth and trapping. Negative results indicating failure of organic amendments and predacious fungi in controlling plant parasitic nematodes have also been published. These failures are mainly due to antagonistic soil microflora and also soil fungistatic against introduced nematode trapping fungi. Addition of organic matter to the soil temporarily suspends the soil fungistatsis. In addition of organic matter to the soil, the nematodes trapping fungi are first stimulated into saprophytic competition with other organisms. Only after intense competition and simultaneous builp up of nematodes population, nematophagous fungi begin to feed on nematodes. Easily decomposible amendments increased the activity of nematophagous fungi. The information so far obtained on the activity of natural enemies of nematodes to prove that the natural antagonists destroy plant parasitic nematodes in amended soils. It is also evident that so far so promising method has been developed to use the nematophagous fungi for controlling plant parasitic nematodes on a field scale. This was due to complex nature of soil, intense antagonism among soil organisms and lack of information on ecological behaviour of the nematophagous fungi. The efficacy of nematode destroying fungal parasites is limited by lack of mobility and inability to seek put the prey or host and low population densities which occur in most of the cultivated soils. Predacious fungi tested do not colonise and exploit soil micro habitat. These fungi are considered to be poor competitive saprophytes and are susceptible to antagonism from other soil organisms. Nevertheless, there are conclusive evidence to show that nematode control on a field scale has been possible through organic amendment of soil without evidence of increased parasitsm and predation. This points to the possibility that the control achieved is dependent on many other factors in addition to parasitism and predation. Some of the possibilities that have been explored are described here.

(h) Toxicity of decomposition products

In recent years the role of toxic decomposition products in the suppression of plant parasitic nematodes in soil has been emphasised by many workers. The microbial activity in amended soil leads to release of a wide variety of chemically different substances which may be directly toxic to nematodes present in the vicinity of the decomposing residues. The metabolites of the microorganisms active during decomposition of organic matter may also be toxic to plant parasitic nematodes.

(i) Osmotic concentration of soil solutes and its effect on nematodes

Use of sugar, dextrose and other carbohydrates although unusual soil amendments have been shown to be effective against many plant parasitic nematodes. Osmotic destruction of nematodes in the presence of high concentration of sugar in soil and solution was attributed for the rapid reduction of plant parasitic nematodes.

(ii) Alteration of physical and chemical conditions in amended soils

Addition of decomposable organic matter to cultivated soils affects soils physical and chemical conditions. The increased aeration and solubilization of nutrients results in better root growth. The adverse effect on are survival of citrus nematode was attributed to high nitrate levels and a decrease in soil pH. The effectiveness of cereal straw decomposition was increased by the addition of ammonium nitrate and super phosphate or potash. It was suggested that inorganic nitrogen hastens the rate and completeness of decomposition of straw and consequent release of substances toxic to nematodes. Crop residue decomposition products are also known to influence physiology of the plant and thereby its growth. It seems probable, that in amended soils favourable conditions for rapid root growth, development and replacement of diseased roots with new ones, and better and prolonged utilisation of nutrients by the enhanced root system helps in marking the effect of nematode injury to plants (Singh and Sitaramaiah, 1973).

It appears from the information available that nematode suppression by organic amendments of soil may not be due to one mechanism but a combination of several mechanisms including that of host nutrition and vigour. There is absolutely no doubt that these amendments increase yields of plants in nematode infested soils whatever the mechanisms may be involved. Discussed management of the antagonistic potential in agricultural ecosystems for the biological control of plant parasitic nematodes with the help of rhizobacteria, mycorrhizal fungi and fungal antagonists of nematode eggs along with organic amendments of soil. He described multitude of management tools that are available to nematologists and phytopathologists who are working in the soil ecosystem for use in manipulating or controlling the behaviour of nematode antagonists. He emphasized that multidisciplinary research is drastically needed to develop alternative strategies to control plant parasitic nematodes.

6. RESISTANT VARIETIES

Soil treatment with chemical and employing available cultural practices are common methods of nematode control. However, in nematode control programme, these methods should not be the first choice for consideration. The most effective and economic method is the use of resistant cultivars. Growing resistant varieties of crops in crop rotation is economical for the farmer in reducing nematode populations gradually in the infested fields. According to Oostenbrink (1960) the cultivations of a resistant variety may suppress nematodes population from 10 to 50 per cent of its harmful density, with reduced multiplication and population density through successive cultivation of resistant varieties for some time, it is possible to grow even a susceptible cultivar at frequent intervals without extra cost on use of nematicides. The use of nematicides can be directed to high value cash crops or where the farmer wants to grow a particular crop continuously in the same field year after year. The use of nematode resistant cultivars and chemical soil treatment should be considered complimentary methods of control and not as competitive measures.

Considerable progress has been made in the U.K., Europe and the U.S.A. in breeding crop varieties resistant to specific sedentary and migratory endoparasitic nematodes. Resistant varieties have been developed in cotton, cowpea, tobacco, bean, soybean, tomato, grapes, and peaches against the root-knot nematodes, in alfalfa for resistance to *Ditylenchus dipsaci*, in potatoes against the cyst nematode and in citrus for resistance to *Tylenchulus semipenetrans*.

Nematode resistance has been discovered in both cultivated and wild plants. In most of the economically important crops like potato, tomato, barley and oats, resistant genes to several endoparasitic nematodes have been identified. Resistance to the potato cyst nematode has been found in *Solanum vernei, Somullidissectum, S. oplocense, S. tuberosum* s.sp *andigena*. These studies were conducted extensively in the U.K. and the Netherlands. Several genes for resistance (R. genes) to many pathotypes of *Globodera rostochiensis* have been found and used for developing commercial cultivars. Resistance to root-knot nematode have been found in *Lycopersicon L. pimpinellifolium*. Gilbert and McGuire (1952 & 1956) studied the inheritance of resistance in tomato to root-knot and found the presence of a single dominant resistant gene Mi which was effective against all species of *Meloidogyne* except *M. hapla*. Subsequently two dominant and one recessive genes for resistance were designated on L Mi, R1 LMi R_2 (cultivars Nematex and Small Ery-1) and L.MiR_3 R_3 (cultivar cold set-1).

High resistance to the cereal cyst nematode *H. avenae* was found in species of *Avena* (*A. sterilis, A. byzantina*). These species possess resistance to several pathotypes of *H. avenae* and were effectively utilized to develop resistance in cultivated oats (*A. sativa*). Resistance in barley against this nematode has been detected in barley varieties primers, svanhals, chevalier, etc. At present there are atleast six resistance genes identified in various barley cultivars which have been designated differently in different countries. High resistance to *Heterodera schachtii* has been found in *Beta petellaris. B. procumbens* and *B. webbiania*. These sources of resistance are useful for incorporation of resistance in the cultivated species (Singh and Sitaramaiah, 1994).

Types and measure of resistance have been described by Sitaramaiah (1989). The resistance in plants to nematodes can be due to hosts reaction to invasion, where in the host does not allow development of Juvenilies to adult egg laying stages. These aspects were described in detail.

7. FUTURE CHALLENGES

Biotechnology

Incorporation of nematode resistance :

Embryo rescue technique

Embryo rescue technique involves hybridization between economic plant species and wild relatives followed by excision and culturing of hybrid zygotic embryo on nutrient media. One of the classical examples of successful application of embryo rescue techniques in plant nematology is the synthesis of tomato cultivar resistant to root-knot nematodes. Cultivated tomato species, *Lycopersicon esculentum* is highly susceptible to attack by *Meloidogyne* spp. but the related wild species *L. peruvianum* has a high degree of resistance against the nematodes. A cross between these two species yielded a hybrid in which case the endosperm does not develop and the embryo aborts. This problem was solved by carefully isolating and culturing the hybrid embryo on tissue culture media (Smith, 1944). This technique thus provides a mechanism to transfer nematode resistance gene from *L. peruvianum* into cultivated species of tomato, *L. esculentum*.

Protoplast Fusion/Somatic Hybridization

Advances in protoplast technology provide a mechanism for overcoming intergeneric and interspecific barriers to hybridization and for introducing new genetic information into plant cells without sexual reproduction. Another advantage of protoplast fusion is that a set of genes that is responsible for specific useful traits can be transferred without knowing the molecular details of genes that confer the traits.

In somatic hybridization, plant tissues

from two parental sources are digested with a mixture of pectinase and cellulase in an osmotically stable solution to produce protoplasts. Protoplasts are marked cells surrounded by a plasma membrane and have the capacity to fuse spontaneously or induced to fuse by way of chemicals or electrofusion. Chemical fusion is a method of inducing cell fusion by use of high concentration of various chemical fusogens such as polyethylene glycol., dextran and polyvinyl alcohol. These fusogens cause cell plasma membrane to adhre to those of any surrounding cells. On the other hand, electrofusion employs high DC electric field pulse(s to induce fusion to cell (protoplast) pairs that are formed by an AC field induced dielectrophoresis or other methods. The fused protoplasts are capable of cell wall regneration. After cell division a single nucleus may be formed containing the chromosomes of each parent cell or more frequently, selective elimination of chromosomes of one of the parental species takes palce during cell division and tissue formation (Gaur and Sirohi, 1998).

This technique of protoplast fusion has been used as a means of transferring root-knot nematode resistance factors from *Solanum sisymbrifolium* into egg plant *S. melongena*, S_1 and S_2 progenies of *S. sisymbrifolium* plant regenerated from callus host resistance to *Meloidogyne javanica* but retained resistance to *M. incognita* (Fassuliotis and Bhat. 1982). Tomato plants cloned from leaf discs of root-knot nematode resistant "Patriot" were completely susceptible to *M. incognita* while sections of stem and leaves rooted in sand in absence of hormones retained resistance. Changes in resistance persisted for 3 generations. It is postulated that exogenous hormone constituents of culture medium modify the expression of genetic resistance.

Regeneration of plants via callus phase can be used to introduce courses of resistance in to useful germplasm, *Larkin et al.*, (1990), studied the introgression of resistance to the cereal cyst nematodes. *H. avenae* from rye into wheat. A line of triticale with high level of resistance to cereal cyst nematode was identified (Fisher, 1982) and this was subsequently shown to be located on the long arm of rye chromosome 6R (Asiedu *et al.*, 1990).

An insertion generates a hybrid or chimeric plasmid or phage consisting in part of the authentic DNA of original genome and in part of the additional "foreign sequence".

Application of genetic engineering made possible the synthesis of transgenic plants. Transgenic plant is one in which foreign DNA (gene a transgene) is stably incorporated in the genome early in embryonic development and so is present in both somatic and germ cells, is expressed in one or more tissues, and is inherited by offspring in a Mendelian fashion. The aim of gene transferring technology is to improve a top variety in one additional character without disturbing the rest of the genome. For gene transfer, three methods are available; vector mediated gene transfer, microinjection by capillaries, laser microbeams or microprojectiles; and direct transfer.

8. APPLICATION OF RECOMBINANT DNA TECHNOLOGY IN NEMATOLOGY

The incorporation of resistance genes into important crop plants epitomizes effective biological disease control. Resistance is heritable and, therefore, inexpensive and permanently available once introduced. Also, genetically detemined resistance is generally effective when plants are grown in a wide range of climates, soil types and cropping regimes. However, problems have been ecountered in obtaining resistance sources against a particular pest/pathogen and emergence of resistance breaking pathogen races or biotype. These problems can be overcome by molecular cloning of numerous resistant genes from many different plants that are targetted against several pests and pathogens (Keen *et al.*, 1993). Then it would be possible to transform various combinations of these genes into susceptible cultivars including species entirely unrelated to the plants from which the resistant genes were originally isolated. This approach would permit the use of previously unavailable resistance specifications in particular crop plants and should lower the frequency with which newly virulent pathogens races become predominant.

Genetic host resistance is the most cost effective and environmentally sound method for management of plant parasitic nematodes. Resistance to thc various species of root-knot nematodes (Fassuliotis, 1979) and cyst nematodes (Cook and Evans. 1987) is available in limited number of host species. Resistant cultivars on infection exhibit activation of number of inducible responses that are thought to be disease resistance mechanism (Raja and Dasgupta, 1986; Sirohi and Dasgupta, 1993, a.b; Bowles, 1990). Hypersensitive response (HR) is one such response in which there is a rapid localized necrosis of cells at the feeding site initiation limiting the parasitic ability of the nematode. It is known that many plant genes are up-regulated during a resistance reaction, including genes encoding proteins such as glucanases, Chitinases and other enzymes commonly referred to as pathogenesis related proteins (Boles, 1990).

Field trials of a transgenic root-knot nematode resistant tobacco plant having tobacco root specific gene Tob RB7 promoter nematode responsive element showed substantial reduction in galling and egg production (Opperman and Conkling, 1995). Tomato plants carrying Mi gene show resistance to *M. arenaria, M. javanica. M. incognita* but not to *M. hapla* (Guilbert and McGuire, 1956). Structural protein genes related to the extensis family have been isolated from root-knot nematode infected tomato roots in a screen designed to identify upregulated genes (Van der Eycken *et al.*, 1992) . One gene (PMRI) with unkwon function has been isolated from potato roots undergoing infection by *G. rostochiensis* (Gurr *et al.*, 1991). For the most past however, few plant genes up-regulated during nematodes infection have thus been characterised.

Nematodes resistant transgenic plants can be designed by various approaches using the available information and knowledge. The simplest would be to introduce a nematode toxin gene under a constitutive promoter. This strategy has been adopted to design transgenic plants carrying the BT gene that are resistant to insect pests (Barton *et al.*, 1987). If a suitable toxin gene could be identified, this very approach can be used to construct nematicidal transgenic plants. However, a number of toxins known to be active against vertebrates also act upon nematodes but are undesirable due to the broad spectrum of activity they exhibit (Opperman and Conkling, 1994). Toxins identified, from invertebrates need to be tested against nematodes like Txp-1 toxin from female mites of the species (*Pyemotes tritici* is extremely potent against wide range of insect species causing immediate paralysis upon exposure to very low doses (Tomalski and Miller, 1991).

A lot of information is being generated with respect to identification of resistance gene local against various nematode pests in different crops of importance. Particular interest has been shown in the chromosomal location of resistance to *G. rostochiensis* in potato (Pireda *et al.*, 1993).

H. schachtii in beet (Jung *et al.*, 1992) and root - knot nematode, *Meloidogyne* spp. (Klein Lankhorst *et al.*, 1991; Messeguer *et al.*, 1991). The identification and cloning of such genes would facilitate their transfer and expression in transgenic crops. In the case of potato, a major dominant locus Gro 1 conferring resistance to some pathotypes of *G. rostochiensis* was mapped by restriction length polymorphism (RELP) to potato linkage group IX; the nearest neighbouring RELP markers are CP 51 and TG 20 (a) (Barone *et al.*, 1990). Other groups are developing high resolution RELP maps around ther Mi genes in tomato. The Mi genes originated from wild tomato species *Lycopersicon peruvianum* and confers resistance to all major species of *Meloidogyne*. The Mi gene is located at position 44 on chromosome 6 and is closely linked to the leaf calyx markers of yv (yellow Virescent) and acid phosphatase -1 (Aps-1). The work of Messequer *et al.*, (1991) led to the production of a more detailed consensus linkage map of chromosome 6. Four co-dominant RELP markers (TG 297, CD 14, Tom 25 and GP 79) can be used as markers for selection of the Mi gene. The extent of *L. peruvianum* DNA segment in different Mi gene containing cultivar was also determined. Three groups of cultivars (ABC) with different amounts of *L. peruvianum DNA* were found. Only in the cultivars of group A there is linkage between Mi and Aps-1. For cultivar of group B the marker Gp 79 can be used and new markers are needed .for nematode resistance in group C cultivars.

Much of the research for possible use of transgenic plants to control plant parasitic nematodes has been concentrated to three genera, *Meloidogyne*, *Heterodera* and *Globodera*. Although feeding sites of root-knot nematode (giant cell complexes) differ in many respects from those of cyst forming nematodes (Syncytia), the end function remains the same i.e., to mediate the effective transfer of the nutrients from the plants to the feeding nematodes. The active involvement of the hosts own genes in the production of giant cells or syncytia has been proposed. The molecular cloning of host genes and promoters involved in the formation of feeding sites could facilitate the production of antisense constructs to inhibit expression of key genes. The antisense RNA is thought to base pair with its target RNA forming double strandad RNA. This duplex formation may inhibit mRNA maturation and/or translation or it may lead to rapid mRNA degradation. Antisense RNA has now been used to regulate the expression of numerous plant genes including chalcone synthetase, polygalacturonase and phosphenotricin acetyl trnasferase. In principle, careful selection and down regulation of the most appropriate host gene or genes may prevent the normal development of nematode feeding sites while minimizing adverse effects on the plants.

The expression of specific intracellular antibodies in plants could confer resistance to particular plant pathogen including plant parasitic nematodes by binding proteins that are essential for disease process. The expression of antibodies that are specific to nematode esophageal gland secretion may be alternative route to engineering resistance. Monoclonal antibodies specific to oesophageal gland products from *M. incognita* and *H. glycines* have been produced. Antibodies such as these will aid out understanding of the nature and role of glands and their products in specific host parasite relationship and expression of these antibodies in transgenic plants may also interfere with the development of feeding sites or disrupt normal feeding processes.

9. TRANSGENIC PATENTS

Many patents have already been made in the area of transgenic plants resistant to nematodes. A transgenic potato with a novel nucleic acid molecule capable to increasing the resistance of a plant to nematode infection of a plant comprising a nucleotide sequence encoding a polypeptide with a collagenase (EC-3, 4.24.3) activity or its precursor has been patented by Stratford and Shields (1995). A lectin from garlic has been patented to confer resistance against wide array of nematode pests (Geoghegan *et al.*, 1995). Similarly, protease from *Paecilomyces lilacinus* (Patented information) can be used to construct nematode resistant transgenic plants (Den-Belder *et al.*, 1994). A transgenic plant with nematode resistance may be prepared by introduction of an RNA ase gene under control of a promoter which induces RNA production in response to nematode attack of root tissue, resulting in prevention of formation of nematode feeding sites. This strategy has been patented by Monsanto, USA for control of plant pathogenic fungi.

10. CONCLUSION

The developments in biotechnology, especially in the area of molecular techniques have shown their immense potential in fundamental and applied nematology. Some of these advances have been imbibed, although yet on a limited experimental stage, for characterization/ differentiation of otherwise difficult to distinguish nematode taxa and to incorporate resistance into plants against economically important parasitic nematodes. The embryo resuce technique has been used to incorporate resistance to wild plant species into cultivated species, such as from tomato, *L. peruvianum* to the cultivated *L. esculentum.* Similarly, recombinant DNA technology has found use in the isolation and incorporation of nematode resistance genes to produce transgenic plants. These techniques have the potential of providing environment friendly and economically desirable plant types to overcome the nuisance caused by nematodes and other pests/pathogenes. However, the research costs and possible risks of undesirable effects on humans and other non-target organisms, whosoever remote, are still the limitations that must be tackled.

11. INTEGRATED APPROACH

Integrated nematode management (INM) includes the research, development, technology transfer, and implementation needed to integrate two or more control measures to manage one or more nematode species (Bird, 1981). The aim of INM is to reduce and stabilize the nematode pest population below damaging levels, resulting in favourable long term socio-economic and environmental monitoring and environmental consequences. The components of INM systems include (1) biological and (2) environmental monitoring (3) agricultural production systems (APS) and pest crop ecosystem models (4) system design and (5) implementation. The proceedings available for INM system design and implementation are based on the principles of exclusion, population reductions and tolerance (Bird, 1981). He discussed in detail the management of plant parasitic nematodes by integrated approach (i.e.,) problem detection, sampling, production system analysis and nematode management optimization and INM case descriptions with *Pratylenchus thornei* on wheat in Mexico, *P. coffea* and mushroom nematodes *Aphelenchoides* spp. in India.

REFERENCES

Asiedu, R., Fisher, J.M. and Driscoll, C.J. 1990. Resistance to *Heterodera quenal* in the ryegenome of triticale TAG 79 : 331-336.

Barone, A., Ritter, E., Schnachtschabel. U., Debner, T., Salamini, I. and C. Gibhardt, 1990. Localization by restriction fragment length polymorphism mapping of a major dominant gene occurring resistance to the potato cyst nematode. *Globodera rostochiensis*. Mol. Gen. Genet 224 : 177-182.

Barton, K.A., Whitely, H.R. and N.S. Yang 1987. *Bacillus thuringiensis* endotoxin expressed in transgenic *Nicotiana tabaccum* provides resistance to lepidopterodes insects. Plant physiol 85 : 1103-1109.

Bird, G.W. 1981. Integrated nematode management for plant protection p. 355-375. In : Plant parasitic Nematodes Vol. III. B.M. Zuckerman and R.A. Rohda Academic Press, New York USA 508 pp.

Bowles, D.J. 1990. Defence related proteins in higher plants. *Ann. Rev. Biochem.* 59 : 873-907.

Boosalis, M.G. and Mankan, R. 1965. Parasitism of soil microorganisms, pp. 374-391. In : K.F. Baker and W.C. Snyder (Eds.) Ecology of soil borne plant Pathogens, Univ. California Press, Berkeley.

Canning, E.W. 1973. Protozoal Parasites as agents for biological control of plant Parasitic nematodes. *Nematologica* : 19 : 342-348.

Cook, R. and K. Evans 1987. Resistance and Tolerance In : Principles and Practice of nematode control in crops Eds. R.H. Brown and B.R. Kerry, Academic Press, Sydney pp : 179-232.

Cook, R.J. and K.F. Bheker 1984. The nature and practice of Biological control of plant pathogens, American phytopathological Society. St. Paul., Minn. 539 pp.

Cooke, R.C. 1968. Relationships between nematodes destroying fungi and soil borne phytonematodes. *Phytopathology* : 58 : 909- 913.

Das Gupta, M.K. 1998. *Phytonematology*, Naya Prokash Pub. Calcutta, India.

Den-Belder, E., P.F.L. Fitteres and C. Waalnijk 1994 PN : EP 62367 09.11.94 Research Institute Plant Protection.

Doncaster, C.C. 1962. Predators of soil nematodes proceedings of the British Society for parasitology. Parasitology : 52 : 19.

Dreschsler, C. 1933. Several more fungi prey on nematodes. *J. Wash. Acad. Sci.* 23 : 355-257.

Dreschsler, C. 1937. Some hyphomycetes that prey on free living terricolous nematodes. *Mycologia* 29 : 447-552.

Dreschsler, C. 1941. Predacious fungi, Biol. Rev. 16 : 265-290.

Duddington, C.L. and C.M.G. Duthoit, 1960. Green manuring and cereal root eelworm. *Plant Path.* 9 : 7-9.

Ellenby, C. 1945. Control of Potato root eelworm, *Heterodera rostochiensis* by allyl isothiocynate, the mustard oil of *Brassica nigra* Ann. *Appl. biol.* 32 : 237-239.

Fisher, J.M. 1982. Towards a constant laboratory assay for resistance to *Heterodera avenae*. *Eppo. Bull.* 12 : 445- 449.

Garrett, S.D. 1965. Towards biological control pp. 4-17 In : Baker, E.F., W.C. Snyder (eds). Ecology of soil borne plant pathogens, University California, California Press, Berkeley.

Geoghegan, I., W. Robertson, N. Birch and A.M.R. Gatehouse 1995. PN : WO 9526634 : 12 : 10-95 Axis Genet, Cambridge, UK.

Gilbert, J.C. and D.C. McGuire 1956. Inheritance of resistance to severe root knot from *Meloidogyne incognita* in commercial tomatoes. *Proc. Am. Hort. Soc.* 68 : 437-442.

Guar, H.S. and A. Sirohi 1998. Biotechnology in plant nematology pp : 393-406. In : Nematode Diseases in Plants. Ed. P.C. Trivedi C.B.S. Publ., New Delhi, India.

Gurr, S., MC Pherson., M.J. Scollan. C., H.J. Atkinson. and D.J. Bovles 1991. Gene expression in nematode infected plant roots, *Mol. Gen. Genet* 226 : 361-366.

Gurr, S., MC Pherson., M.J. Scollan, C., Atkinson, H.J. and Bovles, D.J. 1991. Gene expression in Nematode infected plant roots. *Mol. Gen. Genet.* 226 : 361-366.

Johnson, L.F. 1963. Temperature as a factor in the control of tomato root knot with oat straw. *Phytopathology* 53 : 879.

Jung, C., U. Claussen, B. Harsthemks, F. Fisher and R.G. Herman 1992. A DNA library from an individual *Beta patellaris* chromosome conferring nematode resistance obtained by microdissection of meiotia metaphase chromosome *Pl. Mol. Biol.* 20 : 503-511.

Fassuliotis, G. 1979. Plant Breeding for root knot nematode resistance. In *Root-knot nematodes Meloidogyne* Species systematics, biology and control. *F. Lambarti* and C.E. Taylor Eds. Academic Press London. pp. 425-453.

Fassuliotis, G and D.P. Bhat. 1982. Potential of tissue culture for breeding root-knot resistance into vegetables. *J. Nematol.* 14 : 10-14.

Feldmesser, J. 1971. Soc. Nematol (USA) special Publication No. 1.

Hague, N.G.M. 1980. Nematodes the unseen enemy. A guide to Nematode damage. Du pont de nemocrs & Co. p. 19.

Hemwall, J.B. 1962. *Phytopathology* 52 : 1108-1115.

Homeyer, B., 1971. *Pflschutz-Nachr Bayer* 24 : 48-68.

Keen, N.T., A. Bent. and B. Staskawicz 1993. Plant disease resistance genes. Interactions with pathogens and their improved utilization to control plant disease. In : Biotechnology in plant disease control. Ex. Iochet-wiley-Liss, New York pp. 65-88.

Klein Lankhorst, R., Reitveld., B. Machiels., R. Verlok., R. Weide., C. Gebhardt., M. Koorneef and P. Zabel 1991. RELP markers linked to the root knot nematode resistance in gene Mi in tomato. *Theoritical and applied Genetics*, 81 : 661-667.

Kleinschmidt., G.D. and J.W. Gerdeman 1972. *Phytopathology* 62 : 1447-1453.

Larkin, P.J., Spindler, L.H. and Banks, T.M. 1990. Cell culture of alien chromosome addition lines to induce somatic recombination and gene introgression. In : Progress in plant cellular and molecular biology. Eds. H.J. Nijkemp. I.H.W. Vanderplas and J. Venu A. Artiiik Kluwer pp : 163-168.

Lear, B. 1959. Application of castor pomace and castor cropping to soil to reduce nematode populations. *Plant Dis. Reptr.* 43 : 459-460.

Linford, M.B., F. Yap and J.M. Olivera 1938. Reduction of soil populations of the root-knot nematode during decomposition of organic matter. *Soil science* 45 : 127-140.

Mathur, V.K. 1993. Regulatory measures- Nematodes. In : Hand Book of Economic Nematology pp : 317-327. Singh, R.S. and K. Sitaramaiah (eds), Cosmo Publications, New Delhi, India.

Mathew, V.K., K.C. Sanwal and A. Lal 1986. *Indian J. of Nematology* 16 : 122-123.

Mankan, R. 1980. Biological control of *Meloidogyne populations* by *Bacillus penetrans* in West Africa. *J. Nematol.* 12 : 230.

Messeguer, R., M. Gand., M.C. De Vicente., N.D. Young and S.T. Tanksley 1991. High resolution RELP map around root-knot nematode resistance gene (Mi) in tomato. *Theor. App. genet.* 82 : 529-536.

Mathur, B.N., Handa, D.K. and Swarup, G. 1987. Effect of deep summer ploughings on the cereals cyst nematode, *Heterodera avenae* and yield of wheat in Rajasthan India, *Indian J. Nematol.* 17 : 292-295.

Linford, M.B. and F. Yap 1938. Root knot injury restricted by a nematode trapping fungus. *Phytopathology* 28 : 14-15 (Absn).

Linford, M.B., F. Yap and J.M. Oliveira 1938. Reduction of soil populations of the root knot nematode during decomposition of organic matter. *Soil Sci.* 45 : 127-141.

Lear, B. 1959. Application of Castor pomace and cropping of castor beans to soil reduces nematode populations. *Plant Disease Reporter* 63 : 459-460.

Loewenberg, J.R., Sullivan. T. and Schuster, M.L. 1959. A virus disease of *Meloidogyne incognita* the southern root-knot nematode. *Nature* 184 : 1896.

Opperman, C.H. and Conkling, M.A., 1995. Field evaluations of transgenic root-knot nematode resistant tobacco. *J. Nematol* 27 : 513-514.

Opperman, C.H. and M.A. Conkling 1994. Nematode - induced gene expression and related control strategies, *Fundam. Appl. Nematol* 17 : 211-217.

Patel, R.M. and M.V. Desai 1964. A possible biological control of root-knot, plant dis. rptr. 48 : 167-168.

Prasad, N. and R. Mankau 1969. Studies on a Iporozoan endoparasite of nematodes. *J. Nematol* 1 : 301-302 (Abstr).

Prasadjee, J.K. and K. Sitaramaiah 1992. Effect of plant growth regulators alone and their combination with nematicides on root- knot severity and yield in tomato. Indian *Journal of Nematology* 22 : 77-81.

Pireda, O., M.W. Bonterbale., R.L. Plaisted., B.B. Brodie. and S.D. Tanksley, 1993. Identification of RFLP makers linked to the H gene confering resistance to the potato cyst nematode (*Globodera rostochiensis*) *Genome* : 36 : 152-156.

Raja, A. and D.R. Dasgupta, 1986. Enhanced synthesis of messenger DNA in relation to resistance expression in cowpea (*Vigna unguiculata*) infested with root-knot nematode *Meloidogyne incognita* Kerve *Nematol.* 9 : 35-48.

Rao. V.K. and K. Krishnappa 1995. Integrated management of *M. incognita Fusarium oxysporum* f. sp. *ciceri* wilt disease complex in chickpea. *International journal of pest management* 44 : 234-237.

Rohde, R.A. 1972. Expression of resistance in plants to nematodes, *Ann. Rev. Phytopathol.* 10 : 233-52.

Sayre, R.M. 1971. Biotic influence in soil environment pp. 235- 236 In : B.M. Zuckerman, W.F. Mai and R. Rhde (eds) Vol. I Plant Parasitic Nematodes. Academic Press.

Seshadri, A.R. 1970. *Nematology* pp : 370-411. In : Agriculture year Book. New vistas in crop yields. Indian Council of Agricultural Research, New Delhi.

Singh, R.S. and K. Sitaramaiah 1966. Incidence of root-knot of okra and tomato in oil cake amended soil. *Plant Disease Reports* 50 : 668-672.

Singh, R.S. and K. Sitaramaiah 1967. Effect of decomposing green leaves, sawdust and urea and the incidence of root knot of okra and tomato. *Indian Phytopath.* 20 : 349-355.

Singh, R.S. and K. Sitaramaiah 1970. Control of plant parasitic Nematodes with organic soil amendments. *Pests Articles and New's Summaries* 16 : 287-297.

Singh, R.S. and K. Sitaramaiah 1971. Control of root knot through organic and inorganic amendments of soils. Effect of oil cakes and sawdust. *Indian J. Mycol. Pl. Pathology* 1 : 20-29.

Singh, R.S. and K. Sitaramaiah, 1973. Control of plant Parasitic Nematodes with organic amendments of soil. Final Technical report on effect of organic amendments, green manuring and inorganic fertilizers on root-knot of vegetable crops. G.B. Plant Univ. Agril. & Tech. Bulletin. No. 6, 289 pp.

Singh, R.S. and Sitaramaiah, K. (eds) 1993. Hand Book of Economic Nematology . Cosmo publications, New Delhi, India.

Sitaramaiah, K. 1971. Cultural control of Nematodes pp : 233- 345. In : problems and progress in Economic Nematology Eds. D.S. Bhatti and R.K. Walia. Haryana Agricultral University, Hissar.

Sitaramaiah, K. 1990. Mechanism of reduction of plant parasitic nematode in soils amended with organic materials. pp : 263-295. In : Progress in plant Nematology Eds. S.K. Saxena *et al.*, CBS publishers. New Delhi.

Sitaramaiah, K. 1993. Chemical control of nematodes pp : 298-315 In : Hand Book of Economic Nematology, Singh, R.S. K. Sitaramaiah ed. Cosmo publication, New Delhi.

Sitaramaiah, K. and R.S. Singh 1990. Interaction of Nematodes with fungal and bacterial plant pathogens 195-208. In : Progress in plant Nematology Eds. S.K. Saxena *et al.*, CBS publishers, New Delhi, India.

Sitaramaiah, K. and R.S. Singh 1978a. Role of Fatty acids in nematode control through amendment of soil with margosa cake. *Indian Journal of Agriculture Science* 48 : 270.

Sitaramaiah, K. and R.S.Singh, 1978b. Effect of organic amendment on phenolic content of soil and plant and response of *Meloidogyne javanica* on its host to related compounds. *Plant and Soil* 50 : 671-679.

Sitaramaiah, K. and K.N. Pathak 1979. Effect of phenolics and an aromatic acid on *Meloidogyne javanica* infecting tomato *Nematologica* 25 : 281-287.

Sitaramaiah, K., R.S. Singh and K.P. Singh 1969. Induction of ovoviviparity in *Meloidogyne javanica* by fatty acids *Nematologica* 15 : 163.

Sitaramaiah, K., 1984. Plant Parasitic nematodes of India, Today and Tomorrows Printers and publisers, New Delhi, India.

Sitaramaiah, K. and K.N. Pathak 1981. Effect of growth regulators, phenolics and an aromatic acid on root knot severity *Meloidogyne incognita* and *M. javanica*, on tomato *Z. pflanzenkrankheiten* and *Pflanzuschutz* 88 : 651- 654.

Sitaramaiah. K., 1989. Host parasite relationships of plant parasite nematodes. Perspective in phytopathology, 155-165 eds. V.P. Agnihotri *et al.* Today and Tomorrow's Printers and Publisers, New Delhi, India.

Sitaramaiah, K. and J.K. Prasadjee 1989. Effect of plant growth regulators on *Meloidogyne javanica* infecting tomato. *Indian Journal of Nematology* 19 : 232-245.

Sitaramaiah, K. and K.N. Pathak 1992. Nematode bacterial disease interactions pp : 232-250. In : Nematode interactions. Eds : M.W. Khan, Chapman and Hall England.

Smith, P.G. 1944. Embryo culture of a tomato species hybrid. *Proc. Am. Hort.* Sc. 44 : 413-416.

Schaeffenberg, B., 1951. Lentersuchungam liber die Bedeutung der Enchytraeiden als Nematoden friende Mitteilungen aus cler Biologischen Zentralan - Stalt fur land-und-Forst Wirt Schaft 70 : 55-58.

Sirohi, A. and Dasgupta D.R. 1993. Mechanism of resistance in cowpea to the root-knot nematode. *Meloidogyne incognita* race 1.1 : Early induction of phenylalanine ammonia lyase (E.C. 4.3.1.5) *Indian J. Nematol.* 23 : 42-52.

Stratford, R. and Shields, R. 1995. PN : W09530017, 09.11.95, Uniliven, Rotterdam, The Netherlands.

Thorne, G. 1927. The life history, habits and economic importance of some mononchs. *J. Agric. Res.* 34 : 265-286.

Taylor, A.L. 1967. Introduction to Research on plant Nematology, FAO Rome, p. 133.

Tomalski, M.D. and L.K. Miller 1991. Insect paralysis by baculovirus mimediated express in of a mite neurotoxin gene. *Nature.* 352 : 82-85.

Zwillenberg, L.O. 1953. Anthonie Van Leenwenhoek. *J. Microbial serol.*, 19 : 101-106.

Thorne, G. 1927. The life history habits and economic importance of some mononchs. *J. Agric. Res.* 34 : 265-286.

Webster, J.M. 1972. Nematodes and biological control. pp. 469- 496. In : Economic Nematology, J.M. Webster (Ed.) Academic Press. London.

CHAPTER 6

CROPPING SEQUENCES FOR THE MANAGEMENT OF PLANT PARASITIC NEMATODES

Mansoor A. Siddiqui and M. Mashkoor Alam

INTRODUCTION

Crop rotation is one of the oldest and most important method for managing nematodes that feed on roots of annual crop plants (Nusbaum & Ferris, 1973; Idowu, 1989, Barker, 1991). It is an old but effective and widely used land management practice for the reduction of nematode population in the soil. The concept of rotating crops is a part of the agricultural history of many subsistance level farming operations. Planting crops in special sequences was used by early civilization in Peru to control the potato cyst nematode without knowing the presence of nematode. The resistant crops or non-host crops for the nematode management must be rotated with susceptible crops. Growing a resistant or non host crop for only one year is helpful but generally inadequate. But two or more years in production of a non-host crop resulted in decrease of the soyabean cyst nematode, *Heterodera glycines* to low or undetectable levels with acceptable subsequent yields of soyabean (Schmidt, 1991). Two resistant non-host crops between susceptible crops may be necessary for effective control of nematodes (Sasser, 1989). The optimal crop rotation strategies varied with crop value and initial nematode population level (Ferris & Greco, 1992).

Certain cyst and root knot nematodes attack only a limited number of crops. This situation provides a wider choice of non-host crops available for use in the rotation. For instance, soyabean is the congenial host of soyabean cyst nematode. *Heterodera glycines*. Plant non-host crops like corn, cotton, peanut, tobacco and many other crops in the rotation rapidly decreases the population of this nematode.

The term "cropping system" covers all kinds of crop sequences including continuous monoculture, whereas "crop rotation" implies an inflexible cycle or a fixed series of crops (Nusbaum and Ferris, 1973). Crop rotation allows the effective control of soil nematodes and the spatial diversity can be achieved by mixing varieties, intercropping and using trap crops (Derron, 1988).

On the basis of the experiences of farmers this method was, by and large, adopted to such management practices as a means of improving soil fertility and crop productivity. This method was developed emperically on the basis of life history of nematodes and host ranges and were targeted at the control of a single nematode species on a susceptible main crop. It requires no high cost or high technology impacts. Crop sequences were evaluated on the basis of root disease severity and performance of the main crop at the time of harvesting. It is particularly suitable for use in developing areas of the world. Sequential alternation of crops with varying degrees of susceptibility to the nematode pest can produce striking results in the reduction of crop losses.

Crop rotation is acceptable and adaptable to a wide variety of nematode problems and offer great flexibility in application. In an optimum crop rotation systems, the preceding crop prevents damage to the following crop by the target nematode species and suppress its population without increasing other species of nematodes that may be parasitic to the future crop (Johnson, 1985). In cropping sequences, the crops are grown at different intervals in such a way that the nematode populations remain at safer levels. The tolerance level for most hosts of *Heterodera goettingiana* is less than one egg/g soil. The conventional economic thresholds and optimal crop rotating varied with crop value and initial nematode population level. Generally a host crop can be profitably grown one year in 3 years; at low initial population levels, 2 hosts can be grown in three years (Ferris & Greco, 1992). Nematodes having narrow host range are easy to be managed by cropping. The high nematode population is brought down to a safe level by growing crop that are either non hosts or very poor hosts (Hesling, 1958; Stone, 1960).

A number of studies have been made on the effects of growing non-host crops in different cropping sequences on the single populations of phytoparasitic nematodes (Ivanowa, 1978; Nusbaum & Ferris, 1973; Noe, 1986; Noe *et al.*, 1991).

The effectiveness of cropping sequences in 'Integrated Pest Management' programmes can be enhanced, if the precise knowledge is acquired on the dynamics of nematode populations in relation to epidemiology of root diseases involving these pests. For instance, several plant parasitic nematodes are recognized. Some are sedentary during development and reproductive periods, whereas others are migratory. Endoparasitic nematodes penetrate into the roots, ectoparasites remain outside the roots and others are intermediate. The multiplication rate shows with increasing population density until an equilibrium density is reached at which multiplication just suffices to maintain the population density. The characterization of crop cultivars in terms of 'E' (Equilibrium density) and 'a' (the maximum reproduction rate of the nematodes on that crop) for each nematode species and under various environmental situations then becomes invaluable in planning cropping systems for nematode population management. SInce 'E' and 'a' are not necessarily correlated with 'T' (Threshold of Tolerance i.e. the population density below which no loss in growth and yield occurs) Seinhorst, (1967). Both 'T' and 'E' would be high for a good host and low for a poor host. The general model developed by Seinhorst (1965, 1970), for determination of tolerance limit and yield responses to nematodes is :

$$Y = m + (1\text{-}m)Z^{P\text{-}T}$$

where Y = a ratio between yield at nematode population density P and that in the absence of nematodes; m = the yield at nematode densities in which all available space is occupied by the minimal nematodes (minimal yield), Z = a constant less than 1 (= to the population of plants not attacked at the population density P =1); P = the population density, which is usually ≥ T (Y = 1 where P is equal or less than T); and T = the nematode density below which no detectable loss in yield occurs. The Z value depends on nematode species, host species, and external conditions. Z^T usually ranges from 1.05 to 1.15 (Seinhorst, 1965, 1970). The most profitable management alternative for a particular plant parasitic nematode infestation would be to use the land as pasture for a number of years, thus rotating to a profitable "non-host". At the time of termination of experiment i.e. harvesting is important to note that the population of plant parasitic nematodes had been reduced upto the economic threshold in the future crops.

Control measures may aim at the protection of one crop in a year from damage or it balancing the multiplication of a nematode on one or more hosts in a rotation (Seinhorst, 1973). A cropping system should be selected so that one crop does not produce the population of nematodes larger than the economic threshold density of the succeeding crop in the system.

Crop rotation is one of the options in regulating nematode populations, but its greatest potential undoubtedly lies in the role it may play in integrated control programmes of the future (Van Gundy, 1972). Use of crop rotations can be determined by their relative value in complementing other practices in total management systems that are effective, economical and ecologically sound.

1. AIM OF CROPPING SEQUENCES

The specific aim of this technique (Noe, 1986) are :

(1) To provide farmers with optimum crop rotation/management sequences for limiting crop losses due to plant parasitic nematodes.

(2) To quantify the relationships of nematode numbers to crop yields for a variety crops and plant parasitic nematodes of world wide importance.

(3) To demonstrate the feasibility of optimizing cropping systems for farmers through an 'International co-operative Research' effort.

Bessey (1911) and Birat (1969) found out that certain cropping systems restrict root knot nematode in susceptible vegetables. The selection of appropriate cropping sequences can be an effective tool in the management of plant parasitic nematodes population. Various studies have done on crop rotation to manage plant parasitic nematodes.

Muro (1975 a,b) pointed out that ground nut, maize, wheat, and sorghum reduced the incidence of *Meloidogyne incognita* larvae in soil. Reduction in population of *M. incognita* on tobacco had been observed by rotating it with barley, fescue and orchard grass (Southards & Nichols, 1972). Roy (1978) tested seven different cropping sequences for reducing the infestation of *Meloidogyne graminicola* on rice. He concluded that rotation with poor hosts were found to reduce nematode population but did not eradicate the nematode completely. Reduction in numbers of *M. graminicola* juveniles was highest (85%) in a year when mustard and guizitil (*Guizotia abbysimia*) were mixed sown in winter followed by mixed sowing of deep water rice with sesame. There was 65% reduction when mustard was grown alone followed by deep water rice mixed with millet (Rahman, 1990). Two successive croppings of cowpea and mungbean grown after rice reduced the population density of *Pratylenchus zeae* resulting in the higher yield (Aung & Prot, 1990). Hutton *et al.*, (1982, 1983) successfully controlled *M. incognita* in field trials in a crop rotation using *Hibiscus sabdariffa*/clean fallow/ *Manihot esculenta*/weed fallow/clean fallow/*Vigna sinensis*/*Amarantus viridis*/ *H. sabdariffa*. Amosu (1982 a,b) pointed out that the root knot galling caused by *Meloidogyne* sp. was lowest in sequence with rice, groundnut, maize and cassava. Number of *M. hapla* larvae in soil was poor after three radish crops and highest after carrot with onion (Brodie & Laughlin, 1977). Juvenile population of *M. hapla* increased and carrot yield decreased following 37 months of continuous carrot crops. A two year sequence of cotton wheat cotton, cotton-barley-cotton and maize-onion-peanut were found most effective for lowering root galling and subsequently greater yield (Hashmi & Hashmi, 1990). *Eragrostic curvula* decreased *M. javanica* population in soil (Koen & Grobbelaar, 1965), *Tagetes erecta* significantly reduced *M. javanica* population on tomato (Abid & Maqbool, 1990). Rotation with marigold reduced the *M. javanica* infection in carrot and increased yield (Huang, 1984); and *Paspalum notatum* of *M. incognita* on cotton (Rodgriguez-kabana & Pearson, 1972); while *Arachis prostrata* and *Prosphocarpus palustris* maintained soil populations of *M. incognita* (Wilson & Caveness, 1980). Crop rotation with barley, fescue and orchard grass reduced the population of *M. incognita* on tobacco.

Murphy *et al.*, (1974) found that *M. incognita* population was reduced significantly in *Crotalaria* marigold, bermudagrass or bahia grass. Sinnadurai (1973) claimed that rating of *Meloidogyne* sp., was lowest under groundnut - tomato rotation. The peanut followed by peanut and maize followed by peanut were the best treatments for reducing the population of *M incognita* race and *M. javanica*. Maize followed by maize and *Crotalaria juncea* came in second place. Maize and *C. juncea* apparently hosted the nematodes but their reproduction was very low (Moura, 1991). The survival of *Meloidogyne exigua* was studied in the soil rhizorphere of 9 crops cultivated in rotation on an eradicated coffee area, previously infested with nematode. The second stage larvae population of *M. exigua* was assessed in the soil for 23 months when coffee seedlings were again planted as a field bioassay. The planting of rice, sorghum, makrueni (*Panicum maximum*), bean, soyabean, *Crotalaria spectabilis* and *Stilozobium aterrimum* drastically reduced the number of *M. exigua* larvae in the soil after six months. The absence of second stage larvae on coffee planted after 23 months of rotation confirmed the efficacy of all crops used in the study (Almeida & Campos, 1991). Sharma *et al.*, (1980) obtained several fold increase in the larval population of *Meloidogyne incognita* with tomato monoculture followed by a substantial increase in the tomato-gram and soyabean-tomato rotations. Moreover, larval population was considerably reduced under groundnut mustard rotation. Maize and sorghum reduced the *Meloidogyne arenaria* juveniles population in soil, however, the population recovered quickly when peanut followed either of the two crops (Rodgriguez Kabana & Tochton, 1984, Rodgriguez Kabana *et al.*, 1990, 1991a,b.).

Rotation with a non susceptible crop prevented build up of large population of *Heterodera glycines* on soyabean. However, the population rose to high levels on soyabean followed by cowpea (Ross,

1962; Slack *et al.*, 1981). Soyabean rice cv. IR- 880-C rotation for 2 years highly increased the population of *Tylenchorhynchus martini* in soil, while *Pratylenchus zeae* was the most pathogenic. The soyabean yield was more in nematicidal treated plants than untreated plants (Fernandez *et al.*, 1987). In a three year host rotations for the control of *Heterodera glycines* with soyabean the plots were planted to soyabean or corn (maize) after fumigation with a split application of 1,3-dichloropropene. Nematicidal effects were no longer evident during the 4th year. Yields were most improved by the greatest number of years in non-host crop. Highest yields in descending order were from plants following three years of maize, 2 years of maize and 1 year of maize. Highly significant negative correlations occurred each year between initial nematode population densities and seed yields.

The population of *Meloidogyne incognita* on turnip and maize were reduced following treatment with methyl bromide or DD-MENCS. Application of fenamiphos to cabbage heavily infested with *Heterodera schachtii* resulted in a tenfold increase in plant growth and larval population were reduced by 86% in cabbage cucumber rotation (Rhoades, 1982). According to the findings of Lamberti (1973) and Tacconi *et al.*, (1988) the crop rotations and mineral fertilizers affected *Tylenchorhynchus dubius* and *Pratylenchus thornei* populations in different ways. Low rates of inorganic fertilizers increased the numbers of nematodes. The row crops sugar beet and tomato decreased the population levels whereas an annual forage crop showed an increased effect.

Peas, sugar beet, maize or potato were used as preceding crops for reducing the soil infestation by *Heterodera avenae* (Tenkovtseva & Datsuk, 1975). Resistant barley variety (Anderson, 1975; Graham & Stone, 1975), and maize (Bachthaler & Hien, 1976) have been used against *Heterodera avenae*.

In soil heavily infested with *Heterodera carotae*, carrots were grown once, twice or thrice in different rotations. The nematode population increased in all rotations but least when carrots were grown only once (Abrogioni & Palmisano, 1976).

Behringer (1978, 1979a, b) found that a 2-year rotation rapidly reduce the population of *Heterodera schachtii* in highly infested fields, but he suggested that longer rotations (Upto 6 years) is needed in heavily infested fields. In different crop rotations at 2-3-4-nd 6-year intervals with non host plants soyabean and cereals caused a decreased of the *H. schachtii* infestation in the soil below the population threshold for sugar beet with a consequent increase in sugar beet yield (Tacconi *et al.*, 1990; Tacconi & Sauti, 1991). Number of viable eggs and larvae of *H. schachtii* decreased in fields in a rotation on non Host (NH) e.g. corn, wheat and oats with host crops (H) e.g. cabbage and table beets (Mai & Abawi 1980). *Zea mays* and *Allium cepa* greatly reduced cyst number of *H. schachtii* (Vinduska, 1978). Jones (1956) reported that crucifers in general caused greater increase in the population of *H. schachtii* than the numbers of Chenopodiaceae and the population was reduced by growing onions and beans (Griffin, 1977); and by growing early cabbage (Brzeski & Radzikowska, 1981). The quadrennial rotation (Autumn winter cereals maize autumn cereals sugar beet) produced a marked decline in the numbers of eggs and juveniles of *H. schachtii.* Growing radish, mustard and their resistant varieties in combination with sugar beet rapidly decreases the population of *H. schachtii* (Hincke, 1989). Rotations including 75% root crops out yielded those with 25% root crops by upto 40% on wheat soils and 75% on rye soils. There was no increase in infestation by *H. schachtii* but infestation by *H. rostochiensis* increased in some plots. Soil humus content decreased in the rotations with root crops. The percentage of potatoes could be increased upto 50% but they should only be grown in alternate years (Kus & Gonet, 1990).

A 2-year rotation with bean and pea were not effective in reducing *Heterodera rostochiensis* population. However, a 3-year rotation (potato-bean pea and a 4-year rotation (potato pea pea pea) reduced *H. rostochiensis* population (Menon & Thangaraju, 1973). The population of *Globodera rostochiensis* was reduced following one or two plantings of non host crops e.g., wheat, sugarbeat and azuki (Inagaki, 1978). Magnusson (1981) suggested that crop rotation of resistant potatoes and barley/timothy could be used for *G. rostochiensis* control.

In the plots with tomato monoculture Tiilkikkalak (1991) observed that the tuber yield decreased from 35 to 4.6 t/ha. The population density of the nematode was raised during the first 3 years from 0.1 to 265 larvae per g soil. The population density fluctuated afterwards between 30 and 136 larvae per g. soil. He further used resistant (H1) and susceptible cultivars alternatively decreased the nematode density under economic threshold or even under the level of detection.

Ponin *et al.*, (1976) and Ponin (1980) pointed out that non susceptible crops and resistant potato varieties reduced the intensity of soil infestation by plant parasitic nematodes in potato soils. Guinez (1981), Poskorz & Roszak, 1989) observed that high population of phyto parasitic nematodes in potato soils were maintained by continuous cropping of potato. About 60% cysts of nematodes were reduced by growing non host crop or keeping the land fallow. Cropping sequences of soyabean, millet, wheat, potato and corn controlled *Meloidogyne hapla, Heterodera glycines* and *Paratylenchus* spp. (Nishizawa *et al.*, 1971; Nishizawa 1978 and Raymundo, 1985).

During 2-year rotation of different cropping sequences, the highest decrease in the number of cysts of *Heterodera zeae* was observed in wheat, egyptian clover, mustard, barley and pigeon pea (Siddiqui & Saxena, 1985-1989). The benefits of crop rotation in banana cultivation such as *Canavalia ensiformis, Crotalaria juncea, Mucuna pruriens* and *Desmodium distortum* result from a reduction in root nematodes particularly *Radopholus similis* and increased fertility, which reduced the costs of nematicidal treatments and gave higher crop yields. It was noted that during these studies the population of *Meloidogyne* spp., *Helicotylenchus* spp. and *Hoplolaimus* spp. increases on banana roots (Ternisien, 1989, Ternisien & Melin 1989, Ternisen & Gangrye, 1990). Populations of *Radopholus similis* and *Pratylenchus coffeae* on banana roots were significantly reduced by growing antagonistic crops viz., *Tagetes, lucerne, sunn hemp (Crotalaria juncea)* or coriander (*Coriandrum sativum*) as intercrop for 4 months (Naganathan, *et al.*, 1988).

The use of resistant variety of soyabean (Pickett-71) was found very successful to suppress population of *Rotylenchulus reniformis*. William *et al.* (1983) pointed out that growing a resistant variety considerably managed the population of *Rotylenchulus reniformis* when growing cotton or a succeptible soyabean cultivar in the following year. Pea and mustard were grown either alone or together in alternate rows declined significantly the populations of *R. reniformis* and *Tylenchorhynchus* spp. (Haque & Gaur, 1988).

Oostenbrink *et al.*, (1956) and Winslow (1964) pointed out that cereals generally support the multiplication of migratory nematodes than row crops such as sugar beat and potatoes within a given soil.

Oostenbrink (1961) demonstrated a significant linear relationship between initial population density and the effect on host plant. Beet and marigolds were found to suppress the population of *Pratylenchus penetrans* and he advised that one of these crops should be grown on infested land before potatoes and red clover, which are susceptible. Ferris & Bernard (1967) noted different responses of plants to *Pratylenchus species*, *P. penetrans* and *P. scribneri* increased in corn and soybean plots, whereas *P. neglectus* on wheat. The increase in population of *P. hexincisus* on soyabean was found moderate. Number of mixed *Pratylenchus* spp. communities were most numerous on wheat after maize and wheat monoculture had intermediate numbers. Sugerbeat when integrated as the previous crop decreased the populations of *pratylenchus* (Esmenjaud *et al.*, 1990). Conventional, minimum and zero tillage trials with maize (summer) and oats (winter) for 10 years variably reduced the population of *Pratylenchus*, *Cephalenchus*, *Ditylenchus*, *Plectus*, *Panagrolaimus* etc. (Yeats & Hughes, 1990). Todd (1991) observed the responses of *Belonolaimus* and *Pratylenchus scribneri* populations to 8 cropping regimes (5 continuous; lucerne, maize, sorghum, soyabean, and fallowed soil; and 3 rotations : fallow maize soyabean maize and maize soyabean) were monitored at depth of 30 and 31-60 cm in sandy soil. He concluded that with the exception of lucerne, all crop species examined supported substantial increase in population of both nematodes. Largest nematode population increase in the field occurred in maize plots, whereas lucerne did not allow reproduction factor by either species during the 2 years of observation. Soil populations of both nematodes remained at detectable levels after 2 years of fallow. The distribution of numbers of *Belonolaimus* sp. between soil depths varied with sampling dates where *P. scribneri* were consistantly concentrated in the top 30 cm of soil. Prasad & Rao (1978) reported that the population of *P. indicus* decreased under *Sesamum indicum, Brassica nigra Phaseolus mungo* var. *radiatus*, *P. mungo* and barely and wheat. Under 11 different crops grown in monoculture, with or without weed control the population of *Pratylenchus* sp. were high on cereals and low on lucerne potato and linseed, *Helicotylenchus* spp. numbers were high on witner rye, winter wheat, red clover and low on oats and potatoes while *Tylenchorhynchus* sp. numbers were high on winter rye, winter wheat and weed plots of red clover and lucerene, and low on maize, peas, potatoes, linseed and sugarbeet (Rossner, 1979). Prasad *et al.*, (1983) pointed out that *Phaseolus radiatus* decreased the nematode populations but the population increased in rotations involving *Carthamus tinctorius* or *Nicotiana tabacum*. Scholte & Jacob (1983) showed that continuous cropping of maize resulted in an increase in population of *Pratylenchus* sp. and *Tylenchorhychus* sp. and decrease *Meloidogyne* population.

Johnson *et al.* (1974, 1975) noted that with various rotations having corn, cotton, peanut and soyabean, the monocropping of peanut was most effective. However, no single cropping system suppressed all the plant parasitic nematodes. They suggested that if cotton or corn are to be grown, peanut and soyabean should be in rotation. Mukhopadhyay & Prasad (1968) found out that *Tylenchorhynchus* spp. was abundant in all the cropping sequences tested however, monoculture supported the nematode population more than the rotation. Moreover, the nematode population was higher in later stages of plant growth. The population of *Pratylenchus brachyurus* was affected by different cropping sequences (Egunjobi, 1954). Ferris and Bernard (1971) found that large number of *Helicotylenchus pseudorobustus*, *Tylenchorhynchus martini*, *Xiphinema americanum* and *Paratylenchus projectus* were supported by all or most rotations involving corn, soyabean, oats, wheat and a legume grass forage mixture. The population of *Meloidogyne*, *Pratylenchus* and *Helicotylenchus* was reduced in a rotation of tobacco, maize and ground nut (Norse, 1972).

The population of *Hoplolaimus indicus* was reduced by growing gram, guava, tobacco, watermelon, sugar beet and rapeseed (Gupta and Atwal, 1971); *Trichodorus christiei* by okra (Good *et al.*, 1973); *Pratylenchus alleni* by marigold, castor, bean and chrysanthemum (Hackney & Dickerson 1975), *Xiphinema diversicaudatum* by hop, spring barley, potato, sugar beet, cabbage, winter wheat and winter beans (Cotton, 1977); but the population of *X. diversicaudatum* increased under strawberry and, to a lesser extent in rye grass (McNamara & Pitcher, 1977).

Monoculture of some crops resulted in the build up of nematode populations, but proper crop rotation reduced the nematode population to safe level (Chawla & Prasad, 1973; Khan *et al.*, 1973. Omiyi 1977, Kinolch, 1983 and Kirkpatrick & Sasser, 1984, Alam, 1990).

Brodie & Murphy (1975) found that six-week fallow followed by a cover crop was as effective in reducing the population of *Pratylenchus brachyurus*, *Trichodorus christie* and *Meloidogyne incognita* as a nine month fallow.

Brodie *et al.* (1970a) studied that in a 6 year sequence of cover crop the nematode population varied with different cover crops. Marigold was found best to suppress *Belonolaimus longicaudatus*, *Pratylenchus brachyurus*, *Trichodorus christiei* and *Xiphinema americana*. Brodie *et al.*, (1970b) found that the population of *Tylenchorhynchus claytoni*, *Trichodorus christiei*, *Pratylenchus brachyurus*, *Helicotylenchus dihystera* and *Xiphinema americanum* increased in newly cleared pine land. The population varied with various types of cropping sequences. But susceptible crops with cultivated crops apparently decreased the nematode population upto damaging levels. Brodie *et al.*, (1970c) noted that the population of nematode varied with different row crop sequences. They suggested that corn and cotton should not follow each other in soils infested with *Belonolaimus longicaudatus*.

Shafshak *et al.*, (1974, 1979) found that the populations of *Tylenchorhynchus*, *Pratylenchus*, *Tylenchus*, *Heterodera* and *Aphelenchus* were least in the wheat plots followed by *Trifolium alexandrinum*.

Khan *et al.*, (1975) pointed out that the population of *Meloidogyne incognita* increased extensively in the monoculture of tomato, eggplant, okra, chilli and sponge gourd or in their combinations but decreased when these crops were grown following marigold or when the field was left fallow. *Tylenchorhynchus brassicae* population was found to have increased following eggplant, tomato, okra, spinach and eggplant while *Longidorus bravicaudatus* and *Pratylenchus coffae* were not materially affected by any of the crop combinations tried. Khan *et al.*, (1987) investigated that wheat okra fallow okra wheat fallow onion rotation significantly decreased the *Quinisulus curvus* population; wheat-fallow maize chilli wheat fallow onion sequence reduced the *Hoplolaimus seinhorsti* population. The population of *Logicaudatus bravicaudatus* decreased in both the sequences, while the population of *Helicotylenchus indicus* increased at the end of the cropping sequences.

Castillo *et al.* (1978) pointed out that the cultivation of susceptible crops followed by 2-3 years of non susceptible crops was found to be more effective in reducing the population of *Rotylenchulus reniformis* and

Meloidogyne sp. on rice based cropping. Alam *et al.*, (1977) reported that broad bean and Kochia in addition to marigold were very effective in suppressing the population of most of the phytoparasitic nematodes. In various cropping sequences *Tagetes erecta* and *Zinnia elegans* reduced the population of all the plant parasitic nematodes (Alam *et al.*, 1976, 1980, 1981, 1988 Alam 1990). Saxena *et al.*, (1977) observed that the population densities of *Tylenchorhynchus brassicae* was greater when cauliflower, cabbage and turnip were grown than in raddish, tomato and marigold. Populations were markedly reduced under fallow. In field and green house studies more than 75% nematode populations of *Meloidogyne javanica* and *M. incognita* died during 1 to 2 month fallow and carrot yields in soil which had been fallow for 1, 5, 9 or 13 weeks were 77, 35, 31 and 46% respectively (Huang & Porto, 1988).

Khan *et al*, (1984) noted that cropping sequences including mung, Sesbania, marigold wheat, barley or a fallow period considerably reduced the populations of *Meloidogyne* sp., *Rotylenchulus reniformis* and *Tylenchorhynchus brassicae* to very low levels.

Alam *et al.* (1978, 1979) screened sixty eight crop plants against seven important species of plant parasitic nematodes (Tables I, II). The information could very well help in selecting suitable crop for suppressing prevailing nematode population.

2. DISADVANTAGES

Crop rotation though very effective but has certain limitations :

(1) The amount of damage inflicted on a crop depends on the level of resistance in the crops used in the rotation, and on the number of years of planting these crops between susceptible crops.

(2) The non host crop or resistant crop inhibit significantly a specific nematode may help in the population build up for another species of plant parasitic nematode present in the same field.

(3) Low commercial value of some non host or resistant crops.

(4) Some farmers have limited land to rotate the crop.

3. METHODOLOGY

Management and optimization models can be combined in the short and long term.

The short term method

The returns for single growing seasons may be optimized by reduction of the nematode population to a specific level, but the problem may exacerbated for subsequent crops (Duncan, 1983, Duncan & Ferris, 1983; Ferris, 1978). However, data generated for analytical models of population increase during a single growing season on a given host crop (Duncan, 1983, Duncan & Ferris, 1983) allow prediction from a preplant population estimate of the expected crop value (or crop loss) and the expected final nematode population. Further, data on nematode over wintering surviving (Duncan, 1983, Goodell *et al.*, 1983; Vrain, 1983) allow prediction of the expected nematode population densities at the start of the following cropping season. These relationship allow prediction of expected crop yield or value on a series of possible subsequent crop. Calculations of this type generate the possibility of optimizing cropping sequences for nematode management over multiple years (Duncan & Ferris, 1983). Such approaches are particularly appropriate with plant parasitic nematodes, and perhaps with other soil borne pathogens, because of relatively slow reproductive and dispersal

Table 1. Suitability of some cultivated plants to phyto-nematodes

Latin names	English/common names	Host response to nematodes		
		Rot	Tyl	Mel
Abelmoschus esculentus Moench.	Okra	G	G	G
Allium cepa L.	Onion	P	M	P
A. sativum L.	Garlic	P	G	P
Althaea rosea Cav.	Hollyhock	P	G	P
Antirrhinum majus L.	Antirrhinum	M	G	G
Avena sativa L.	Oats	P	M	P
Benincasa hispida Cogn.	White gourd/petha	M	G	M
Beta vulgaris L.	Table beet	M	M	G
Brassica campestris L.	Mustard	P	M	P
B. rapa L.	Turnip	G	G	M
B. oleracea botrytis L.	Cauliflower	P	G	P
B. oleracea capitata L.	Cabbage	M	G	M
B. oleracea caulocarpa Pasq.	Knol khol	M	G	P
Cajanus cajan Milsp.	Pigeon pea/arhar	P	M	P
Calendula officinalis L.	Calendula	M	M	M
Capsicum annuum L.	Chilli	M	G	G
Celosia cristata L.	Cock's comb	P	P	M
Centaurea cyanus L.	Corn flower	M	M	M
C. moschata L.	Sweet sultan	P	M	P
Cheiranthus cheiri L.	Wall flower	P	G	M
Chrysanthemum crinatum L.	Chrysanthemum	P	M	P
Cicer arietinum L.	Bengal gram	G	G	G
Coreopsis tinctoria Nutt.	Tickseed	P	P	P
Coriandrum sativum L.	Coriander/dhania	P	P	M
Cosmos bipinnatus Cav.	Cosmos	P	P	P
Crotalaria juncea L.	Sann hemp/phulsan	P	P	P
Cyamopsis tetragonoloba Taub.	Cluster bean/ guar	P	P	G
Dahlia pinnata Cav.	Common or garden dahlia	P	M	P
Daucus carota L.	Carrot	G	M	G
Dimorphotheca sinuata DC.	Cape marigold/African dasy	P	P	P
Dolichos biflorus Roxb.	Horse gram/khurti	P	M	M
Eruca sativa Mill.	Rocket salad/duan	P	P	P
Gossypium Sp. L.	Cotton	G	G	G
Hibiscus cannabinus L.	Brown hemp/patsan	M	M	G
Hordeum vulgare L.	Barley	P	G	P
Impatiens balsamina L.	Balsam	P	P	P
Kochia scoparia Schrad.	Kochia	G	G	P
Logenaria leucantha Rusby.	Bottle gourd/lauki	M	M	G
Lens culimaris Medic.	Lentil/masoor	G	G	P
Linum usitatissimum L.	Linseed	P	M	G

Table 1 contd...

Table 1 contd...

Latin names	English/common names	Host response to nematodes		
		Rot	Tyl	Mel
Luffa acutangula Roxb.	Ridge gourd/torai	M	G	G
Lycopersicon lycopersicum (L.) Karsten.	Tomato	G	G	G
Nicotiana tabacum L.	Tobacco	G	M	G
Papaver somniferum L.	Poppy	G	M	M
Pennisetum typhoides Stapf.	Pearl millet/bajra	M	G	P
Phaseolus aconitifolius Jacq.	Kidney bean/moth	M	M	P
P. aureus Roxb.	Green gram/moong	G	G	G
P. mungo radiatus L.	Black gram/urad	M	M	M
Phlox drummondii Hook.	Phlox	G	G	P
Pisum sativum L.	Pea	M	M	G
Portulaca oleracea L.	Portulaca/kulfa	P	G	G
Raphanus sativus L.	Radish	P	G	P
Ricinus communis L.	Castor	G	M	M
Sacchrum officianum L.	Sugarcane	M	G	P
Sesamum indicum L.	Sesame/til	P	P	P
Sesbania aculeata (Willd.) Poir.	Dhaincha	P	M	P
Solanum melongena L.	Eggplant	G	G	G
S. tuberosum L.	Potato	G	M	G
Sorghum vulgare Pers.	Sorghum/jowar	M	G	P
Spinacia oleracea L.	Spinach	G	P	G
Tagetes erecta L.	Marigold	P	P	P
Trigonella foenum graecum L.	Fenugreek/methi	G	G	M
Triticale Sp.	Triticale	M	G	P
Triticum aestivum L.	Wheat	M	P	P
Vigna catjang Walp.	Cowpea/lobia	M	M	P
Vicia faba L.	Broad bean/bakla	P	M	M
Zea mays L.	Corn/maize	P	M	P
Zinnia elegans Jacq.	Zinnia	P	G	P

P, M, G = Poor, moderate good hosts
Rot = *Rotylenchus reniformis* Linford & Oleviera, *Tyl* = *Tylenchorhynchus brassicae* Siddiqi,
Mel = *Meloidogyne incognita* (Kofoid & Wite) Chitwood.

rates and yield loss predictability from simple analytical models.

The selection of a cropping sequence involves some forecasting of crop values over a series of years and an expectation that average weather conditions will prevail. However, the commencement of a particular cropping sequence does not lock the user into that programme should condition drainage over the period for which the prediction were made. The optimal solution may involve reassessment of the system through biological or economic monitoring at the beginning of each growing season and selection of the most profitable route from that point forward. Linear programme technique are adaptable to this technique.

Table 2. Suitability of sixty cultivated plants to phytonematodes

Latin names	English/common names	Host response to nematodes			
		Hop	Hel	Tyl	Tri
Abelmoschus esculentus Moench.	Okra	G	M	M	M
Allium cepa L.	Onion	P	P	G	G
A. sativum L.	Garlic	P	P	P	G
Althaea rosea Cav.	Hollyhock	P	M	G	P
Antirrhinum majus L.	Antirrhinum	M	P	P	M
Avena sativa L.	Oats	P	P	P	M
Benincasa hispida Cogn.	White gourd/petha	M	M	M	P
Beta vulgaris L.	Table beet	G	G	M	M
Brassica campestris L.	Mustard	M	P	P	P
B. rapa L.	Turnip	P	P	M	P
B. oleracea botrytis L.	Cauliflower	M	M	G	M
B. oleracea capitata L.	Cabbage	P	P	G	M
B. oleracea caulocarpa Pasq.	Knol khol	P	P	M	P
Cajanus cajan Milsp.	Pigeon pea/arhar	G	P	P	P
Calendula officinalis L.	Calendula	M	M	P	M
Capsicum annuum L.	Chilli	M	M	M	M
Celosia cristata L.	Cock's comb	P	P	P	P
Centaurea cyanus L.	Corn flower	M	M	P	P
C. moschata L.	Sweet sultan	P	P	P	P
Cheiranthus cheiri L.	Wall flower	P	P	P	P
Chrysanthemum crinatum L.	Chrysanthemum	P	G	P	P
Cicer arietinum L.	Bengal gram	P	P	M	P
Coreopsis tinctoria Nutt.	Tickseed	P	P	P	P
Coriandrum sativum L.	Coriander/dhania	G	P	M	P
Cosmos bipinnatus Cav.	Cosmos	G	M	P	P
Crotolaria juncea L.	Sann hemp/phulsan	M	P	P	M
Cyamopsis tetragonoloba Taub.	Cluster bean/ guar	G	M	P	P
Dahlia pinnata Cav.	Common or garden dahlia	P	P	P	P
Daucus carota L.	Carrot	M	M	M	M
Dimorphotheca sinuata DC.	Cape marigold/African dasy	P	P	P	P
Dolichos biflorus Roxb.	Horse gram/khurti	P	P	M	M
Eruca sativa Mill.	Rocket salad/duan	P	P	P	P
Gossypium Sp. L.	Cotton	G	M	G	M
Hibiscus cannabinus L.	Brown hemp/patsan	G	M	M	M
Hordeum vulgare L.	Barley	G	M	P	M
Impatiens balsamina L.	Balsam	P	M	P	P
Kochia scoparia Schrad.	Kochia	P	P	G	P
Logenaria leucantha Rusby.	Bottle gourd/lauki	P	G	M	M
Lens culinaris Medic.	Lentil/masoor	P	P	M	P
Linum usitatissimum L.	Linseed	P	M	M	M

Table 2 contd...

Table 2 contd...

Latin names	English/common names	Host response to nematodes			
		Hop	Hel	Tyl	Tri
Luffa acutangula Roxb.	Ridge gourd/torai	P	M	G	M
Lycopersicon lycopersicum (L.) Karsten.	Tomato	G	M	M	P
Nicotiana tabacum L.	Tobacco	M	G	P	M
Papaver somniferum L.	Poppy	M	P	M	P
Pennisetum typhoides Stapf.	Pearl millet/bajra	M	G	M	M
Phaseolus acontifolius Jacq.	Kidney bean/moth	M	G	M	P
P. aureus Roxb.	Green gram/moong	M	P	G	P
P. mungo radiatus L.	Black gram/urad	M	M	P	P
Phlox drummondii Hook.	Phlox	P	M	P	P
Pisum sativum L.	Pea	G	G	P	M
Portulaca oleracea L.	Portulaca/kulfa	G	M	G	M
Raphanus sativus L.	Radish	P	P	P	G
Ricinus communis L.	Castor	G	M	M	M
Sacchrum officianum L.	Sugarcane	G	G	M	G
Sesamum indicum L.	Sesame/til	G	P	P	P
Sesbania aculeata (Willd.) Poir.	Dhaincha	G	M	M	P
Solanum melongena L.	Eggplant	P	P	M	P
S. tuberosum L.	Potato	G	M	G	M
Sorghum vulgare Pers.	Sorghum/jowar	P	P	M	P
Spinacia oleracea L.	Spinach	P	P	G	M
Tagetes erecta L.	Marigold	P	P	P	P
Trigonella foenum-graecum L.	Fenugreek/methi	M	M	G	G
Triticale Sp.	Triticate	M	M	M	P
Triticum aestivum L.	Wheat	G	M	M	M
Vigna catjang Walp.	Cowpea/lobia	M	M	M	M
Vicia faba L.	Broad bean/bakla	P	P	P	M
Zea mays L.	Corn/maize	M	P	P	P
Zinnia elegans Jacq.	Zinnia	G	M	G	G

P. M. G = Poor, moderate good hosts
Hop = *Hoplolaimus indicus*. Hel = *Helicotylenchus indicus*
Tyl = *Tylenchus filiformis*. Tri = *Trichodorus mirzai*.

REFERENCES

Abid, M. & M.A. Maqbool 1990. Effect of intercropping of *Tagetes erecta* on root knot disease and growth of tomato. *Int. Nematol. Network Newsl.* 7 : 41-42.

Alam, M.M. 1990. Effect of different cropping sequences on ornamental plants on the population of plant parasitic nematodes. *Current Nematology* 1 : 21-24.

Alam, M.M., A.M. Khan & S.K. Saxena 1978. Control of plant parasitic nematodes by Management practices : crop rotation : *Proc. Nat. Acad. Sci.*, India 46B(I & II) : 308-312.

Alam, M.M., A.M. Khan & S.K. Saxena 1988. Management of plant parasitic nematodes by different cropping sequences. *Indian J. Pl. Pathol.* 6 : 102-109.

Alam, M.M., S.K. Saxena, & A.M. Khan 1977. Influence of different cropping sequences on soil population of plant parasitic nematodes. *Nematol. medit.* 5 : 65-72.

Alam, M.M., S.K. Saxena & A.M. Khan 1978. Suitability of crops to certain ectoparasitic nematodes. *Acta Bot. Indica* 6 (Suppl.) : 205-207.

Alam, M.M. S.K. Saxena & A.M. Khan 1979. Suitability of crops to the reniform, stunt and root knot nematodes. In 'Recent Researches in Plant Sciences.' Ed. S.S. Bir, Kalyani Publishers, New Delhi, India : 631-634.

Alam, M.M., S.K. Saxena, & A.M. Khan 1981. Soil population of plant parasitic nematodes under various cropping sequences. *Nematol. medit.* 9 : 40-41.

Alam, M.M., Z.A. Siddiqui, A.M. Khan & S.K. Saxena 1980. Effect of different cropping sequence on the population of plant parasitic nematodes. *Indian J. Nematol.* 10 : 35-39.

Almeida, V.F. De & V.P. Campos 1991. Crop rotation and survival of *Meloidogyne exigua* in eradicated and infested coffee areas. Altermancia de cultures e sobrevivencia de Meloidogyne exigua em areas de cafezal infestado e errdicao. *Nematol Brasileira* 15 : 30-42.

Ambrogiaoni, L. & A.M. Palimisano 1976. The effect of crop rotations on *Heterodera carotae* (Nematoda : Heteroderidae) and on carrot yield infested soil. Effecto di avvicendamenti culturali su Heterodera carotae (Nematoda : Heteroderidae) e sulla produzione di carota in terrano infestato. *Radia* 5 : 355-368.

Amosu, J.O. 1982a. Cropping system for control of root knot nematodes. In Proc. 3rd Res. Plan. Conf. on root-knot nematodes *Meloidogyne* spp. regions IV & V *Int. Institute of Tropical Agriculture* : 254-258.

Amosu, J.O. 1982b. Control of root knot nematode by cultural practices. In *Proc. 3rd Res. Plan. Conf. on root knot nematodes Meloidogyne* spp. regions IV & V. 16-20 Nov., Ibadan, Nigeria (IPM) *Int. Institute of Tropical Agriculture* : 259-264.

Aung, T. & J.C. Prot 1990. Effect of crop rotations on *Pratylenchus zeae* and on yield of rice cultivar UPL Ri 5. *Rev de Nematol.*, 13 : 445-447.

Bachthaler, G. & L. Hien 1976. Results of long term cereal rotation with varying proportion of maize. *Soil & Fertilizers* 40 : 1397.

Barker, K.R. 1991. Rotation and cropping systems for nematode control : the North Carolina experience - Introduction. *J. Neamatol.* 23 : 342-343.

Behringer, P. 1978. Nematodenkontrollierter Riibenanbau. Festellung des Ertragsausfalles sowie Entwicklung der Populations dynemic. *Deutsche Zuckerribenzeitung* 15 : 4-5.

Behringer, P. 1979a. Gafahrlose Fruchtfolgen and Zwischenfriichte Gute Fruchtfolge erspart oft chemieeinsatz gegen Nematoden. *Deutsche Zuckerribenzeitung* 15 : 2-3.

Behringer, P. 1979b. Erfassung und Bekampfung des Rubennematodesn (*Heterodera schachtii*) in Bereich der Pramken - Zuckernach den Grundsatzen des integrierten Pflanzenschutzes. *Gesunde Pflanzen* 31 : 185-188.

Bessey, E.A. 1991. Root knot and its control. U.S.D.A. Bur Plant Indua Bull. No. 219 : 89 pp.

Birat, R.B.S. 1969. Influence of cropping pattern on the population of *Meloidogyne javanica* (Treub, 1885) Chitwood, 1949. *Indian Phytopath.* 22 : 391-393.

Brodie, B.B., J.M. Good & C.A. Jaworski 1970a. Population dynamics of plant nematodes in cutlviated soil Effect of summer crops in old agricultural land. *J. Nematol.* 2 : 147-151.

Brodies, B.B., J.M. Good & C.A. Jaworski 1970b. Population dynamics of plant parasitic nematodes in cultivated soil Effect of symmer crops in newly cleared land. *J. Nematol.* 2 : 217-222.

Brodie, B.B., J.M. Good & W.H. Marchant 1970c. Population dynamics of plant nematodes in cultivated soil : Effect of soil based rotations in Tifton sandy loam. *J. Nematol.* 2 : 135-138.

Brodie, B.B. & W.S. Murphy 1975. Population dynamics of plant nematodes as affected by combinations of fallow and cropping sequence. *J. Nematol.* 7 : 91-92.

Brody, J.K. & C.W. Laughlin 1977. The effect of vegetable cropping sequences on population development of *Meloidogyne halpa. Revista de Agricultura* 52 : 13-18.

Brzeski, M.W. 1973. [Control of *Meloidogyne halpa* Chitw. on carrot by crop rotation] Zwalczanie matwika potnocnego (*Meloidogyne hapla* Chitw.) na marchwi. *Materiaty Ogolnopolskiego Ziazdu Warzwyniezego, Skiermiewice.* : 85-86.

Castillo, M.B., M.S. Alejar and R.R. Harwood 1975. Nematode in cropping patterns II. Control of *Meloidogyne incognita* through cropping patterns and cultural practices. *Philippine Agricultrist* 59 : 295-312.

Castillo, M.B., M.S. Alejar and J.A. Litsinger 1978. Effect of crop rotation on nematode populations in dryland rice based cropping systems of Batangas, Philippines. Int. *Rice Res. Newsletter* 3 : 18-19.

Chawla, M.L. & S.K. Prasad 1973. Multiple cropping and nematodes. I-Effect of fertility and crop yield. *Indian J. Nematol.* 3 : 34- 39.

Cotten, J. 1977. Effect of annual and perennial cropping regions in microplots on the population density and vertical distribution of *Xiphinema diversicaudatum* and on soil porosity. *Ann. Appl. Biol.* 86 : 397-404.

Derron, J.O. 1988. Crop rotation : an important component of integrated pest management. [La rotation des cultures : un element essential de la lutte integree). *Schweizerische Land Wirt Schaftliche Forschung* 27 : 91-101.

Duncan, L.W. 1983. Preceding effects of plants parasitic nematode communities on crop growth. Ph.D. Thesis. Univ. Calif. Riverside, 160 pp.

Duncan, L.W. & H. Ferris 1983. Effects of *Meloidogyne incognita* on cotton and cowpeas in rotation. PP22-26. In proc. Beltwide Cotton Prod. Res. Conf. 347 pp.

Egunjobi, O.A. 1984. Effects of intercropping maize with grain legumes and fertilizer treatment on population of *Pratylenchus brachyurus* Godfrey (Nematoda) and on the yield of maize (*Zea mays* L.) *Protection Ecology* 6 : 153- 167.

Esmenjaud, D., R. Rivoal & H. Marzin 1990. Numbers of *Pratylenchus* spp. (Nematoda) in the field on winter wheat in different cereal rotations. *Nematologica* 36 : 217-226.

Fernandez, M., J. Ortega & A. Berbe 1987. Soybean - rice rotation and the role of plant parasitic nematodes (Posibilidad de la rotacion soya arrez respecto a los fitonematode). *Cienias de la Agricultura* No. 31, 14-18.

Ferris, H. 1978. Nematode economic thresholds : derivation, requirements and theoretical considerations. *J. Nematol.* 10 : 341-350.

Ferris, H. & N. Grec 1992. Management strategies for *Heterodera goettingiana* in a vegetable cropping system in Italy. *Fundamental & Applied Biology* 15 : 25-33.

Ferris, V.R. & R.L. Bernard 1967. Population dynamics of nematodes in fields planted to soybeans and crop grown in rotation with soybeans. I. The genus *Pratylenchus* (Nemata : Tylenchida). *J. Econ. Ent.* 60 : 405-410.

Ferris, V.R. & R.L. Bernard 1971. Crop rotation effects on population densities of ectoparasitic nematodes. *J. Nematol.* 3 : 119-122.

Good, J.M., W.S. Murphy & B.B. Brodie 1973. Population dynamics of plant nematodes in cultivated soil. Length of rotation in newly cleared and old agricultural land. *J. Nematol.* 5 : 117-122.

Goodell, P.B., H. Ferris & N.C. Goodell 1983. Overewintering population dynamics of Meloidogyne incognita in cotton. *J. Nematol.* 15 : 480 (Abstr.)

Graham, T.W. & L.E.W. Stone 1975. Field experiment on the cereal cyst nematode (*Heterodera avenae*) in South East England 1967-72. *Ann. Appl. Biol.* 80 : 61-73.

Griffin, G.D. 1977. Influence of non host cultivars on the population dynamics of Heterodera schachtii and growth of sugarbeats. *J. Nematol.* 9 : 269-270.

Griffin, G.D. 1980. Effect of non host cultivars on Heterodera schachtii population dynamics. *J. Nematol.* 12 : 53-57.

Guinez, S.A. 1981. [Fluctuations in nematode populations due to crop rotation]. Fluctuation en la populacion de nematodes mediante la rotacion de cultivos. *Agricultura Tecnica* 41 : 57-59.

Gupta, J.C. & A.S. Atwal 1971. Biology and ecology of Hoplolaimus indicus. II. The influence of various environmental factors and host plants on the reproductive potential. *Nematologica* 17 : 277-284.

Hackney, R.W. & O.J. Dickerson 1975. Marigold, castor, bean, and chrysanthemum as controls of *Meloidogyne incognita* and *Pratylenchus allenl. J. Nematol.* 7 : 84-90.

Haque, M.M. & H.S. Gaur 1988. Effect of intercropping of pea and mustard on the population growth of plant parasitic nematodes. *Indian J. Nematol.* 18 : 351-353.

Hashmi S. & G. Hashmi 1990. Crop rotation and Meloidogyne javanica in vegetable production. Int. Nematol. *Network Newsl.* 7 : 19-21.

Heincke, D. 1989. [Integrated contol of sugarbeet neamtodes]. Rubennematoden integrient bekampfen. *Pflanzenschutz - Prazis*, No. 1, 31-33.

Hesling, J.J. 1958. *Heterodera* major O'Schmidt 1930 - Population changes in field and in plots of fallow soil. *Nematologica* 3 : 274-282.

Huang, S.P. 1984. Cropping effects of marigolds, corn, and okra on population levels of *Meloidogyne javanica* and on carrot yields. *J. Nematol.* 16 : 396-398.

Huang, S.P. & M.V.F. Porto 1988. [Effect of fallow on population of root knot nematodes and on carrot yield]. Efeito do alqueive na poplacao dos nematoides das galhas na producao de canoura. *Fitopatologia Brasileira* 13 : 377-381.

Hutton, D.G., P.L. Coates Beckford & S.A.F. Eason Health 1982. Three year crop rotation for control of *Meloidogyne incognita* on the South central plains of Jamaica (interim report). In *Proc. 3rd Res. Plan. Conf. on root knot nematodes, Meloidogyne* spp. Region I : 136-146.

Hutton, D.G., P.L. Coates-Beckford & S.A.F. Eason Heath 1983. Management of *Meloidogyne incognita* population by crop rotation in a small scale field trial and nematode pathogenic effects on selected cultivars. *Nematropica* 13 : 153-163.

Idowu, A.A. 1989. Assessing advantages of intercropping in nematode management. *Int. Nematol. Network Newsl.* 6 : 31-33.

Inagaki H. 1978. Decrease of *Globodera rostochiensins* population by resistant potato varieties and non host crops in green house trials. *Japanese J. Nematol.* 8 : 11-15.

Ivanova, I.V. 1978. The role of crop rotation in the control of *Ditylenchus dipsaci*. Byulletin' Vsesoyuznogo Institute Gel' mintologii im K.I. Skrybina No. 23 (Fitogel' mintolgiya) : 8-14.

Johnson, A.W. 1985. Specific crop rotation effects combined with cultural practices and nematodes In "An Advanced Treaties on *Meloidogyne* Vol. I. Biology and Control, IMP (Ed. J.N. Sasser and C.C. Carter), North Carolina Univ. Grpahics, Raleigh : 283-301.

Johnson, A.W., C.C. Dowler & E.W. Hauser 1974. Seasonal population dynamics of selected plant parasitic nematodes on four monocultured crops. *J. Nematol.* 6 : 187-190.

Johnson, A.W., C.C. Dowler & E.W. Hauser 1975. Crop rotation and herbicide effects on population densities of plant parasitic nematodes. *J. Nematol.* 7 : 158-168.

Jones, F.G.W. 1956. Soil populations of beet eelworm (*Heterodera schachtii* Schm.) in relation to cropping II. Microplots and field plots results. *Ann. Appl. Biol.* 44 : 25-26.

Khan, A., A.H. Khanzada, F.M. Bilqees & M. Aslam 1987. Influence of crop rotation on the population of some plant parasitic nematodes. *Pak. J. Nematol.* 19 : 447-450.

Khan, A.H., S.K. Saxena & I. Mahmood 1984. Effect of different cropping sequences on the population of plant parasitic nematodes. *Pak. J. Nematol.* 2 : 29-36.

Khan, A.M., S.K. Saxena, Z.A. Siddiqi & R.S. Upadhyay 1975. Control of nematodes by crop rotation. *Indian J. Nematol.* 5 : 214-221.

Khan, A.M., Z.A. Siddiqi, M.M. Alam & S.K. Saxena 1976. Note on the effect of different cropping sequences on the population of plant parasitic nematodes. *Indian J. Agric. Sci.* 46 : 439 - 441.

Khan, M.W., A.M. Khan & S.K. Saxena 1973. Fungi and nematodes in vegetable fields with different cropping pattern. *Indian Phytopath.* 26 : 437-445.

Kinolch, R.A. 1983. Influence of maize rotations on the yield of soybean grown in *Meloidogyne incognita* infested soil. *J. Nematol.* 15 : 398-405.

Kirkpatrick, T.L. & J.N. Sasser 1984. Crop rotation and races of *Meloidogyne incognita* in cotton root knot management. *J. Nematol.* 16 : 323-328.

Koen, H. & N. Grobbelaar 1965. The development effect of *Eragrostis curvula* on *Meloidogyne javanica* population of soil. *Nematologica* 11 : 573-580.

Kus, J. & Z. Gonet 1990. The effect of root crop percentage on productivity of crop rotations and soil fertility. Whpyw uzdialu roslin Opokowyh na produckcyjnose zmianowan i zyznose gleby. *Pamietnik Pulawski* No. 95 : 125-143.

Lamberti, F. 1973. Comparison between rotation and fertilization in the Mediterranean environment. Note II : The effect on populations of an important parasitic nematode, *Tylenchorhynchus dubius*. Confronto tre rotazioni e coneimazioni in ambiente mediterraneo. II. Nota : Influenza sulle popolazioni di un nematode parassita di importanti colture agrarie, *Tylenchorhynchus dubius* (Butschli) Filipjev. Revista di Agronomica 7 : 27-32.

Magnusson, M.L. 1981. Vaxfoljd - ett bra alternative till kemick bekampning av potatiscystnematoden. In Nordisk Vaxtsky ddskonferens Skokloster 17-19 mars. Del. II. Zodogisk och nematologisk sektion. Uppasala, Sweden, Sverigen Lant bruk suniversitet : 88-89.

Mai, W.F. & G.S. Abawi 1980. Influence of crop rotation on spread and density of *Heterodera schachtii* on a commercial vegetables farm in New York. *Plant Disease* 64 : 302-305.

McNamara, D.G. & R.S. Pitcher 1977. The long term effects of four monocultural regimes on two field populations of the nematodes *Xiphinema diversicaudatum* and *Longidorus* spp. *Ann. Appl. biol.* 86 : 405-413.

Menon, P.P.V. & D. Thangaraju 1973. Effect of crop rotation on the population of the potato golden nematode (*Heterodera rostochiensis* Woll.) in the Nilgiris. *Indian J. Agric. Sci.* 43 : 304-306.

Moura, R.M. De 1991. Two years of crop rotation on sugar cane field to control root knot. Part I : effect of the treatment on the nematode population. Dois anos de rotacao de cultura em campos de cana de acucar para controle da Meloidoginose. Efeito dos trtementos na populacao de namatoide. *Nematologia Brasileira* 15 : 1-7.

Mukhopadhyay, M.C. & S.K. Prasad 1968. Population dynamics of *Tylenchorhynchus*. *Nemtologica* 14 : 404-418.

Muro, A. Di 1975a. Root knot nematode (*Meloidogyne incognita*) control by two and three-year crop rotation. *Annali dell. Istituto Sperimentale per il Tobacco* 2 : 89-91.

Muro, A.Di. 1975b. Effetto di varie piante sui nematodi galligeni del tobacco in rotazionii biennali, *Annali dell. Istituto Sperimentale per il Tobacco 2 : 93-99.*

Murphy, W.S., B.B. Brodie & J.M. Good. 1974. Population dynamics of plant nematodes in cultivated soil : Effects of combinations of cropping systems and nematicides. *J. Nematol.* 6 : 103-107.

Naganathan, T.G., R. Arumugam, M. Kulasekaran & S. Vadivellu 1988. Effect of antagonistic crops as intercrops on the control of banana nematodes. *South Indian Horticulture* 36 : 268- 269.

Nishizawa, T. 1978. Annual changes of soil nematodes in the field of continuous cropping or rotation. *Kasetsart J.* 12 : 3- 13.

Nishizawa, T., Y. Ohshima & H. Kurih ara 1971. Survey of the nematode population in the experimental fields for successive or rotational cropping. *Proc. Kanto Tosan Plant Protect Soc* No. 18 : 121-122.

Noe, J.P. 1986. Cropping systems analysis for limiting losses due to plant parasitic nematodes : guide to research methodology. North Carolina State Univ. Graphics, Raleigh, North Carolina, USA, 1-18 pp.

Noe, J.P., J.N. Sasser & J.L. Imbriani 1991. Maximizing the potential of cropping system for nematode management. *J. Nematol.* 23 : 353-361.

Norse, D. 1972. Nematode populations in a maize groundnut tobacco rotation and the resistance of maize varieties to *Meloidogyne javanica. Tropical Agric.* 49 : 355-360.

Nusbaum, C.J. & H. Ferris 1973. The role of cropping systems in Nematode population management. *Ann. Rev. Phytopath.* 11 : 423-440.

Omiyi, P.I. 1977. Influence of some cropping systems on nematode populations in the soil. *Samaru Agric. Newsl., Nigeria 19 : 87-89.*

Oostenbrink, M. 1960. Population dynamics in relation to cropping, manuring and soil disinfection. In "Nematology" (Ed. J.N. Sasser & W.R. Jenkins), Univ. Carolina Press Chapel, Hill : 439-442.

Oostenbrink, M. 1961. Nematode in relation to plant growth II. The influence of crop on the nematode population. *Neth. J. Agric. Sci.* 9 : 55-60.

Oostenbrink, M., J.J. Jacob & K. Kuiper 1956. An interpretation of some crop rotation experiences based on nematode surveys and population studies. *Nematologica* 1 : 202-215.

Piskorz, B. & W. Roszak 1989. Potato cultivation in crop rotation and monoculture. Part II. Effect of increased share of potatos in crop rotation on occurrence of *Globodera rostochiensis* and other disease. Uprawa ziemmiaka w zmianowanue i w monokulturze. Cz. II. Wplyv zwiekszonego matwika ziemmiac zanego i chorob. Roczniki Nauk Rolniczych Seria A, Produckja Roslinna 108 : 41-53.

Ponin, I.Y.A. 1980. The potato nematode and crop rotation. Zaschita Rastenii, Minsk USSR No. 5, 116-20.

Ponin, I.Y.A., B.P. Ivanova & N.N. Timofun 1976. Nematode diseases in maximal saturation of crop rotation with potato. Zashchita Rastenii Minsk, USSR Vypusk 1 : 62-70.

Prasad, J.S., B.T.S. Moorty & Y.S. Rao 1983. Cultural control of the root nematodes *Pratylenchus indicus* in rice through crop rotations.. *Beitrage zur Tropischen Landwirtschft* and *Veterina medizin* 21 : 231-234.

Prasad, J.S. & Y.S. Rao 1978. The influence of crop rotations on the population densties of the root lesion nematode, *Pratylenchus indicus* in rice and rice soils. *Ann. de Zool. Ecol. Animale.* 10 : 627-633.

Rahman, M.L. 1990. Effect of different cropping sequences on root knot nematodes. *Meloidogyne graminicola* and yield of deep water rice. *Nematol medit* 18 : 213-217.

Raymundo, S.A. 1985. Cropping systems research and root knot nematode control. In : "An Advanced Treatise on *Meloidogyne*" Vol. I. Biology and Control. IMP (Ed. J.N. Sasser & C.C. Carter) North Carolina Univ. Graphics, Raleigh : 277-281.

Rhodes, H.L. 1982. Effect of fenamiphos on populations of *Heterodera schachtii* and subsequent plant injury in a cabbage cucumber rotation. *Nematropica* 12 : 289-293.

Rodgriguez-Kabana, R., D.B. Weaver, D.G. Robertson, R.W. Young, & E.L. Carden 1990. *Sorghum* in Rotation with soybean for the management of cyst and root knot nematodes. *Nematropica* 20 : 111-119.

Rodgriguez-Kabana, R., D.G. Robertson, C.F. Weaver & L. Wells 1991a.Rotations of bahiagrass and castorbean with peanut for the management of *Meloidogyne arenaria. J. Nematol.* 23 : 658- 661.

Rodgriguez-Kabana, R.D.G. Robertson, L. Wells, C.F. Weaver & P.S. King 1991b. Cotton as a rotation crop for the management of *Meloidogyne arenaria* and *Sclerotium rolfsii* in peanut. *J. Nematol.* 23 : 652-657.

Rodgriguez Kabana, R. & J.J. Touchton 1984. Corn and Sorghum as rotational crops for management of *Meloidogyne arenaria* in peanut. *Nematropica* 14 : 26-36.

Rodgriguez-Kabana, R. & R.J. Collins 1979. Relation of fertilizer treatments and cropping sequence to populations of two plant parasitic nematode species. Nematropica 9 : 151-166.

Rodgriguez-Kabana, R. & R.J. Collins 1980. Relation of fertilizer treatments and cropping sequence to populations of *Pratylenchus scribneri. Nematropica* 10 : 121-129.

Rohde, W.A., A.W. Johnson, C.C. Dowler & N.C. Glaze 1980. Influence of climate and cropping patterns on the efficacy of ethoprop, methyl bromide, and DD-MENCS for control of root knot nematodes. *J. Nematol* 12 : 33-39.

Ross, J.P. 1962. Crop rotation effects on the soybean cyst nematode population and soybean yields. *Phytopathology* 52 : 815-818.

Rossner, J. 1979. Einfluss langjahringer Monokulturen und unkrautbekampungsmassnahmen auf die Populationsdichte Wandernder Wurzelnematoden. *Zeitschrift fur pflanzenkran Kheiten und pflanzenschutz* 86 : 257-265.

Roy, A.K. 1978. Effectiveness of rotation with non host or fallow on reducing infestation of *Meloidogyne graminicola. Indian J. Nematol.* 8 : 156-158.

Saxena, S.K., M.M. Alam & A.M. Khan 1977. Population dynamics of nematodes. I. Soil population of *Tylenchorhynchus brassicae* in relation to cropping. *Acta Bot. Indica* 5 : 180-182.

Schmitt, D.P. 1991. Management of *Heterodera glycines* by cropping and cultural practices. *J. Nematol.* 23 : 348-352.

Scholte, K. & J.J. Jacob 1983. The influence of continuous cropping and free living root lesion nematodes on yield of fodder maize. *Netherlands J.Pl. Pathol.* 89 : 127-139.

Seinhorst, J.W. 1965. The realtionship between nematode density and damage to plants. *Nematologica* 11 : 137-154.

Seinhorst, J.W. 1970. Dynamics of population of plant parasitic nematodes. *Ann. Rev. Phytopath.* 8 : 131-156.

Sceinhorst, J.W. 1973. The relation between nematode distribution in a field and loss in yield at different average nematode densities. *Nemotologica* 19 : 421-427.

Shafshak, S.E., A.S. El-Debabay, G.A. Sary & M.S. Salem 1974. Effect of preceding crops on the growth and yield of two succeeding crops, maize and wheat. *Zeitschrift für Acker und Pflan.* 140 : 44-53.

Shafshak, S.E. & M.S. Salem 1979. Effect of preceding crops on the spread on weeds and nematode in cotton. *Zeitschrift für Ackerund Pflan* 148 : 318-326.

Sharma, S.K., I., Singh & P.K. Sakhuja 1980. Influence of different cropping sequences on the population of root knot nematodes, *Meloidogyne incognita* and the performance of the subsequent mungbean crop. *Indian J. Nematol.* 10 : 53-58.

Siddiqui, M.A. & M.M. Alam 1984. Studies on the relative efficacy of different cropping sequences and mode of ploughing for the management of plant nematodes (Abstr.). 1st Int. Cong. Nematol., Univ. of Guelph, Ontario, Canada (5-10 Aug.); 94.

Siddiqui, M.A. & S.K. Saxena 1985-89. Seasonal fluctuation in the populations of *Heterodera zeae* in fields under different cropping sequences. *J. Environ. Res.* 6 : 77-79.

Sinnadurai, S. 1973. Crop rotation to control nematodes in tomatoes. *Ghana J. Agric. Sci.* 6 : 137-139.

Slack, D.A., R.D. Riggs & M.L. Hamblem 1981. Nematode control in soybeans : rotation and population dynamics of soybean cyst and other nematodes. Arkansas Exp. Sta. Rept. Series No. 263:36 pp.

Southards, C.J. & B.C. Nichols 1972. Population dynamics of *Meloidogyne incognita* in a tobacco monoculture : effect of winter cover crops. *J. Nematol.* 4 : 234-235.

Stone, L.E.W. 1960. Field experiments on cereal root eelworm. *Ann. Appl. Biol.* 48 : 681-686.

Tacconi, R., E. Biancardi & R. Olimpieri 1990. [Effect of crop rotations and intercrops on Heterodera schachtii (2nd cvontribution)]. Effeto di a v vicendamenti colturali e di colture intercalari su Heterodera schachtii (II contribution). 40 : 47-51.

Tacconi, R., G. Giordani and G. Toderi 1988. Effect of fertilization and rotations on a population of *Pratylenchus thornei.* Informatore *Fitopatologicao* 38 : 47-50.

Tacconi, R. & R. Sauti 1991. Effect of crop rotations in *Heterodera schachtii* (2nd contribution). Effetto di avvicenda meenti colturali su *Heterodera schachtii* (II contributo). Informatore *Fitopatologicao* 41 : 57-59.

Ten'Kovtseva, E.S. & N.M. Datsuk 1975. Preceding crops and the numbers of cereal root nematodes. *Zashchita Rastenii* No. : 9-14.

Ternisien, E. 1989. [Study of crop rotations in banana plantations. II. Impact of rotated crops on banana production and the health of the soil]. Etude des rotations culturales en bananeraie. II. Impact des cultures de rotation sur la production bananiere et 1 etat sanitaire du sol. *Fruits (Paris)* 44 : 445-454.

Ternisien, E. & J. Gangry 1990. [Crop rotation in intensive banana cultivation]. Rotations culturales en culture bananiere intensive. *Fruits* (Paris) Special Issue : 98-102.

Ternisien, E. & P. Melin 1989. [Study of crop rotation in banana plantations I. Assessement of crops for rotation]. Etude des rotations culturales en bananeraie. I. Bilan des cultures de rotation. *Fruits* (Paris) 44 : 373-383.

Tikkikkala, K. 1991. Effect of crop rotation on *Globodera rostochiensis* and on potato yield. Bulletin OEPP 21 : 41-47.

Todd, T.C. 1991. Effect of cropping regime on populations of *Belonolaimus* sp. and *Pratylenchus scribneri* in sandy soil. *J. Nematol* 23 : 646-651.

Vinduska, L. 1978. Ovlivneni populace hadatka repnenho *Heterodera schachtii* Schmidt nekterymi druhy rosthn. Ochrama Rostlin 14 : 135-138.

Vrain, T.C. 1978. Influence of chilling and freezing temperature on infectivity of *Meloidogyne incognita* and *M. hapla. J. Nematol.* 10 : 177-180.

William, C., D.F. Gilman, J.E. Jones & W. Birchfield 1983. Cotton Soyabean rotation for reniform nematode control. Bull. *Louisana Agr. Exp. Sta.* No. 741 : 1-22.

Williams, T.D., J. Beane, M.M. Berry & R.M. Webb 1979. The rotation fumigation experiment, Woburn Experimental Farm, 1969- 1977. In Rothamsted Exp. Sta. Rep. 1978, Part 2, 47-66.

Wilson, G.F. & F.E. Caveness 1980. The effects of rotation crops on the survival of root knot, root lesion and spiral nematodes. *Nematropica* 10 : 56-61.

Winslow, R.D. 1964. Soil nematode population studies. I. The migratory root Tylenchida and other nematodes of the Rothamsted and Woburn six course rotation. *Pedobiologica* 4 : 65-78.

Yeates, G.W., & K.A. Hughes 1990. Effect of three tillages regimes on plant and soil nematodes in an oats/maize rotation. *Pedobiologia* 34 : 379-387.

CHAPTER - 7

PLANTS AND THEIR PRODUCTS ACT AS MAJOR NEMATODE INHIBITORY AGENTS

Rakesh Pandey, R.A. Sikora, A. Kalra, H.B. Singh and S. Pandey

INTRODUCTION

Nematodes are multicellular, bilaterally symmetrical, unsegmented and colourless organisms. Being highly diverse in their habitats occurring in soil, fresh water, ocean, lakes, hills and animals, nematodes feed on bacteria and other micro-organisms, decaying organic matter, plants, animals and human beings. Some species of nematodes involved in the biological complex of the soil environment are parasitic to higher plants causing considerable damage. Most of the soil and plant parasitic nematodes are 0.2 to 1 mm. long but occasionally over 3 mm., vermiform but female become lemon or pear shaped. According to an estimate one acre of agricultural land may contain 3×10^8 nematodes.

Plant parasitic nematodes constitute a small group of nematode species which infect plant roots, bulbs, rhizomes, stems, leaves, buds, flowers, seeds etc. and cause severe damage to the large number of agriculture crops. Phytonematodes cause damage to the plants directly or indirectly. In direct damage, they feed upon the root surface (ectoparasitic nematodes) or partially enter into the roots with their anterior part of the body (semi endoparasitic) or they enter inside the root (endoparasitic). Indirect damage by phytonematodes include their involvement in plant disease complex, making the roots more prone to secondary infections by bacteria and fungi and their role as vectors of many plant viruses.

Reduction in crop yield and loss in a quality of marketable produce results in decline in net profits to the agriculturists. According to an estimate an average loss caused by plant parasitic nematode on large number of crops world wide was more than 12% or more than 77 billion dollars.

The chemicals, which were used earlier to control plant parasitic nematodes, were usually fumigants. Wright (1981) reviewed the mode of action of fumigant nematicides. Their action includes alkylation (Castro and Belser, 1978) and oxidation (Castro and Thompson, 1971). Such materials possess broad biocidal activity. Later, non-fumigant organophosphate and carbamoyl oximes were used for the control of plant parasitic nematodes. The amount of fumigant and non fumigant nematicides applied to control plant parasitic nematodes on various agricultural crops are not only expensive but also caused severe environmental hazards, adversely affect public health, contaminated our ground resources or depletion. Several fumigant nematicides are banned for environment protection or withdrawn from the world market due to their cost of meeting new regulations. There is now pressing demand all around the world for minimising the use of chemical nematicides and search for safer and environment freindly methods of phytonematodes management.

Several methods other than chemicals are known for the management of plant parasitic nematodes on number of agricultural crops. These include biological, physical, cultural, use of resistant varieties etc. Due to high pressure on agricultural land for feeding more mouths of world population, it becomes essential to look out for some cheaper, ecofriendly and safer ways of phytonematode management on agricultural crops. Many plants are known to cause reduction in the nematode population below the damaging level. In few cases plants have been found to be actually antagonistic towards nematodes and some have been shown to produce toxic materials inhibitory to nematodes. Studies pertaining to use of plant organs products, extracts etc. as nematicides have demonstrated promising results. The use of natural plant products to manage phytonematode population has a number of possible advantages over synthetic products. Plant products apart from providing an effective way of managing nematodes are also easy to apply. Such compounds also being biodegradable are unlikely to cause environmental problems than conventional pesticides and may be more readily available and therefore less costly.

Many plants possess nematicidal and nematostatic properties in their roots, shoots, leaves, flowers, seeds etc and their extracts, essential oil, oilseed cakes and products have been successfully tested against variety of phytonematodes (Omidvar 1961,62, Gommers, 1981; Grainge and Ahmad 1988; Sasanelli and Di Vito, 1991; Basu and Majumdar 1998, Sreeja and Charles 1998). Nematicidal properties of the plants and plant products are discussed herein.

1. COMPOUNDS OF PLANT ORIGIN POSSESSING NEMATICIDAL ACTIVITIES

Many plant based chemical compounds like polhythienyls, acetylenes, fatty acids, phenols, alkaloids, phenolic glucosides tannins, terpenoids are known to be toxic to variety of nematodes affecting various crops (Valette *et al.*, 1953; Tarjan and Cheo, 1956; Husain and Masood, 1975). For example *Asparagus officinalis* contains in its stem, leaves and roots a glycoside with low molecular weight toxic to stubby root nematode, *Trichodorus christiei* (Rohde and Jenkins, 1958). Marigold (*Tagetes* spp.) was among the first few plants, which has been examined for nematode killing compound because different species of *Tagetes* often reduce the nematode population in soil. The roots of *Tagetes patula* and *T. erecta* contain a terthieny; and derivatives of bithienyl (Uhlenbroek and Bijloo 1959) which reduce poplulation of *Meloidogyne* sp. and *Pratylenchus* sp. (Oostenbrink *et al.*, 1957; Winoto, 1969). Kanagy and Kaya (1996) reported that a terthienyl at a concentration of 20 and 40 ppm significantly reduces the nematode number that infects insect host. The alpha terthienyl, which is isolated from the root of *Tagetes* sp., is responsible for the reduction of phytonematode population in soil (Varma *et al.*, 1978; Caswell *et al.*, 1991; Zavaleta Meija *et al.*, 1993). For a-terthienyl at concentration of 5 μg/ml was active against stem and bulb nematode, *Ditylenchus dipsaci*, while for *Anguina tritici* it was at 0.5 μg/ml and for *Globodera rostochiensis* it was at 0.1-0.2 μg/ml (Uhlenbroek and Bijloo, 1958).

Several other plants also contain nematicidal compounds effective against variety of plant parasitic nematodes (Nene and Thapliyal 1966; Nene and Krishnakumar, 1967; Swarup and Sharma, 1967; Allen and Feldmesser, 1970; Prot and Kornprobst, 1983; Tada *et al.*, 1988; Stephan *et al.*, 1989). Onda *et al.*, (1965, 1970) reported the activity of three tetracyclic alkaloid viz. chelerythrine, sanguinarine and bocconine isolated from *Bocconia cordata* against *Rhabditis* and *Panagrolaimus* at the concentration of 50-100 μg/100 ml. Fassuliotis and Skucag (1969) reported that monocrotaline isolated from *Crotalaria spectabilis* inhibited movement of *M. incognita* at 10 μg/ml Grainge and Ahmad (1988) reported that alkaloids and tannins isolated from *Ricinus communis* are toxic to *Heterodera schachtii.*

Nematicidal activities of steroidal saponins, asparanin I, A and B isolated from seed and follicle of *Asparagus adescendens* against *M. incognita* have been noticed (Meher *et al.*, 1988). Similar activities have been reported by them in triterpenic saponins, sinusoid, sonumins III isolated from pod walls of *Acacia concinna* and in albichinin II as well as in albichinoside isolated from heart wood of *Albizzia chinensis.* Nematicidal properties of medicarpin and 4-hydroxymedicarpin isolated from *Taverniera abyssinica* against *C. elegans* have also been observed (Stadler *et al.*, 1991). Nematicidal activites of colchicine isolated from *Gloriosa superba*, cyclocurcumin, curcuminoids, curcumin and bisdemethoxy curcumin from *Curcuma longa*, five diphyenylheptanoids, one new and four known compounds from *Curcuma cosmosa* against *M. incognita* have earlier been noticed (Kiuchi *et al.*, 1993; Nidiry *et al.*, 1993; Jurgens *et al.*, 1994). Chandravadana (1994) reported a major compound serpentine in the roots of *Catharanthus roseus* possessing nematicidal activity against *M. incognita.* Further it was stressed that the activity is highly dependent on concentration of compounds and their exposure period as well. Few free amino acids like leucine, aspartic acid, methionine and threonine exhibited nematicidal activity against root knot nematode, *M. incognita* (Nidiry *et al.*, 1994). Some other compounds possessing nematicidal activities against number of plant parasitic nematodes have been listed in Table 1.

2. ESSENTIAL OILS AS ALTERNATE SOURCE OF NEMATICIDES

Essential oils are important for their use in perfumery, pharmaceutical, cosmetics and food industries. Several reports concerning their strong inhibitory activity against a number of agricultural pests including phytonematodes suggest their possible use in pesticides industry. Carvone and limonene, both the major components of *Anethum sowa* Roxb. oil was found to be nematicidal in nature (Saxena *et al.*, 1987). In palmarosa oil the nematicidal activities are reported to be due to geraniol (Abbot, 1925) and linalool (Sangwan, *et al.*, 1985) while in *Nardostachys jatamansi* oil the nematicidal

Table 1. Nematicidal activity of different compounds isolated from various plant species

S.No.	Plants name	Active ingredients	References
1	*Acacia auriculiformis*	Acaciaside A and B	Roy *et al.*, 1993; Ray *et al.*, 1990
2	*Allium fistulosum* var. *caespitosum*	Dipropyl thiosulfinate	Tada *et al.*, 1988
3	*Allium grayi*	1-octanol methyl 4-hydroxybenzoate, methyl 4-hydroxycinnamate, allygrin	Tada *et al.*, 1988
4	*Allium sativum*	Allicin Dially disulphide	Nath *et al.*, 1982 Gupta and Sharma, 1993
5	*Angellica pubescens*	Heptadeca-1, a-diene-4, 6-diyne-3, 8-dios	Munakata, 1983
6	*Arachis hypogaea*	Di-n-butyl succinate	Kimura et al., 1981
7	*Artemisia anma*	Atremisnin, Artemisnic acid, Arteannuin B, Arteether	Bhatacharya *et al.*, 1999
8	*Asparagus officinalis*	Asparaganic acid	Prakash and Rao, 1997
9	*Azadirachta indica*	Azadirachtin, cuercetin, limonoides, nimbidin, nimbidic acid, thionimone	Devakumar *et al.*, 1985, Siddiqui and Alam, 1993
10	*Bocconia cordata*	Bocconine, Chelerythrine, Sanguinarine	Rhode, 1972
11	*Brassica nigra*	Allyl isothiocynates	Ellenby, 1945a, b
12	*Carthamus tinctorius*	3-cis, 11-trans-and & 3-trans, 11 trans-trideca, 1,3,11 triens, 5,7,9-triyene	Munakata, 1978, Kogiso *et al.*, 1976
13	*Cirsium japonicum*	Tridec-1-en-3,5,7,9,11-pentayne, 9,10-epoxyheptadec-16-en-4,6 diyn-8-ol	Kawazu *et al.*, 1980
14	*Coreopsis lanceolata*	1-phenyl hepta-1, 3,5-triyne 2-phenyl-5-(1-propynyl) - thiopene	Kawazu et al., 1980
15	*Cymbopogon caesius*	Linalool	Sangwan et al., 1990
16	*Daphne odora*	Odoracin, Odoratrin	Munakata, 1978
17	*Datura stramonium*	Atropine, Nicotine, Scopolamine	Uhlenbrock & Bijloo, 1959a Winoto, 1959; Kumari et al., 1986
18	*Eclipta alba*	alpha-tetra thienyl dithiophene	Prasad and Rao, 1979
19	*Erigeron philadelphicus*	Methyl 2-trans, 8-cis-deca 2, 8-diene-4, 6-diynoate (2 trans, 8-cis-matricaria esters	Kimura et al., 1981
20	*Gloriosa superba*	Colchicine	Nidiry et al., 1993
21	*Hannoa undulata*	Quassinoids, Chaparrinone, klarneanone, Glaucarbolone	Prot and Komprobs, 1983, 1985
22	*Helenium (hybrid)*	1-tridecanene 3,5,7,9,11 pentanyn 2,3-dihydro-2 hydroxy-3 methlene-6-menthyl benzofuran	Gommers, 1971
23	*Inula helenium*	Alanto lactone	Mahajan et al., 1986
24	*Lawsonia inermis*	Palmitic acid, linoleic acid and oleic acid	Badami and Patil, 1975
25	*Lycopersicon esculentum*	a tomatine	Prakash and Rao, 1997
26	*Ocimum basilicum*	Linalool, Methyl chavicol	Gokte *et al.*, 1991
27	*Ocimum sanctum*	Eugenol	Gokte et al., 1991

Table 1 contd...

Table 1 contd...

S.No.	Plants name	Active ingredients	References
28	*Piper marginatum*	Phenyl propanoids, transan ethole, Estragole	Husain and Doveda, 1990
29	*Solidago canadensis* (= *S. altissima*)	Cis-dehydromatricaria ester	Kawazu *et al.*, 1980
30	*Sophora flavescens*	Cystime, (-) N-methyl cystine	Matsuda *et al.*, 1991
31	*Tagetes erecta*	a terthinenyl 5-(3-butene-1-ynyl)-2, 2-bithienyl	Uhlenbrock & Bijloo, 1959
32	*Tagetes minuta*	(E) - tagetenone, Jasmolins I and II, Limonene, Myrcene, beta-caryophyuenc, cinerins I and II	Kamal and Mangla, 1987, Green et al., 1991
33	*Tagetes patula*	Theophene, a tetrathienyl	Uhlenbrock & Bijloo, 1959, Winoto, 1969, Kumari *et al.*, 1986
34	*Tephrosia purpurea*	Rotenoids, Flavonoids, Chalcones	Kiuchi et al., 1989a, b
35	*Xiphinema strumarium*	Oleic acid, 3-4-dihydroxy cinnamic acid	Ghosh and Sukul, 1992

activities are due to patchouli alcohol (Masood *et al.*, 1980). In 1982 Chatterjee and co workers reported nematicidal activities of eugenol and linalool against *M. incognita*. Sangwan and co workers in 1985 reported nematicidal properties of the major components of essential oils of aromatic grasses viz. geraniol, citral, citronellol and citronellal in essential oil of aromatic grasses against *Anguina tritici, Heterodera avenae, Meloidogyne javanica* and *Tylenchulus semipenetrans*. Nematicidal properties in essential oils of *Acotium calaum, Allium sativum, Anethium sowa, N. jatamansi, Eucalyptus citrodera, C. martini* and *Xanthium straumarium* against various nematodes have also been reported (Malik, *et al.*, 1987; Saxena *et al.*, 1987; Gupta and Sharma, 1991).

Essential oils of *Acorus calamus* (rhizome), *Foreniculum vulgare* (herb) and *Tribulus terrestis* (seed), *Octimum canum* (herb), *Ocimum sanctum* (herb) and *Nicotiana tabacum* (leaf) were found to be toxic to *M. incognita* (Singh *et al.* 1991). Pandey *et al.*, (1998) reported the nematicidal properties in the essential oil of eucalyptus, lemon grass, spear mint, methol mint, peppermint, jamarosa, palmarosa and their actives. Details of nematicidal activities in essential oils have been given in Table 2.

3. PLANT EXTRACTS AS ROUTINE TOOLS OF NEMATICIDAL PRODUCTS

Large number of plants belonging to family Asteraceae suppressed nematode population to a large extent. These plant contain 1,2 dithins in their roots. Root exudates and extract of seedlings of sesame (*Sesamum orientale*) inhibited hatching of eggs and penetration of roots by *M. incognita* (Tada *et al.* 1989). Similarly root exudates of several other plants also inhibited hatching of eggs of variety of phytonematodes (Giebel, 1974; Ellenby 1945 a,b,c; Rohde and Jenkins, 1958, Tanda *et al.*, 1988). Husain and Saxena (1969) reported inhibition of hatching of *M. incognita* eggs in the fruit extracts of *Duranta repen, Solanum xanthocarpum* and seed extracts of *Argemone maxicana, Cucurbito pepo* and *Cucurbito maxima*. These extracts also possess nematicidal activity against variety of other plant parasitic nematodes. Pandey and Haseeb (1988) and Pandey (1990) reported strong nematicidal properties in root and shoot extracts of *Aloe barbadense, Ammi majus,*

Table 2. Nematicidal action of some oils extracted from different plant parts.

S.No.	Plants name	Oil extracted from	References
1	*Acrous calamus*	Seed, Rhizome, Plant	Singh et al., 1991, Saxena, *et al.*, 1987
2	*Allium sativum*	Bulb	Gupta and Sharma, 1991
3	*Anethum sowa*	Plant	Saxena *et al.*, 1987
4	*Argemone mexicana*	Seed	Das and Sukul, 1988
5	*Azadirachta indica*	Seed	Devakumar *et al.*, 1985
6	*Carthamus tinctorius*	Seed	Miller, 1979
7	*Cymbopogon flexuosus*	Leaf	Gokte et al., 1991
8	*Cympobogon flexuosus*	Herbage	Pandey *et al.*, 1998
9	*Cympobogon martini*	Leaf	Pandey *et al.*, 1998
10	*Cympobogon nardus* var. *conferiflorus xjyworancusa*	Leaf	Gokte et al. 1991
11	*Cympobogon wintrianus*	Leaf	Gokte et al., 1991
12	*Datura stramonium*	Seed	Kumari, et al., 1986
13	*Eucalyptus citrodera*	Leaf	Pandey, et al., 1998
14	*Eucalyptus hybrid*	Leaf	Pandey et al., 1998
15	*Foeniculum vulgare*	Herbage	Singh *et al.*, 1991
16	*Guizotia obyssinica*	Seed	Miller 1979
17	*Lawsonia alba*	Seed	Kumari et al. 1986
18	*Mentha arvensis*	Herbage	Pandey et al., 1998
19	*Mentha piperita*	Herbage	Gokte et al., 1991
20	*Mentha spicata*	Herbage	Pandey et al. 1998
21	*Myristica fragrans*	Herbage	Gokte *et al.* 1990
22	*Nardostachy jatamansi*	Plant	Saxena et al., 1987
23	*Nicotiana tabacum*	Seed	Singh et al., 1991
24	*Ocimum basillicum*	Herbage	Gokte et al., 1991; Pandey *et al.*, 1998
25	*Ocimum canum*	Seed	Singh et al., 1991
26	*Ocimum sanctum*	Leaf, Seed	Singh. *et al.*, 1991; Gokte *et al.*, 1991
27	*Palargoneum graveoleans*	Herbage	Pandey et al., 1998
28	*Tagetes minuta*	Leaf	Mukherjee, 1983
29	*Tribulus terrestris*	Seed	Singh et al., 1991
30	*Xanthium strumarium*	Leaf	Ghosh and Sukul, 1992, Malik *et al.*, 1987

A. peryii, Artemisia annua, A. pallens, Lactuca sativa, Gloriosa superba and *Scillia indica* against *M. incognita*. They further reported that these extracts reduced the hatching of *M. incognita* and were dependent on concentration and exposure time. Toxicity of number of plant extracts belonging to family Compositae against several nematodes has been also reported (Tyagi *et al.*, 1988). *Tagetes lucida* was found toxic to variety of phytonematodes in an *in vitro* study and maximum toxicity was reported in flowers followed by seeds, leaves and root extracts (Siddiqui and Alam, 1988). Mohanty and Das (1988) found that leaf extract of *Erythrina indica* at 1:1 and 1:5 concentration was highly toxic to *M. incognita* and *Tylenchorhynchus mashoodi*. Per cent mortality of both the nematodes increased with exposure time. Sarosh *et al.* (1989) observed nematicidal and hatch inhibitory effect of aqueous and ethyl acetate extracts of root, leaves and pods of *Cassia occidentalis* against *M. incognita*. The aqueous extract was found to be more effective than ethyl acetate extract. The compunds were identified as coumestan and flavone. Gokte *et al.*, (1990) tested nematicidal activities in different extract of *Myristica fragrans* against *M. incognita*, *MacKeen et al.*, (1997) tested nematicidal activity in the methonolic extract of 79 Malayasian plants belonging to 42 plant families. They reported strong activity in *Sauropus androgynus, Eugenia polyantha, Areca catachu, Piper betel, and Piper nigrum* against pine wood nematode, *B. xylophilus* at minimum effective dose (0.625 mg/ball). Strong activity was also found in the extract of *Spondias cyntherea, Codiageum variegatum, Eliodia glabra* and *Cicca acida* while moderate nematicidal activity was recorded in *Carica papaya, Ipomoea aquatica, Ocimum basilicum, Leea gigantea, Pithecellabium girringa, Crypteronia paniculata, Myristica fragrans, Murraya koenigii, Leucaena leucocephala, Melostoma malabaricum and Morinda citrifolia.* Weak nematicidal activity were also found in the plant extract of *Ipomoea batatas, Cymbopogon citratus Gancinia atroviridis, Psophocarpus tetragonolobus, Tamarindus indica, Allium odorum* and *Stenchalaena palustris* (MacKeen *et al.*, 1997). Nematicidal activities in other plant extracts have been presented in Table 4.

Over past few years use of variety of cover crops or inter crops releasing nematicidal compounds in to the soil have been used for managing nematode infestation. Marigold (*Tagetes* spp.) has been successfully used as inter crop for the management of different types of nematode pest (Hackney and Dickerson, 1975). Oostenbrink *et al.* (1957) reported that *Tagetes patula* reduces 90% population of *Pratylenchus* sp. and further suggested that *Tagetes* also prevent damage to new planting site. Similarly Good (1965) reported that American marigold as promising nematode reducing cover crop for the control of *M. incognita* population. Their study showed that *Crotalaria*, Marigold, beggarweed, hairy indigo, sudan grass, millet, coastal bermuda grass were effective for reducing phytonematode population. Suatmadji (1969) suggested that simultaneous cultivation of *Tagetes* with trees and ornamental would be more effective against the nematode infection.

Basu and Majumdar (1998) reported nematicidal activities of five neem based compound viz. neem seed kernel, neem seed coat, achook, neemmark and nimbecidine against second stage larvae of *M. incognita*. Mortality of nematodes was found to be directly correlated with the concentration of the extract and exposure period. Sreeja and Charles (1998) screened various plant extracts for their nematicidal properties against *Radopholus similis* and reported that leaf extract of *Piper betel, Kalachoe pinnata*, root extract of *Indigofera tinctoria* and *Solanum indicum* as well as rhizome extract of *Acorus calamus* and *Cyperus rotundus* have high degree of nematostatic properties. Similarly the nematicidal activities were also reported by number of research investigator in roots of large number of plant.

4. PLANTS AND THEIR OIL SEED CAKES AS NEMATICIDAL PRODUCTS

As various agricultural plants possess nematicidal activities in their different parts, their parts as such or after chopping if incorporated in to the soil would reduce the nematode population and severity of root knot nematode infection (Khan, 1990).

The incorporation of plant materials in soil has been demonstrated as a satisfactory method for the control of plant parasitic nematodes on large number of agricultural crops. This has advantages of low cost and easy applicability. Apart from achieving a good control of nematodes, addition of organic amendments also improves the nutrient status of the soil, and soil structure. Oil seed cake are widely used to suppress nematode population in soil

Table 3. Chopped/Amendments of plant parts in soil for nematode inhibition on agricultural crops

S.No.	Plants Name	Plant parts Used/ Amendments	Target nematode species	Reference
1	*Achyranthes aspera*	Plant material	*M. incognita*	Chattopadhyay, 1991
2	*Adhatoda vasica*	Leaf powder	*M. incognita*	Pandey, 1995
3	*Aloe barbadensis*	Leaf	*M. incognita*	Mahmood *et al.*, 1979
4	*Anacardium occidentale*	Plant material	M. incognita	Chattopadhyay, 1991
5	*Andrographis paniculata*	Dried plant	*M. incognita*	Goswami & Vijailakshmi, 1985
6	*Ananus comosus* L.	Leaves	*Meloidogyne* sp.	Linford *et al.*, 1938
7	*Anthocephalus chinensis*	Leaf	*M. incognita*	Chatterjee & Sukuk, 1980
8	*Argemone mexicana*	Chopped leaf, Plant material	*M. incognita*	Reddy, *et al.*, 1990, Chattopadhyay, 1991
9	*Artocarpus heterophyllus*	Chopped shoot	*Meloidogyne sp. R. reniformis Tylenchorhynchus brassicae*	Siddiqui *et al.*, 1987
10	*Azadirachta indica*	Chopped leaf, Leaf powder Kernel powder	*M. incognita, M. javanica, A. tritici, H. indicus, Helicotylenchus indicus, T. brissicae, T. filliformis*	Lall & Hameed, 1969, Bhatnagar et al., 1980, Gokte & Swarup, 1988 Singh & Sitaramaiah, 1967; Singh & Kataria, 1991
11	*Boehmerea nivea*	Leaf amendment	*M arenaria*	Mian & Rodriguez, 1982
12	*Caesalpinia crista*	Leaf	*M. incognita*	Krishnamurthy & Murthy, 1990
13	*Calendula officinalis*	Dried plant Amendment	*M. incognita*	Goswami & Vijailakshmi, 1985
14	*Calotropis gigantea*	Leaf amendment Leaf powder Flower and Plants	*M. incognita, R. reniformis, H. indicus*	Kumar & Nair, 1976, Rao & Pandey, 1982, Bhatnagar et al., 1980 Vijailakshmi & Goswami, 1985 Sivakumar & Marimuthu, 1986, Chattopadhyay, 1991
15	*Calotropis procera*	Chopped leaves	*M. incognita*	Reddy et al., 1990
16	*Canabis sativa*	Plant amendment	*M. incognita*	Goswami & Vijailakshmi, 1986
17	*Carica papaya*	Chopped shoot	*Meloidogyne sp. R. reniformis Tylenchorhynchus brassicae*	Siddiqui et al. 1987
18	*Cassia fistula*	Leaf amendment	*M. javanica*	Singh and Sitaramaiah, 1967
19	*Cassia tora*	Plant	*Meloidogyne sp.*	Verma, 1993
20	*Catharanthus roseus*	Leaf and Plant amendment	*M. incognita, M. javanica, R. reniformis T. vulgaris*	Patel, 1995, Patel *et al.*, 1987, Krishnamurty & Murthy, 1990, Chattopadhyay, 1991
21	*Chenopodium anbrossoides*	Plant amendment	*M. incognita*	Garciaespinosa, 1980
22	*Clerodendrum inerme*	Plant amendment, Leaf powder	*M. incognita, M. javanica*	Patel, 1995, Patel et al., 1987
23	*Crotalaria* sp.	Leaf	*M. arenaria, M. hapla*	Yuhara, 1971b Mian & Rodriguez Kabana, 1982

Table 3 contd...

Table 3 contd...

S. No.	Plants Name	Plant parts Used/ Amendments	Target nematode species	Reference
24	*Datura metel*	Leaf amendment	*M. javanica*	Kaliram & Gupta, 1980, 1982
25	*Datura stramonium*	Plant materials	*M. incognita*	Chattopadhyay, 1991
26	*Eupatorium odoratum*	Leaf amendment	*M. incognita*	Kumar *et al.*, 1976, Kuriyan, 1988, Subramanium, 1985
27	*Ficus carica*	Chopped shoot	*Meloidogyne sp. R. reniformis, Tylenchorhynchus brassicae*	Siddiqui et al., 1987
28	*Ficus elastica*	Stem	*M. incognita, R. reniformis, T. brassicae*	Siddiqui *et al.*, 1987, 1992
29	*Ficus glomerata*	Chopped shoot	*Meloidogyne sp. R. reniformis Tylenchorhynchus brassicae*	Siddiqui *et al.*, 1987
30	*Giliricidia sepium*	Chopped leaf	*M. incognita R. similis*	Kuriyan, 1988; Jasy & Koshy, 1992
31	*Holarrhena antidysenterica*	Powdered bark	*M. Incognita*	Chattopadhyay, 1991
32	*Ipomoea fistulosa*	Chopped shoot	*Meloidogyne sp. R. reniformis Tylenchorhynchus brassicae*	Siddiqui *et al.*, 1987
33	*Lantana camara*	Chopped leaf	*M. incognita*	Reddy et al., 1990
34	*Leucaena leucocephala*	Chopped leaf	*M. incognita*	Paruthi *et al.*, 1987
35	*Mangifera indica*	Leaf	*M. incognita Criconemoides annulatum*	Kumar & Nair, 1976, Jain & Saxena, 1993
36	*Melia azeridach*	Dried leaf Chopped leaf	*M. javanica, M. incognita*	Singh & Sitaramiah, 1967, Goswami & Vijailakshmi 1986, Verma, 1993
37	*Nerium odoratum*	Chopped shoot	*Meloidoyne sp. R. reniformis Tylenchorhynchus brassicae*	Siddiqui et al., 1987
38	*Ocimum sanctum*	Plant material	*M. incognita*	Chattopadhyay, 1991
39	*Parthenium hyterphorus*	Chopped leaf	*M. incognita*	Reddy *et al.*, 1990
40	*Pongamia glarba*	Mulching of leaf	*M. incognita*	Govindiah *et al.*, 1989
41	*Prosopis juliflora*	Chopped leaf	*M. incognita*	Sunderbabu *et al.*, 1990
42	*Ricinus communis*	Leaf powder and Chopped leaf	*M. incognita H. indicus, H. dihystera, T. vulgaris, M. javanica, R. reniformis, T. brissicae*	Bhatnagar *et al.*, 1980, Ravidutt & Bhatti, 1986, Goswami & Vijailakshmi, 1986 Janesh & Lamberti, 1983, Zaki & Bhatti, 1986
44	*Sesbania aculeata*	Leaf amendment	*M. javanica*	Singh & Sitaramaiah, 1967; Goswami & Vijailakshmi, 1986
44	*Sesamum indicum*	Chopped leaf	*M. incognita*	Reddy *et al.*, 1990
45	*Tabernaemontana coronaria*	Chopped shoot	*Meloidogyne sp. R. reniformis Tylenchorhynchus brassicae*	Siddiqui et al., 1987
46	*Tagetes erecta*	Chopped leaf	*M. incognita, H. indicus, H. dihystera, T. vulgaris*	Varma and Anwar, 1995
47	*Typhonium trilobatum*	Corn powder	*M. incognita*	Chattopadhyay & Mukhopadhyay, 1989
48	*Vernonia cineraria*	Plant material	*M. incongita*	Chattopadhyay, 1991

Table 4. Oil seed cakes as nematode inhibitory product

S.N.	Plant name/ Common name of Oil cakes	Target nematode species	Host plant	References
1	*Arachis hypogaea* Linn. (Ground nut oil seed cake)	*Aphelenchoides sacchari, Aphelenchus avenae, Ditylenchus cypei, Helicotylenchus erythrine Helicotylenchus sp. Hoplaimus indicus, M. exigua, M. incognita, M. javanica, Pratylenchus delattrei, R. reniformis, T. brassicae*	Brinjal, Citrus, Coffee, Guava, Mulberry, Okra, Tobacco, Tomato	Khan et al., 1967; Singh & Sitaramaiah, 1971; Gaur & Prasad, 1970; Desai *et al.*, 1973; Bhatnagar *et al.* 1980; Kaliram & Gupta, 1980; Khan & Husai 1988, Poornima & Vadivelu, 1990
2	*Azadirachta indica* Juss. (Neem oil seed cake)	*Aphelenchoides sacchari, Aphelenchus avenae, Ditylenchus cypei, Helicotylenchus dihystera Helicotylenchus erythrine Helicotylenchus sp. Hemicriconemoides Hopolaimus indicus Meloidogyne javanica, Meloidogyne incognita, Pratylenchus delattrei, Pratylenchus zeae, R. reniformis, Radopholus similis, T. brassicae, T. vulgaris, Tylenchorhynchus elegans*	Arecanut, Soyabean, Brinjal, Citrus, Davana, Guava, Mint Okra, Tobacco Mulberry, Betelvine, Tomato, Wheat, Mungbean, Cowpea	Sing and Sitaramaiah, 1967; Khan *et al.*, 1967; Lall & Hameed, 1969; Egunjobi & Afolami, 1976, Singh & Sitaramaiah, 1969, 1970; Sitaramaiah and Singh, 1977; Bhatnagar *et al.*, 1978; Sikdar *et al.*, 1986; Jagadale *et al.*, 1985a; Bhattacharya & Goswami, 1987; Misra and Prasad, 1989, Gnanaprogosam *et al.*, 1993, Pandey 1995, 1997, Singh & Kataria, 1991; Pandey *et al.*, 1992; Poornima & Vadivelu, 1990
3	*Boehmeria nivea* (Ramie oil cake)	*M. arenaria*	—	Mian & Kabana, 1982
4	*Brassica campestris* L. (Mustard cake)	*Aphelenchus avenae Hoplolaimus indicus Meloidogyne arenaria Meloidogyne incognita Meloidogyne javanica Heterodera avenae Tylenchus filiformis T. brassicae, T. vulgaris, Heterodera rostochinensis*	Okra, Potato, Tomato, Maize, Sunhemp	Hameed, 1968; Singh and Sitaramaiah, 1971; Misra & Prasad, 1973; Khan *et al.*, 1966; Alam *et al.*, 1977, Ellenby, 1945a, b Patel, *et al.*, 1993
5	*Carthamus tinctorius* Linn. Safflower cake	*Meloidogyne incognita M. javanica*	Tomato	Hameed, 1968; Goswami, & Vijailakshmi, 1986
6	*Eruca vesicaria* L. Rocket Salad cake	*M. incognita*	Okra	Bhatnagar et al., 1978
7	*Gossypium hirsutum* Linn. Cotton oil seed cake	*M. incognita*	Vegetable crops	Lanjeswar & Shukla, 1986.
8	*Guizotia obyssinica* cass	*M. incognita*	Tomato	Gowda, 1972
9	*Helianthus annuus* Linn. Sunflower cake	*M. javanica, M. incognita*	Brinjal	Singh & Singh 1988
10	*Linum usitatissimum* Linseed cake Linn.	*M. incognita, M. javanica*	Brinjal Okra, Tomato	Singh & Sitaramaiah, 1969; Srivastava, 1971, Singh & Singh, 1988

Table 4 contd...

Table 4 Contd...

S.N.	Plant name/ Common name of Oil cakes	Target nematode species	Host plant	References
11	*Madhuca indica* Gmel. Madhua cake	*Apelenchus avenae, Ditylenchus cypei, Pratylenchus delattrei M. incognita, M. javanica, Hoplolaimus indicus, R. reniformis, Tylenchorhynchus brassicae*	Okra, Tomato, Brinjal, Citrus, Guava, Menthol mint, Davana	Singh and Sitaramaiah, 1967; Mobin and Khan, 1969; Khan *et al.*, 1974; Alam and Khan, 1974; Pandey, 1989, 1995; Pandey *et al.*, 1992; Poornima and Vadivelu, 1993.
12	*Pongamia glabra* Karanja Cake	*M. incognita, M. javanica*	Tomato, Tobacco, Betelvine, Mulberry	Hameed, 1968; Singh and Sitaramaiah, 1967; Desai *et al*, 1979; Jagadale *et al*, 1985a, b
13	*Ricinus communis* Linn Castor cake	*Aphelenchus avenae Ditylenchus cypei Hoplaimus indicus Helicotylenchus dihystera Meloidogyne incognita Meloidogyne javanica Heterodera avenae Tylenchus filiformis T. brassicae, T. vulgaris*	Okra, Tomato Citrus, Brinjal, Mulberry, Papaya, Menthol mint, Davana	Singh and Sitaramaiah, 1967; 1976; Khan et al., 1973; Alam *et al.*, 1977; Mukherjee 1983; Sikdar *et al.*, 1986; Reddy *et al.*, 1990; Pandey, 1989, 1995; Pandey *et al.*, 1992
14	*Shorea robusta* Gaertn. Sal cake	*Meloidogyne incognita*	Wheat	Sharma et al. 1985

(Khan *et al.*, 1966, Singh and Sitaramaiah, 1970). Generally amendments of oil seed cakes in soil suppressed population of phytonematodes through toxic decomposed products like phenolics, ammonia or increased predacious and parasitic activity of soil biota. Similarly the use of cotton waste, alfalfa pellet, alfalfa hay, castor pomace, olive pomace, dried crop residues, corn and soybean meal, chitin have been found effective. Different types of oil seed cakes like neem, castor, mahua, mustard, linseed, peanut etc. not only reduces the nematode infestation and galling but also reduce nematode population in roots and soil and their egg laying capacity. Details have been given in Table 4.

In conclusion research on use of different plant materials, their by products, essential oils and other organic materials in last one decades has been carried out tremendously due to hazardous effect of chemical nematicides on environment and human health. In coming era it could be one of the best substitute for nematicides. However, a more extensive investigations are needed for identification of nematicidal principles in plants and their mode of action against variety of plant parasitic nematodes. The cost benefit ratio should always be kept in mind before using or applying these plants or their products for the phytonematode management.

Table 5. Nematicidal activities in the extracts of different parts of various plants

S.No.	Name of the Plant	Extracted part of the plant	Reference
1	*Acacia auriculiformis*	Funicle	Roy et al., 1993
2	*Acacia concinna*	Fruit	Vijailakshmi,et al., 1979
3	*Achillea millefolium*	Seed	Vijailakshmi, et al., 1979
4	*Acorus calamus*	Seed rhizome	Kiuchi et al., 1989
5	*Adhatoda vasica*	Leaf	Pandey, 1995; Hussaini, et al., 1996
6	*Ageratum conzoides*	Plant and leaf	Mukherjee, 1983, Vijailakshmi et al., 1979, Nath *et al.*, 1982
7	*Allium cepa*	Bulbs	Nirdhy et al., 1994
8	*Allium sativum*	Bulbs	Gupta and Sharma, 1991, Gupta et al., 1995
9	*Aloe barbadensis*	Plant & root	Pandey and Haseeb, 1988
10	*Aloe peryii*	Plant & root	Pandey and Haseeb, 1988
11	*Alpina galenga*	Rhizome	Kiuchi et al., 1989
12	*Amaranthus gracilis*	Plant	Singh et al., 1955, Haider, 1993
13	*Anagallis arvensis*	Leaf	Singh et al., 1955, Haider, 1993
14	*Andrographis paniculata*	Leaf & Plant	Mukherjee & Shukul, 1978
15	*Angelica pubscens*	Plant	Katsura, 1981
16	*Annona squamosa*	Leaf	Hussain and Masood, 1975, Mahmood *et al.*, 1979
17	*Anthocephalus chinensis*	Root	Sarra-Guzmann, 1984
18	*Antigonon leptopus*	Leaf & flower	Ali *et al.*, 1991, Ahmad and Khan, 1991
19	*Antipogon cephalus*	Leaf	Ali *et al.*, 1991
20	*Arachis hypogaea*	Leaf	Majumdar and Misra, 1991
21	*Araucaria cookii*	Leaf	Jain and Shukul, 1988
22	*Argemone mexicana*	Leaf, root, shoot & seed	Das and Shukul, 1988 Majumdar and Misra, 1991; Hussaini et al., 1996
23	*Artobotrys odoratissimus*	Leaf	Chattopadhyay and Mukhopadhyay, 1989
24	*Artemisia absinthium*	Leaf	Vijailakshmi, et al., 1979
25	*Artemisia annua*	Root	Pandey, 1990
26	*Artemisia pallens*	Root & shoot	Pandey, 1990
27	*Artemisia vulgaris*	Root	Sarra-Guzmann, 1984, Sherif et al., 1987
28	*Artocarpus heterophyllus*	Shoot	Siddiqui et al., 1987
29	*Asparagus officinalis*	Root	Das, 1983
30	*Asparagus racemosus*	Root	Swarup and Sharma, 1967
31	*Azadirachta indica*	Leaf	Qamer, et al., 1989, Vats and Nandal, 1995, Akhtar, 1995
32	*Balanites aegyptica*	Plant	Ibrahim, 1992
33	*Bauhinia variegata*	Bark	Vijailakshmi, et al., 1979
34	*Berberis asiatica*	Stem	Vijailakshmi, et al., 1979
35	*Boerhaavia diffusa*	Leaf	Vijailakshmi et al., 1979
36	*Bongainvillea spectabilis*	Leaf	Sunderbabu et al., 1990
37	*Caesalpenia cristata*	Leaf	Vijaylakshmi et al., 1979
38	*Calendula officinalis*	Leaf & Shoot	Sharma and Trivedi, 1991 Tyagi et al., 1988
39	*Colotropis gigentea*	Plant, Flower	Vijaylakshmi, et al., 1979, Vijailakshmi and Goswami, 1985, Hussaini *et al.*, 1996
40	*Colotropis procera*	Leaf	Nazar and Nath, 1989 Jain et al., 1986
41	*Cannabis sativa*	Leaf, root and shoot	Vijaylakshmi et al., 1979, Nazar and Nath, 1989 Haseeb, *et al.* 1982

Table 5 contd...

S.No.	Name of the Plant	Extracted part of the plant	Reference
42	*Carica papaya*	Shoot and fruit	Siddiqui *et al.*, 1987, Hoan & Dravid, 1979
43	*Carthamus oxycantha*	Leaf and Plant	Mukherjee 1983
44	*Carthamus tinctorius*	Root and flower	Katsura, 1981
45	*Carum capticum*	Seed	Vijailakshmi *et al.*, 1979
46	*Cathranthus roseus*	Leaf	Hussaini *et al.*, 1906
47	*Cassia angustifolia*	Leaf	Vijailakshmi *et al.*, 1979
48	*Cassia carandas*	Leaf	Haseeb *et al.*, 1982
49	*Cassia fistula*	Leaf	Haseeb *et al.*, 1982, Vijailakshmi *et al.*, 1979
50	*Cassia occidentalis*	Leaf & root	Sarosh *et al.*, 1989
51	*Centrosema pubescens*	Leaf	Hoan & David, 1979
52	*Chenopodium album*	Plant	Nandal & Bhatti, 1983
53	*Chenopodium ambrosoides*	Leaf, root and shoot	Husain & Masood, 1975 Haseeb *et al.*, 1978
54	*Chromolaena odorata*	Leaf	Hoan & David, 1979, Muller & Sturhan, 1994
55	*Chrysanthemum coronarium*	Shoot	Tyagi *et al.*, 1988
56	*Chrysanthemum indicum*	Leaf	Sharma and Trivedi, 1991
57	*Chugsan themium*	Shoot	Tyagi *et al.*, 1988
58	*Cichorium intybus*	Leaf	Vijailakshmi et al., 1979
59	*Cineraria martina*	Leaf & root	Sasanelli and Adabbo, 1993
60	*Cirsium arvense*	Plant	Mukherjee 1983
61	*Citrus reticulata*	Leaf, fruit & peel	Mukherjee, 1983
62	*Cleome viscosa*	Leaf	Mani *et al.*, 1986
63	*Clerodendron infortunatum*	Leaf	Haider, 1993
64	*Clerodendron inerme*	Leaf	Patel, 1995
65	*Celosia antigarum*	Leaf	Haseeb *et al.*, 1982
66	*Commiphora wightii*	Leaf	Sharma and Trivedi, 1991
67	*Cosmus bipinnatus*	Shoot	Tyagi *et al.*, 1988
68	*Crotalaria spectabilis*	Leaf & seed	Hoan and Kosy, 1992
69	*Crotalaria juncea*	Leaf & plant	Jasy and Kosy. 1992 Mian and Rodriguez, 1982 Dar, 1983
70	*Croton sparsiflorum*	Leaf	Islam 1983
71	*Cuscuta reflexa*	Stem, Flower	Haider, 1993, Vijailakshmi and Goswami, 1985
72	*Cycas circinalis*	Leaf	Sharma and Trivedi, 1991
73	*Cymbopogon flexuosus*	Leaf	Gokte *et al.*, 1991
74	*Cymbopogon citratus*	Leaf	Kumar and Air, 1976, Hoan & David, 1979
75	*Cynodon dactylon*	Leaf & Plant	Hoan & David, 1979
76	*Cyperus rotundus*	Plant	Hoan & David, 1979
77	*Dahalea pinnata*	Shoot	Tyagi *et al.*, 1988
78	*Daphne odora*	Root	Munakata, 1978, Katsura, 1981
79	*Datura alba*	Leaf	Mukhtar, *et al.*, 1994
80	*Datura metel*	Leaf	Vijailakshmi *et al.*, 1979, Majumdar and Misra, 1991, Mukherjee, 1983
81	*Datura stramonium*	Leaf & seed	Kumari *et al.*, 1986, Mani *et al.*, 1986, Chatopadhyay, 1991
82	*Derris elleptica*	Root & seed	Hoan & David, 1979
83	*Desmodium zongelicum*	Root	Surra-Guzman, 1984
84	*Digitaria decumbens*	Root and Plant	Haroon & Smart, 1983, Haroon et al., 1982

Table 5 contd...

S.No.	Name of the Plant	Extracted part of the plant	Reference
85	*Dimorphotheca sinuata*	Shoot	Tyagi *et al.*, 1988
86	*Dalbergia sissoo*	Leaf	Haseeb, *et al.*, 1982
87	*Eclipta alba*	Plant, leaf, shoot and root	Prasad and Rao, 1979, Vijailakshmi *et al.*, 1979, Majumdar and Misra, 1991, Tyagi *et al.*, 1988, Bano *et al.*, 1986
88	*Eichhornia crassipes*	Leaf and root	Jain *et al.*, 1986, Sarra-Guzmann, 1984, Fazal and Husain, 1991
89	*Emblica officinalis*	Leaf	Vijayalakshmi *et al.*, 1979
90	*Embelica vibes*	Seed	Vijailakshmi *et al.*, 1979
91	*Embelica officinalis*	Leaf	Haseeb *et al.*, 1982
92	*Eragrostis amabillis*	Leaf	Hoan & David, 1979
93	*Erythrina indica*	Leaf	Qamar *et al.*, 1984 Mohanty and Das, 1988
94	*Esthascholpzia caliphorni*	Leaf	Sharma and Trivedi, 1991
95	*Eucalyptus globulus*	Leaf	Hussaini *et al.*, 1996, Jain and Gupta, 1995
96	*Eucalyptus tereticornis*	Leaf	Vats and Nandal, 1993
97	*Eupatorium odoratum*	Plant material	Subramaniyam, 1985
98	*Euphorbia cyparissias*	Plant	McIndoo, 1982
99	*Euphorbia pilulifera*	Leaf	Hussaini *et al.*, 1996
100	*Ficus glomerata*	Shoot, Flower	Vijailakhmi *et al.*, 1979, Siddiqui *et al.*, 1987
101	*Fleurya interrupta*	Leaf	Mukherjee & Shukul, 1978, Mukherjee, 1983
102	*Foeniculum vulgare*	Plant	Singh *et al.*, 1991
103	*Gaillardia picta*	Leaf, flower & shoot	Tyagi *et al.*, 1985
104	*Gliricida sepium*	Leaf	Jasy & Koshy, 1992
105	*Gliricida maculata*	Leaf	Jasy and Koshy, 1992
106	*Gloriosa superba*	Root & Shoot	Pandey and Haseeb, 1988
107	*Glycurrhiza glabra*	Bark	Vijailakshmi *et al.*, 1979
108	*Gliricida maculata*	Leaf	Mani *et al.*, 1986
109	*Hannoa kalaineana*	Seed	Prot and Komprobst, 1985
110	*Hannoa undulata*	Root and seed	Prot and Komprobst, 1983, 85
111	*Helianthus annuus*	Root, flower & seed	Dar, 1983
112	*Holarrhena antidysenterica*	Leaf, root & bark	Majumdar and Misra, 1991, Mukherjee, 1983
113	*Hydrocarpus laurifolia*	Flower, fruit & seed	Mukherjee, 1983
114	*Hyssopus officinalis*	Leaf	Vijailakshmi *et al*, 1979
115	*Imperata cylindrica*	Leaf	Hoan & David, 1979
116	*Ipomoea palmata*	Leaf & shoot	Kumari *et al.*, 1986
117	*Ipomoea carnea*	Leaf	Siddiqui and Husain, 1990
118	*Ipomoea fistulera*	Leaf & Shoot	Siddiqui *et al.*, 1987 Patel *et al.*, 1987
119	*Iris japonica*	Leaf	Katsura, 1981
120	*Jatropha curcas*	Leaf	Qamar *et al.*, 1989
121	*Jatropha gossypifolia*	Leaf	Sharma & Trivedi, 1995
122	*Lantana comara*	Leaf	Chandel and Mehta, 1990
123	*Lawsonia alba*	Plant & Seed	Kumari *et al.*, 1986
124	*Lawsonia inermis*	Leaf	Kumari *et al.*, 1986 Vijailakshmi *et al.*, 1979
125	*Leucaena leucocephala*	Root & leaf	Hoan & David, 1979 Grewal & Sohi, 1988
126	*Leucas aspera*	Leaf	Husaini et al., 1996
127	*Linum usitatissimus*	Leaf	Sharma and Trivedi, 1991

Table 5 contd...

S.No.	Name of the Plant	Extracted part of the plant	Reference
128	*Malva sylvertris*	Leaf	Vijailakshmi *et al.*, 1991
129	*Mangifera indica*	Leaf	Vijailakshmi *et al.*, 1979 Jain & Saxena, 1993
130	*Melia azedarach*	Leaf root and shoot	Haseeb *et al.*, 1978, Vijayalakshmi *et al.*, 1979
131	*Mentha piperita*	Plant	Gokte et al. 1991 Vijailakshmi *et al.*, 1979 Haseeb *et al.*, 1978
132	*Mimosa pudica*	Leaf, Seed	Hoan & Davids, 1979, Vijaylakshmi, *et al.*, 1979
133	*Mimusops elengi*	Bark	Vijayalakshmi *et al.*, 1979
134	*Momordica charantia*	Plant	Mukherjee, 1983
135	*Moren bryanthemum*	Leaf	Mukhtar *et al.*, 1979
136	*Moringa pterygosperma*	Leaf	Hoan & Davids, 1979 Sarra-Guzman, 1984
137	*Moringa oleifera*	Leaf	Krishnamurthy and Murthy, 1990
138	*Myristica fragrans*	Plant	Gokte *et al.*, 1990
139	*Nardostachys jatamansi*	Bark	Vijailakshmi *et al.*, 1979
140	*Nasturtium officinalis*	Leaf	Ellenby, 1945a
141	*Nelumbium spaciosium*	Leaf	Vijailakshmi, *et al.*, 1979
142	*Nerium odorum*	Shoot	Siddiqui *et al.*, 1987
143	*Nicotiana tabacum*	Root, shoot and seed	Singh and Tomar, 1991 Haseeb *et al.*, 1978
144	*Ocimum canum*	Leaf, shoot, root and seed	Singh and Tomar, 1991, Chaterjee *et al.*, 1982 Sharma and Trivedi, 1995, Husaini *et al.*, 1996
145	*Ocimum sanctum*	Seed, Leaf Flower	Vijailakshmi et al, 1979, Singh and Tomar, 1991, Fazal and Husain, 1991
146	*Oxoxylon indicum*	Leaf	Vijaylakshmi *et al.*, 1979
147	*Oryza sativa*	Leaf	Hoan & Davids, 1979
148	*Papaver rhoeas*	Root & shoot	Haseeb *et al.*, 1978
149	*Parthenium hysterophorus*	Plant, leaf	Hasan & Jain, 1984, Mani *et al.*, 1986
150	*Petunia vialacea*	Leaf	Sharma and Trivedi, 1991
151	*Phaseolus lunatus*	Seed	Hussain and Massod, 1975
152	*Phlox paniculata*	Leaf	Sharma and Trivedi, 1991
153	*Plantago ovata*	Seed	Vijailakshmi *et al.*, 1979
154	*Polyalnthea longifolia*	Leaf	Sharma & Trivedi, 1995
155	*Pongamia glabra*	Leaf	Majumdar and Misra, 1991
156	*Populus deltoides*	Leaf	Grewal & Sohi, 1988
157	*Portulaca oleracea*	Leaf and shoot	Hoan & Davids, 1979
158	*Prosopsis juliflora*	Leaf	Mani *et al.*, 1986
159	*Psoralea coryfolia*	Seed	Vijayalkshmi *et al.*, 1979
160	*Pterocarpus santalinus*	Wood	Vijailaksmi *et al.*, 1979
161	*Puerasia lobata*	Leaf & root	Mian and Rodriguea-Kabana, 1982
162	*Raphanus sativus*	Root	Dar, 1983
163	*Rauwolfia serpentina*	Root	Vijayalakshmi, et al., 1979
164	*Ricinus communis*	Leaf	Mani *et al.*, 1986, Akhtar and Mahmood, 1994
165	*Ruta graveolens*	Leaf & root	Sasanelli and Addabbo, 1992
166	*Scillia indica*	Root & shoot	Pandey and Haseeb, 1988
167	*Semecarpus anacardium*	Fruit	Vijailakshmi et al., 1979
168	*Sesbania sesban*	Plant	Ibrahim, 1992
169	*Sida cordifolia*	Leaf	Mahmood *et al.*, 1982

Table 5 contd...

S.No.	Name of the Plant	Extracted part of the plant	Reference
170	*Solanum hispidium*	Leaf & Root	Haseeb *et al.*, 1978
171	*Solanum sucrense*	Root	Jacabson, 1958
172	*Solanum pampasense*	Root	Jacobson, 1958
173	*Solanum xanthocarpum*	Stem	Vijailakshmi et al., 1979
174	*Sonchus oleracens*	Shoot	Tyagi *et al.*, 1988
175	*Sphaeranthus indicus*	Stem	Vijayalakshmi *et al.* 1979
176	*Sphenocloea zeylanica*	Leaf and Plant	Mohandas *et al.*, 1982
177	*Sapindus emerginatus*	Leaf	Mani *et al.*, 1986
178	*Swertia chirata*	Plant	Vijayalakshmi *et al.*, 1979
179	*Symplocas racemoses*	Bark	Vijayalakshmi *et al.*, 1979
180	*Syndrella nodiflora*	Leaf	Hussaini et al., 1996
181	*Tagetes erecta*	Leaf & root	Sasanelli and Addabbo, 1993, Hussaini *et al.*, 1993
182	*Tagetes lucida*	Leaf, root, flower & seed	Siddiqui and Alam, 1988 Mukherjee 1983
183	*Tagetes patula*	Plant	Kumari *et al.*, 1986
184	*Tagetes tenuifolia*	Root	Mukherjee, 1983
185	*Tamarix gallica*	Leaf	Vijayalakshmi, *et al.*, 1978
186	*Thuja orientalis*	Root, stem, leaf, fruit	Jain *et al.*, 1986, Fazal and Husain, 1991
187	*Tinospora cordifolia*	Stem	Vijayalakshmi *et al.*, 1979
188	*Tithonia diversifolia*	Flower	Tyagi, *et al.*, 1985
189	*Tribulus terrestris*	Seed	Singh and Tomar, 1991
190	*Tylophora asthmatica*	Bark	Vijayalakshmi, *et al.*, 1979
191	*Ulmus macrophora*	Leaf & flower	Tsaudo *et al.* 1991
192	*Vernonia anthelmintica*	Leaf & root	Mukherjee, 1983
193	*Vernonia cinerea*	Seed	Vijayalakshmi *et al.*, 1979
194	*Withania somnifera*	Root & shoot	Haseeb *et al.*, 1978
195	*Xanthium strumarium*	Leaf	Nandal and Bhatti, 1983
196	*Zennia elegans*	Plant	Bano *et al.*, 1986

REFERENCES

Abbott, W.S. 1925. A method of computing the effectiveness of an insecticide. *J. Eco. Entomol.* 8 : 265-267.

Ahmad, S. and Khan, A. A. 1991. Nematicidal action of *Antigonon leptopus* against *Meloidogyne incognita* Racel. *Ind. J. Nematol.* 2 (1) : 34.

Akhtar, M. and Mahmood, I. 1994. Prophylactic and therapeutic use of oilcakes and leaves of neem and Castor extracts for the control of root knot nematode on chilli. *Nematologia mediterranea* 22 (2) : 127-129.

Alam, M.M. and A.M. Khan 1974. Control of phytonematodes with oil cake amendments in spinach field. *Ind. J. Nematol.* 4 : 239-240.

Alam, M.M.; A.M. Khan and S.K. Saxena 1977. Persistent action of oil cake on the population of nematode in field. *Botyu Kagitu* 42 : 119-124.

Ali, M.A., Mikage, M.; Kiuchi, F., Tsuda, Y. and Kondo, K. 1991. Screening of crude drugs used in Bangladesh for nematicidal activity on the larvae of *Toxocara canis. Shoyakugaku Zusshi* 45 (3) : 206-214.

Allen, E.H. and Feldmesser, J. 1970. Nematicidal effect of alpha tomatine *on Panagrellus redivivus. Phytopathology* 60 : 10- 13.

Badami, R.C. and Patil, K.P. 1975. Minor seed oils, physico chemical characteristics and components of acids of four seed oils. *J. Oil Tech. Assoc.* 7 (3) : 79-81.

Bano, M.; S. Anver, S.A. Tyagi, and M.M. Alam, 1986. Evaluation of nematicidal properties of some members of compositae. *Int. Nematol. Network newslett.* 3 (1) : 10.

Basu, R. and Majumdar, V. 1998. Effect of aqueous extract of neem based pesticides on the mortality of second stage juveniles of *Meloidogyne incognita. Annals of Plant Protection* 6 (1) : 53-55.

Bhattacharya, A.K, Pandey, R., Jain, D.C., SIngh H.N., Kumar, S. and R.P. Sharma 1999. Artemisinin and its biogenetic precursors as possible nematicidal agents. In national Conf. On Plant Biotec. : Towards Strategic Agiculture and Drug development, 15-17 March, 1999 at CIMAP Lucknow : 66.

Bhattacharya, D. and Goswami, B.K. 1987. Comparative efficacy of neem and ground nut oil cakes with aldicarb against *Meloidogyne incognita* in tomato. *Revue de Nematologie* 10 (4) : 467-470.

Bhatnagar, A.; T.K. Mukerjee and B. Taigi 1980. Effect of leaves of some plants on *Meloidogyne incognita. The Andhra Agric. J.* 27 : 312-314.

Bhatnagar, A.; T.K. Mukerjee and Taigi, B. 1978. Control of root knot nematode on Okra by application of water soluble fraction of oil cakes to soil. *Acta Botanica Indica* 6 : 188-189.

Castro, C.E. and Belser, N.O. 1978. Intoxication of *Aphelenchus avenae* by ethylene bromide. *Nematologica* 24 : 37-44.

Castro, C.E. and Thomson, I.J. 1971. Mode of action of nematicides In Zukermann, B.M., Mai W.F. and Rohde, R.A.. (eds.) Plant parasitic nematodes Vol. II New York and London Academic Press. 289-296.

Caswell E.P., Tang, C.S., Defrank, J. and Apt. W.J. 1991. The effect of root exudates of *Chloris gayana* and *Tagetes patula* on *Rotylenchulus reniformis Revue de Nematologie* 14 : 581-587.

Chandel YS and Mehta P.K. 1990. Nematicidal properties of leaf extact of wild sage *Lantana camara* Indian *Journal of Agricultural sciences* 60 (11) : 781

Chandravadana, M.C.; Nidiry, E.S.J., Khan, RM and Rao M.S. 1994. Nematicidal activity of serpentine against *Meloidogyne incognita* incognita. *Fundamental and Applied Nematology.* 17 (2) : 185-186.

Chatterjee, A. and Shukul, N.C. 1980. Nematicidal action of three wild plants. *Nematologica,* 26 : 500-502.

Chatterjee, A, N.C., Shukul, S. Lashar, and S. Ghoshmajumdan 1982. Nematicidal principles from two species of Lamiaceae. *Journal of Nematology* 14 (1) : 118-120.

Chattopadhyay, P.R. 1991. Nematicidal properties of some plant materials on tomato (*Lycopersicon esculentum* var. local) infested with *Meloidogyne incognita. Geobios* 18 (5/6) : 232-236.

Chattopadhyay, P.R. and Mukhopadhyay, M.C. 1989. Comparative studies on the nematicidal properties of *Typhonium trilobatum* and *Melia azedarach. Indian Journal of Nematology* 19 (1) : 59.

Chattopadhyay, P.R. and Mukhopadhyay, M.C. 1989a. Effect of leaf extracts of *Artabotrys odoratissium* R. (Fam. Annonaceae) on the hatching of the eggs of *Meloidogyne incognita* (Kofoid and White, 1919) Chitwood, 1949. *Ind. Journal of Nematology* 19 (1) : 29-31.

Dar, D.W. 1983. Plant species reportedly possessing pest control properties. An EWC/UH DATA BASE Univ. of Hawaii : 249 pp.

Das, S.N. 1983. Department of Nematology, O.U.A.T. Bhuneswar, India Pers. Commn. In Michael *et al.*, 1985. Plant species reportedly possessing pest control properties. An EWC/UH DATA BASE Univ. of hawaii : 249 pp.

Das, S. and Shukul, N.C. 1988. Nematicidal effect of the oil from the seed of *Argemone maxicana. Environment and Ecology* 6 (1) : 194-197.

Desai, M.V., Shah, H.M. and Pillai, S.N. 1973. Nematicidal properties of some plant species. *Indian J. nematol.* 3 : 77-79.

Desai, M.V., Shah, H.M. Pillai, S.N. and Patel, S.N. 1979. Oil cakes in control of root knot nematodes. *Trop. Res.* 5 (1) : 105- 108.

Devakumar C; B.K. Goswami and Mukherjee, S.K. 1985. Nematicidal principles from neem (*Azadirachta indica* A. Juss.) I - Screening of neem kernel fractions against *Meloidogyne incognita. Ind. J. nematol.* 15 : 121-124.

Egunjobi, O.A. and Afolami, S.O. 1976. Effect of neem (*Azadirachta indica*) leaf extract on population of *pratylenchus brachyrus* and on growth of maize. *Nematologica* 22: 125-132.

Ellenby, C. 1945a. The influence of crucifers and mustard oil on the emergence of larvae of the potato root eelworm, *Heteroderra rostochinensis* Wollenw. *Ann. Appl. Biol.* 32 : 67-70.

Ellenby, C., 1945b. Control of potato root eelworm, *Heterodera rostochinensis* Wollenw, By allyl-isothiocyanate, the mustard oil of *Brassica nigra* Linn. *Ann. Appl. Biol.* 32 : 237-239.

Fassuliotus, G. and Skugas, G.P. 1969. The effect of pyrrolizidine alkaloid ester and plants containing pyrrolizidine on *Meloidogyne incognita. J. Nematol.* 1 : 287-288.

Fazal, M. and Husain S. I. 1991. Studies on nematicidal effect of *Ocimum sanctum* and *Thuja orientalis. New Agriculturist*, 1 (2) : 111-112.

Garciaespinosa, R. 1980. *Chenopodium ambrossiodes* Linn. A potential use in combating phytoparasitic nematodes. *Agricultural Trop* 2 : 92-97.

Gaur, H.S., and Prasad S.K. 1970. Effect of organic matter and inorganic fertlizer on soil and plant nematodes. *Indian J. Entomol.* 32 (1) : 186-189.

Ghosh, T. and Shukul, N.C. 1992. Nematicidal principles in the leaves of *Xanthium strumarium Environment and Ecology*. 963-968.

Giebel, J. 1974. Mechanism of plant resistant to nematode : A review. *J. nematol.* 6 (4) : 175-184.

Gnana progosam, N.C.; Monottim, Suresh Kumar B and Udamulla G.P. 1993. Effect of "Javan" a neem based natural pesticide in controlling nematode pests of tea. Srilanka Journal of Tea Science 62 (2) : 47-52.

Gokte, N. and Swarup G. 1988. Effect of neem (*Azadirachta indica*) seed kernel powder treatment on penetration and development of *Anguina tritici* on wheat. *Ind. J. Nematol.* 18 (1) : 49.

Gokte, N., Maheshwari, M.L. and Mathur, V.K. 1990. Nematicidal activity of *Myristica fragrans* against *Meloidogyne incognita. Indian Perfumer* 34 (2) : 105-107.

Gokte, N; Maheshwari, M.L. and Mathur VK 1991. Nematicidal activity of new essential oils against root knot and cyst nematode species. *Ind. Journal of Nematology* 21 (2): 123-127.

Gommers, F.J. 1971. Nematicidal active principles isolated from the root of *Helenium* hybrid plants. *Phytochem.* 10 : 1945.

Gommers, F.J. 1981. Biochemical interaction between nematodes and plants and their relevance to control. Helminthological Abstract Series B *Plant Nematology* 50 : 9-24.

Good, J.M. Minton, N.A. and Jaworski, C.A. 1965. Relative susceptibility of selected crops and coastal bermudagrass to plant nematodes. *Phytopathology* 55 : 1026-1030.

Goswami, B.K. and Vijailakshmi, K. 1985. In Vitro and pot culture studies on the effects of dry products of *Andrographis paniculata, Calendula officinalis, Enhydra fluctuans* and *Solanum khasianum* on *Meloidogyne incognita. Ind. J. Nematol.* 15 (2): 264.

Goswami, and B.K. and Vijailakshmi, K. 1986. Efficacy of some indigenous plant materials and non edible oil seed cakes against *Meloidogyne incognita* on tomato. *Indian J. nematol.* 16 : 280-281.

Govindaiah, N.S., Sharma, D.D. and Gargi 1989. Effect of mulching of green leaves for the control of root knot nematode in mulberry. *Indian J. Nematol.* 19 (1) : 25-28.

Gowda, D.N. 1972. Studies on comparative efficacy of oil cakes on control of root knot nematode, *Meloidogyne incognita* Chitwood on tomato. *Mysore Journal Agric. Sci.* 6 (4) : 524-525.

Grainge, M. and Ahmed, S. 1988. Handbook of plants with Pest Control Properties (J. Wiley and Sons. Eds.), New York : 238-248.

Green, M.M., Singer, J.M., Sutherland, D.J. and Hibben, C.R. 1991. Larvicidal activity of *Tagetes minuta* (marigala) towards *Aedes aegypti. Journal of the American mosquito control Association* 7 (2): 282-286.

Grewal, P.S. and Sohi, H.S. 1988. Toxicity of some plant extracts to *Aphelenchoides compositicola. Indian J. Nematol.* 18 (2) : 354-355.

Gupta R. and Sharma N.K. 1991. Nematicidal properties of garlic, *Allium sativum. Indian Journal of Nematology* 21 (1) ; 14- 18.

Gupta, R. and Sharma N.K. 1993. A study of the nematicidal activity of allicinan activity of allicinan active principle in garlic, *Allium sativum* against root knot nematode, *Meloidogyne incognita* (Kofoid & White, 1919) Chitwood, 1949. *International Journal of Pest Management*, 39 : 390-392.

Gupta, R. Sharma, NK and Gupta R. 1995. Action of garlic (*Allium sativum* L.) extract on the Juveniles of *Meloidogyne incognita* (Kofoid and white) Chitwood. Current Research University of Agricultural Sciences Bangalore 24 : 5 : 91-92.

Hackney, R.W. and Dickerson, O.J. 1975. Marigold, Castor Bean and Chrysanthemum as control of *Meloidogyne incognita* and *Pratylenchus alleni. J. Nematol.* 7 (1) : 84-90.

Haider, MG 1993. Effect of plant extracts and incorporation of plant parts in soil on population of root knot nematode (*Meloidogyne incognita*) and root knot development in tomato (*Lycopersicon esculentum*) *Indian Journal of Agricultural Sciences* 63 (2): 136-139.

Hameed, S.F. 1968. Relative efficacy of different oil cakes on the incidence of root knot nematodes. *Madras Agric Journal* 55 : 168-171.

Haroon, S.A. and Smart G.C.Jr. 1983. Root extracts of *Pangola digitgrass* affect Egg Hatch and larval survival of *Meloidogyne incognita. J. Nematol* 15 (4):

Haroon, S.A., Nation, J.N. and Smart Jr. G.C. 1982. Isolation of natural nematicides from pangola digitgrass. *Digitaria decumbens* that effect egg hatch and larval survival of *Meloidogyne incognita. J. Nematol.* 14 : 444.

Hasan, N. and Jain R.K. 1984. Biotoxicity of *Parthenium hysterophorus* extract against *Meloidogyne incognita* and *Helicotylenchus dihystera. Nematol. Medit.* 12 : 239-242.

Haseeb, A., Singh, B., Khan, A.M. and S.K. Saxena, 1978. Evaluation of nematicidal property of certain Alkaloid Bearing plants. *Geobios* 5 : 116-118.

Haseeb, A., Khan, A.M. and S.K. Saxena 1982. Toxicity of leaf extracts of plants to root knot and reniform nematode. *Ind. J. Parasitol.* 6 (1) : 119-120.

Hoan, L.T. and Davide, R.G. 1979. Nematicidal properties of root extracts of seventeen plants species on *Meloidogyne incognita. Philippine Agriculturist* 62 : 282-295.

Husain,. S.I. and Saxena, S.K. 1969. Studies on nematicidal action of certain anthelmintic substances. All India nematol. Symp. New Delhi 69.

Husain, S.I. and Masood, A. 1975. Effect of some plant extracts on larval hatching of *Meloidogyne incognita* (Kofoid and White) Chitwood. *Acta Botanica Indica* 3 : 142-146.

Husaini S.S., Prasada, Rao, R.V.V. and Pandu, H.K. 1996. Toxicity of water soluble leaf extracts against larvae and egg masses of three *Meloidogyne* species. *Ind. J. Nematol.* 26 (1) : 23- 31.

Ibrahim, A.M. 1992. Antihelmintic activity of some Sudanese medicinal Plants. *Phytotherapy Research.* 6 (3) : 155-17.

Islam, B.N. 1983. Department of Entomology, Bangladesh Agric Univ. Pers. Commn. In Michael *et al.*, 1985. Plant species reportedly possessing pest control properties. An EWC/UH DATA BASE Univ. of Hawaii : 249 pp.

Jacobson, M. 1958. Insecticides from plants. A review of literature (1951-1953). USDA Agric. Handbook, Govt. Printing Office, Washington DC 219 pp.

Jaehn Anario, Levyferrcira and Angelo Catanlo 1988. Nemat, Brasileina Vol XII 1988. Eficiencia Do Abamectin M.K. 936 (1.8% CE) No control De *Meloidogyne incognita* Em Tomateiro, ALFACE E CENOURA.

Jagadale, G.B., Pawar, A.B. and Darekar, K.S. 1985a. Effect of organic amendments and antagonistic plants on control of the root knot nematode infesting Betelvine, *Piper betel* Linn. Presented in IV nematol Symp. At Udaipur, India 35.

Jagadale G.B., Pawar, A.B. and Darekar K.S. 1985b. Effect of organic amendments and antagonistic plants on control of the root knot nematode infesting Betelvine. *Indian J. Nematol.* 15 (2) : 264.

Jain, R.K. and Gupta, D.C. 1995. Effect of a few plant leaf extacts on hatching of root knot nematode, *Meloidogyne javanica* egg. *Ind. J. Nematol* 25 (1): 107-108.

Jain, S.K. and Saxena, R. 1993. Evaluation of the nematicidal potential of *Mangifera indica. Ind. J. Nematol.* 23 (2) : 131-132.

Jain, U., Dutta, S., Trivedi, P.C. and Tiagi, B. 1986. Allelochemic effects of some plants on hatching of *Meloidogyne incognita. Indian J. Nematol.* 16 (6) : 275-276.nbm

Janeh, A and Lamberti, F. 1983. Use of castor oil seed cakes as nematicides in coffee nurseries. *Revista Cafelatera* 232: 28-29.

Jasy, T. and Koshy, P.K. 1992. Effect of certain leaf extracts and leaves of *Gliricidia maculata* (H.B. and K) Steud. As green manure on *Radopholus similis*. *Ind. J. Nematol.* 22 (2) : 117-121.

Jurgens, T.M; Frazler, E.G.; Schaeffer, J.M.; Jones, T.E.; Zink, D.L.; Borris, R.P. Nanakorn, W; Beck, H.T.; and Balick, M.J. 1994. Novel nematicidal agents from *Curricula commas*. *Journal of Natural Products*, 57 (2) : 230-235.

Kaliram, and D.C. Gupta 1980. A note on the efficacy of fresh neem extracts in the control of *Meloidogyne javanica* infesting chickpea (*Cicer arietinum*) *Ind. J. Nematol.* 10 (1) : 96-98.

Kaliram, and D.C. Gupta 1982. Efficacy of plant leaves, nematicides and fertilizers alone and incombination against *Meloidogyne javanica* infesting chickpea. *Ind. J. Nematol.* 12 (2) : 221-225.

Kamal, R. and Mangla, M. 1987. Toxicity of natural pyrethrins against mesoyclops ieuckarti seusu. 1 ato carrier of dranculiasis (sic). *PyrethrumPost* 16 : 4, 125-127.

Kanagy, J.M.N.and H.K. Kaya 1996. The possible role of marigold roots and a Terthienyl in mediating host finding by Steinernematid nematodes. *Nematologica* 42 : 220-231.

Katsura, M. 1981. Nematicidal natural products. ICIPE, Conf. May. 1980 dept. Agric. Chem. Nagoyo Univ. Japan 23 : 1.

Kawazy, K., Nishi Y., Ishii, K. and Tada, M. 1980. A convenient screens method for nematicidal activity. *Agricultural and biological Chemistry* 44 (3) : 631-635.

Khan Fasahat, A. 1990. Nematicidal potentials of some Naturally growing medicinal plants against *Pratylenchus zeae*. *Revue nematol.* 13 (4) : 463-465.

Khan, A.M., Adhami, A., Siddiqui Z.A. and Saxena, S.K. 1966. Effect of different oil cakes on hatching of larvae and on the development of root knot caused by *Meloidogyne incognita* (Kofoid and White) Chitwood, 1949. Proc. Ist Int. Symp. Plant Pathol. New Delhi. on Plant Dis. Prob. 582-588.

Khan, A.M., M.M. Alam, Z.A. Siddiqui and Saxena, S.K. 1967. Effect of differnt oil cakes on hatching of larvae and on the development of root knot caused by *Meloidogyne incognita* (Kofoid and white) Chitwood 1949. Proc. 1st. Symp. Plant. Pathol. New Delhi on Plant Dis. Prob. : 37.

Khan, A.M., Alam, M.M. and Rais, A. 1974. Mechanism of control of plant parasitic nematodes as a result of the application of oil cakes to the soil. *Indian J. Nematol* 4 : 93-96.

Khan, T. A. and Husain, S.I. 1988. Studies on the efficacy of seed treatments with pesticides, oil cakes neem leaf and culture filtrate of *Paecaelomyces lilacinus* for the control of disease caused by the presence to *Rotylenchulus reniformis*, *Meloidogyne incognita* and *Rhizoctonia solani* occurring either individually or concomitantly on cowpea. *Indian J. Nematol.* 18 (2) : 192-198.

Kimura, Y., Mor, M., Suzuki, A., Kobayashi, A. 1981. *Agric. Biol. Chem.* 45 : 2915.

Kiuchi F., Hayashi, M.; Nakamura N and Mikage, M. Tsudy, Konde and K. Namiba T. 1989a. Screening of crude drug used in Nepal for nematicidal activity on the larvae of *Toxocora canis*. Shoyakugaku Zasshi 43 (4) : 294-299.

Kiuchi, F; Nakamura N; Miyashita, N., Nishizawa S., Tsuda, Y. and Kondo K. 1989b. Nematocidal activity of some antihelmintics, traditional medicines and spices by new assay method using larvae of *Toxocora canis*. *Shoyokogaku Zasshi*, 43 (4) : 279-287.

Kiuchi, F; Chen, X; Tsuda Y; Kondok and Kumar V. 1989c. Studies on crude drugs effective on visceral larva migraines, VI, identification of nemtocidal principles in *Tephrosia purpurea* pers. *Shoyakugaku Zasshi*, 43 (1) : 42-49.

Kiuchi, F.; Goroy, Sugimoto N. Akau, N; Kando K and Tsuda 1993. Nematicidal activity of turmeric: synergistic action of Cur cuminoids. Chemical and Sharma sciences. Kanazawa university, 131 Takaramachi, Kanazawa 120 Japan, 1640-1643.

Kogiso, S., Wada, K. and Munakata, K. 1976. Isolation of nematicidal polyacetylened from *Carthamus tinctorius*. *Agricultural and Biological Chemistry* 40 (10): 2085-2089.

Krishnamurthy, G.V.G. and P.S.N. Murthy, 1990. Further studies with plant extract on root knot nematode larvae, *Meloidogyne javanica*. In National Symp. On problems and prospects of Botanical Pesticides in Integrated pest management at C.T.R.I., Rajmundary, India 21-22nd January 1990 : 43-44.

Kumar, T.P. and M.R.G.K. Nair 1979. Effect of some green leaves and inorganic wastes on root knot nematode infestation in okra (Bhindi). *Agric. Res. Journal Kerala* 14 (1) : 64-67.

Kumari, R., K.K. Varma, K.S. Dhindsa and D.S. Bhatti 1986. *Datura, Ipomea, Tagetes* and *Lawsonia* as control agents for *Tylenchulus semipenetrans* and *Anguina tritici. Ind. J. Nematol.* 16 (2) : 236-240.

Kuriyan, K.J. 1988. Organic amendment for control of root knot nematode *Meloidogyne incognita* on bhindi. In Integrated Pest Control Progress and prospective. Eds. Mohandas, N. and Koshy, G. Trivendrum, *Assoc. Adv. Entomol.* 133-134.

Lall, S.B. and Hameed, S.F. 1969. Studies on the biology of the root knot nematodes, *Meloidogyne* sp. with special reference to host resistance and manuring. 1st All India Nematol. Symp., New Delhi : 6.

Lanjeswar, R.D. and Shukla, V.N. 1986. Vulnerability of larvae and eggs of *Meloidogyne incognita* to some oil cakes and fungicides. *Indian J. Nematol.* 16 (1) : 69- 73.

Linford, M.B., Yap, F. and Oleveira J.M. 1938. Reduction of soil populations of root knot nematode during decomposition of organic matter. *Soil Sci.* 45 : 127-141.

Mackeen, M.M., Ali, A.M. Abdullah, M.A. Nasir, R.M., Mat, N.B., Razak, A.R. and Kawazy, K. 1997. Antinematodal activity of some Malaysian plant extracts against pine wood nematode. *Pesticide Science* 51 (2) : 165-170.

Mahajan, R., Singh P., Bajaj, K.L. and Kalsi, P.S. 1986. : Nematicidal activity of some sesquiterpenoids against root knot nematode (*Meloidogyne incognita*) *Nematologica* 32 : 119- 123.

Mahmood, I., Masood, A., Saxena, S.K. and Husain S.I. 1979. Effect of plant extracts of poisonous plants of Iraq on mortality of citrus nematode, *Tylenchulus semipenetrans* Cobb. *Acta Botanica Indica* 9 : 198-200.

Mahmood, I., Saxena, S.K. And Zakiruddin, R. 1982. Effect of certain plant extracts on the mortality of *Rotylenchulus reniformis* (Linford and Olivers, 1940) and *Meloidogyne incognita* (Kofoid and White, 1919) Chitwood, 1949. *Bangladesh J. Bot.* 11 (2) : 154-157.

Majumdar, V. and Misra, S.D. 1991. Nematicidal efficacy of some plant products and management of *Meloidogyne incognita* in pulse crops by soaking seeds in their aqueous extracts. *Current Nematology* 2 (1) : 27-32.

Malik, M.S., Sangwan, N.K. Dhindsa, K.S. and Bhatti, D.S. 1987. Nematicidal activity of extracts of *Xanthium strumarium. Pesticides* 21 (10) : 19-20.

Mani, A., Ahmed, S.N., Kameshwar Rao, P. and Dakshinmurti, V. 1986. Plant products toxic to the citrus nematode. *Tylenchulus semipenetrans* Cobb. *Int. Nematol. Network Newsl.* 3 (2) : 14-15.

Matsuda, K., Yamada, K., Kimura M. and Hamada M. 1991. Nematicidal activity of matrine and its derivatives against pine wood nematode. *J. Agric. Food Chem.* 39 : 189-191.

McIndoo, I. 1982. USDA, Bur. Entomol. And plant Quar. Pers. Commn. In Michael *et al.*, 1985. Plant species reportedly possessing pest control properties. An EWC/UH DATA BASE Univ. of Hawaii : 249 pp.

Meher, H.C., Walia, S. and Sethi, C.L. 1988. Effect of steroidal and triterpenic saponins on the mobility of juveniles of *Meloidogyne incognita Ind. J. Nematol.* 18 (2) : 244-247.

Mian, I.M. and Rodgriguez Kabana R. 1982. Survey of nematicidal properties of some organic materials avaialble in Alabama as amendment to soil for control of *Meloidogyne arenaria. Nematropica* 12 (2) : 235-246.

Miller, P.M. 1979. Vegetable oils as proctyents against nematode infestation. *J. Nematol.* 11 (4) : 402-403.

Miller, P.M., Turner, L.C. and Tomlinson, H. 1973. Toxicity of leaf and stem extracts of plants to *Tylenchorhynchus dubius. J. Nematol.* 5 : 173-177.

Misra, S.D. and H.S. Gour 1989. Control of nematodes infesting mung with nematicidal treatment and field application of aldicarb and neem cake. *Ind. J. entomol.* 5 (14) : 422-426.

Misra, S.D. and S.K. Prasad, 1973. Effect of water extracts of oil seed cakes on the second stage larvae of*Meloidgoyne incognita* at different concentration and exposure times. *Ind. J. Entomol.* 35 (2) : 101-106.

Mobin, M.P. and Khan, A.M. 1969. Effect of organic amendment on the population of rhizosphere fungi and nematode around roots of some fruit trees. All India Nematol. Symp. New Delhi. 11-13.

Mohandas, C., Rao, Y.S. and Sahu, S.C. 1982. Cultural control of root knot nematodes, *Hirschmanniella* spp. with *Sphenochelea zeylanica. Proc. Ind. Acad. Sci.* 90 : 373-376.

Mohanty, K.C. and Das, S.N. 1988. Nematicidal properties of *Erythrina indica* against *Meloidogyne incognita* and *Tylenchorhynchus mashoodi Indian J. Nematol.* 18 (1) : 138.

Mukherjee, S.N. and N.C. Sukul 1978. Nematicidal action of three species on wild herbs. J. Res. 2(2) : 12.

Mukherjee, S.K. 1983. Plant species reportedly possessing pest control properties. ICAR, New Delhi Michel *et al.*, 1985, An EWC/UH/DATA BASE, Univ. Hawaii : 249 pp.

Mukhopadhyay, M.C. and P.R. Chadopadhyay, 1979. Control of root knot nematode, (*Meloidogyne incognita*) Kofoid and white Chitwood with *Typhonium trilobatum* an annual herb. *Ind. J. Nematol* 9 (1) : 53-59.

Mukhtar, T., Ahmod, R., Haq, M. and Jowed, N. 1994. Effect of leaf extracts of some plants on *Meloidogyne incognita. Pakistan Journal of Phytopathology.* 6 (1) :35-37.

Muller, J. and Sturhan D. 1994. Studies on effects of *chromolaena odorata* roots on root knot nematodes (*Meloidogyne incognita*) *Nematologia mediterranea*, 22 (2) : 273-274.

Munakata, K. 1978. Nematicidal substances from plants. Advances in pesticides Sciences Part 2 Pergamon Press 25 : 295-302.

Munakata K. 1983. In Natural products for Innovative Pest management, Whitehead, D.L., Bowers, W.S. Eds. Peagmon Press Oxford : 288-310.

Nakamura, N; Kiuchi F,: Tsuda, Y; Kondo, K. Satot, T. 1991. Nematocidal and bursting activities of essential oils on the larvae of *Tooxocara canins*. *Shoyakugaku Zasshi*, 44 (3) : 183-195.

Nandal, S.N. and D.S. Bhatti 1983. Preliminary screening of some weeds and shrubs for their nematicidal activities against *Meloidogyne javanica*. *Indian J. Nemato.* 13 : 123-127.

Nath, A.; N.K. Sharma, S. Bhardwaj and C.N. Thapa 1982. Nematicidal property of garlic. *Nematologica* 28 (2) : 253-255.

Nazar, M.Z.H. and Nath, R.P. 1989. Nematicidal effect of certain weed on root knot nematode *Meloidogyne incognita* and *M. javanica* mixture. *Journal of Researcharajendra Agricultural university*, 7 (12) : 95-96.

Nene, Y.L. and Thapliyal, P.N. 1966. Nematicidal properties of *Anagallis arvensis* L. extracts. *Indian Phytopath.* 19 : 26-29.

Nene, Y.L. and Kumar K. 1967. Nematostatic properties of *Erigeron linifelins* extracts. *Indian Phytopath.* 19 : 26-19.

Nidiry, E.S.J., Chandravadana, M.C., Khan, R.M. and Rao, M.S. 1994. In vitro nematicidal activity of extract of bulbs and seeds of onion against rootknot nematode *Meloidogyne incognita*. *Nemato Medit.* 22 (1) : 37-40.

Nidiry, E.S.J., Khan, R.M. Reddy, P.P. 1993. In vitro nematicidal activity of *Gloriosa superba* seed extract against *Meloidogyne incognita*. *Nematologia mediterranea*, 21 (2) : 127-128.

Omidvar, A.M. 1961. On the effect of root diffusates from *Tagetes* spp. On *Heterodera rostochiensis* Woll. *Nematologica* 7 : 62-64.

Onda, M.; Takiguchi, K., Hiratura, M., Kukushima, H., Kagawa, M. and Naoi, F. 1965. The constituents of *Maclea cordata* nematicide alkaloids. *Nippon nogeikagaku kaisbi* 39 : 168-174.

Onda, M., Abe, K., Yonezawa, K., Esums, N. and Suzuki, T. 1970. *Chem. Pharm. Bull.* 18 : 435.

Oniffode, A.K. and Egurjobi, D.A. 1994. Suppression of *Meloidogyne incognita* populations with water lettuce in Nigeria. *Afro Asian Journal of Nematology*, 4 (1) : 96-100.

Oostenbrink M, Kuiper, K. and S'Jacob, J.J. 1957. Tagetes als feindpflanzen von *Pratylenchus* Arten. *Nematologica* 2 : 424-433.

Pandey R 1990. Studies on phytonematotoxic properties in the extract of some medicinal and aromatic plants. *Int. Nematol. Network Newsl.* 7 (3) : 19-20.

Pandey R. 1995. Application of oil seed cake, pesticides and dry leaf matter on the reproduction potential of *Meloidogyne incognita* and yield of Japanese mint. *Indian J. Nematol.* 25 (1) : 12-13.

Pandey R. 1997. Ecofriendly approaches for the management of rootknot nematode on camphor basil. *Indian J. Plant Pathol.* 15 (1 & 2) : 74-81.

Pandey, R. and Haseeb, A. 1988. Studies on the toxicity of extracts of certain medicinal plants to root knot nematode *Meloidogyne incognita* (Kofoid and White) Chitwood *Indian Journal of plant Pathology* 6 (2) : 184-186.

Pandey R., Haseeb, A. and Husain, A. 1992. Distribution, pathogenicity and management of *Meloidogyne incognita* on *Mentha arvensis* cv. MAS-1. *Afro Asian J. Nematol.* 2 (1) : 27-34.

Pandey, R.; S. Tandon, N. Mehrotra, S. Sharma, H.N. Singh and S. Kumar 1998. Essential oils and their actives as potent source of nematicidal compounds in National Symposium on rational approaches in nematode management for sustainable agriculture at G.A.U. Anand from 23-25 Nov. 1998.

Paruthi, I.J. Jain R.K. and Gupta D.C. 1987. Effect of different periods of degradation of subabool leaves alone and incombination with nematicide on root knot nematode incidence in okra. *Indian J. Nematol.* 17 (1) : 30-32.

Patel, H.R. 1995. Inhibitory effect of *Clerodendron enermi* on root knot of okra. The *Madras Agric. Journal* 72 : 470-472.

Patel, H.R., Thakur, N.A. and Patel G.C. 1987b. Larval emergence and infestation of *Meloidogyne incognita* as influenced by periwinkle, *Catharanthus roseus. Ind. J. Agric. Sci.* 57 : 863-866.

Patel, D.J.; Patel H.U., Patel, S.K., and Patel, B.A. 1993. Nematicidal properties of some plant materials in tomato nursery. *Indian Journal of Plant Protection* 21 (2) : 242-244.

Poornima K. and Vadivelu S. 1990. Comparative efficacy of nematicides, oil cakes and plant extracts in the management of *Meloidogyne incognita, Pratylenchus delattrei, Rotylenchulus reniformis* on brinjal. *Indian Journal of Nematology* 20 (2) : 170-173.

Prakash, A. and Rao J., 1997. Botanical pesticides in agriculture. CRC., Lewis Publishers, Boca Raton, New York, London, Tokyo, 461 pp.

Prasad, J.S., and Rao, Y.S. 1979. Nematicidal properties of the weed, *Eclipta alba* Hassk (Compositae) revista di parasitologia XL. 87-90.

Prot. J.C., and J.M. Komporobst. 1983. Effect of *Azadirachta indica, Hannoa undulata* and *H. klaineana* seed extract on the activity of *Meloidogyne javanica* juveniles to penetrate tomato roots. *Revi de Nematol.* 6 (2) : 330-332.

Prot. J.C. and J.M. Komprobst. 1985. Effect of quassinoids extracted from *Hannoa undulata* seed on the penetration and reproduction of *Meloidogyne javanica* on tomato. *Revi de Nematol* 8 (4) : 383-389.

Qamar, F., Saeed, M; Kapadia, Z. Seema, N. and Badar Y 1989. Nematicidal properties of crude extracts of some indigenous plants. Part I. *Pakistan Journal of Scientific and Industrial Research,* 32 (9) : 600-602.

Rao, M.S. and Pandey, M. 1982. Comparative efficacy of Karanja leaf and carbofuran on the management of *Aphelenchoides composticola. Indian J. Nematol.* 12 (2) : 158-159.

Ravidutta, B. and Bhatti, D.S. 1986. Effective dose and time of application of nematicides and castor leaves for controlling. *Meloidogyne incognita* in Tomato. *Indian J. Nematol.* 16 (1) : 8-11.

Ray, D., Sinha Babu, S.P., Shukul, N.C. and Mahato S.B. 1990. Nematicidal principle from the funicles of *Acacia curiculiformis, Indian Journal of Nematology,* 1990, 20 : 2, 152-157.

Reddy, V.R., Chitra, K.C. and Rao, K.P. 1990. Efficacy of plant extracts in the control of brinjal spotted leaf beetle, *Henosepilachna vigintiotopunctata* Fabr. In nat. Symp. Prob. & Prospect. Botanical pesticides in IPM at CTRI, Rajmundry, India : 18-19.

Rohde, R.A. 1972. Expression of resistance in plants to the nematodes. *Ann. Rev. Phytopathol.* 10 : 233-252.

Rohde, R.A. and Jenkins, W.R. 1958. Basis for resistance of *Asparagus officinalis* var. *altilis* L. to the stubby root nematode, *Trichodorus christiei* Allen, 1957. *Md. Agric. Exp. Stn. Bull.* A-97 : 1-19.

Roy, D., Sinhababu, S.P., Shukul, N.C. and Mahato, S.B. 1993. Nematicidal principle from the funicles of *Acacia auriculiformis. Indian J. Nematol.* 23 (2) : 152-157.

Sangwan, N.K. Verma, B.S. Verma, K.K. and Dhindsa, K.S. 1990. Nematicidal activity of some essential plant oils. *Pesticide Science* 28 : 3, 331-335.

Sangwan, Naresh, K. Kailash K. Verma, Brahams, Verma, Mangel S. Malik and Kuldip Singh Dhindso 1985. Nematicidal activity of essential oils of *Cymbopogon grasses.* Nematologia 13 (1985) 93-99.

Sarra Guzman, R. 1984. Toxicity screening of various plant extracts against *Meloidogyne incognita* Chitwood and *Radopholus similis* Cobb and characterisation of their nematicidal components. Ph.D. Thesis. Univ. of Philippines at Los Banos College, Laguna, Philippines, 139 pp.

Sarosh, Husain, S.I. and Nisar, S. 1989. Isolation of anti nematode prohibitions from *Cassia occidentalis* and their effect on hatching and mortality of *Meloidogyne incognita* (Kofoid & White) Chitwood. *J. Phytological Res.* 2 (2) : 233-235.

Sassanelli, N. and D'Addabbo, T. 1992. Nematicidal activity of aqueous extracts from leaves of *Ruta graveolens* on *Xiphinema index. Nematol. Medit.* 20 : 53-55.

Sasanelli, N. and D'Addabbo, T. 1992. The effect of *Cineraria maritima, Ruta graveolens* and *Tagetes erecta* extracts on the hatching of *Heterodera schachtii. Nematol. medit.* 20 : 49-51.

Sasanelli, N. and D'Addabbo, T. 1993. Effect of *Cineraria maritima, Ruta graveolens* and *Tagetes erecta* leaf and root extracts on Italian populations of *Meloidogyne* species. *Nematol medit.* 21 (1) : 21-25.

Sasenelli, N. and Di Vito, M. 1991. The effect of *Tagetes* spp. extract on the hatching of an Italian population of *Heterodera rostochensis. Nematol. Medit.* 19 : 135-137.

Saxena, D.B., Goswami, B.K. and Tomar, S.S. 1987. Nematicidal activity of some essential oils against *Meloidogyne incognita. Indian Perfumer* 31 (2) : 150-154.

Sharma, H.K. and Prasad, D. 1995. Evaluation of *Argemone maxicana* plant extract against root knot nematode, *Meloidogyne incognita* on Soybean and brinjal. In national symposium on nematode prob. of India, IARI, New Delhi. 56.

Sharma, R. and Trivedi P.C. 1991. Nematicidal properties of some leaf extracts against *Meloidogyne incognita. Journal of Phytological research* 4 (2) : 131-137.

Sharma, R. and P.C. Trivedi 1995. Efficacy of some botanicals against *Heterodera avenae* on barley. National Symposium of nematode problems in India - An appraisal of the nematode management with ecofriendly approaches and bio-components, IARI, New Delhi, Jan. 1990 : 4.

Sherif, A., Hall, R.G. and El Amangi, M. 1987. Drugs, insecticides and other agents from *Artemisia.* Medical Hypothesis 23 (2) : 187-193.

Siddiqui, MA; and Alam, M.M. 1988. Toxicity of different plant-parts of *Tagets indica* to plant parasitic nematodes. *Indian J. Nematol* 18 (2) : 181-185.

Siddiqui, MA; and Alam, M.M. 1993. Evaluation of nematicidal potential in neem allelochemicals. World neem Conf., Banglore, 24 - 28 Feb., 1993, India : 39.

Siddiqui, MA, Haseeb, A. and Alam, M.M. 1987. Evaluation of nematicidal properties in some latex bearing plants. *Ind. Journal Nematology* 17 (1) : 99-102.

Sikdar, A.K., Govindaiah, Y.R., Madhava, Rao, Jolly, M.S. and Giridhar, K. 1986. Control of root knot nematode disease of mulberry by temik 10 G-oil cakes. *Indian J Sericulture* 25 (1) : 33-35.

Singh, A., Kohli, J.D. and Parihar, D.B. 1955. Search for antihelmintics indigenous remedies I. Action of acetyl saponin from *Anagallis arvensis* L. on annelids and helminths. *Indian J. Vet. Sci.* 25 : 25-29.

Singh RP and Kataria, PK 1991. Insects nematodes and fungi evaluated with neem *Azadirachta indica* A. juss.) in India. *Neem Newsletter* (1) ; 3-10.

Singh, R.P. Tomor, S.S., Devakumar, C., Goswami B.K. and Saxena, D.B. 1991. Nematicidal efficacy of some essential oils against *Meloidogyne incognita. Indian Perfumer*, 35 : 35-37.

Singh, R.S. and K. Sitaramaiah 1967. The effect of decomposition of green leaves, saw dust and urea on incidence of root knot of okra and tomato. *Indian Phytopath.* 20 : 349-355.

Singh, R.S. and K. Sitaramaiah, 1971. Control of root knot nematode through organic and inorganic amendments in soil II. Effect of oil cake and saw dust. *Ind. J. Mycol. Pl. Pathol.* 1 (1) : 20-29.

Singh, R.S. and K. Sitaramaiah 1976. Incidence of root knot of okra and tomato in oil cake amended soil. *Plant Dis. Reptr.* 50 (9) : 668-672.

Singh R.S. and Sitaramaiah K 1970. Control of plant parasitic nematodes with organic amendments of soil. *PANS* 16 : 287-297.

Singh, S.P. and V. Singh 1988. Evaluation of oil cakes and nematicides for the control of *Meloidogyne incognita* infecting egg plants. *Ind. J. Nematol.* 18 (2) : 338-342.

Sitaramaiah K. and R.S. Singh 1977. Response of plant parasites and soil nematodes to plant extracts in amended soil. *Pantnagar J. Res.* 2 (2) : 153-157.

Sivakumar, M. and Marimuthu, T. 1986. Efficacy of different organic amendments against phytonematodes associated with betelvine (*Piper betel* L.). *Indian J. Nematol.* 16 (2) : 278.

Sreeja, P. and Charles, J.S.K. 1998. Screening of botanicals against burrowing nematode, *Radopholus similis* (Cobb, 1893) Thorne, 1949. *Pest Management in Horticulture Ecosystem* 4 (1) : 36-39.

Srivastava, A.S. 1971. Application of organic amendments for control of root knot nematode, *Meloidogyne javanica* Treub. Labbev. *J. Sci. and Tech.* 9B (3/4) : 203-205.

Stadler, M; Dagne, E. and Anke H. 1991. Nematicidal activity of medicorpin 8 4 hydroxymedicorpin from *Taverriera abyssinica. Planta Medica* 57 (supplement 2) : 13.

Stephan, Z.A., AL-Askari, A.A. and Antoon, B.G. 1989. Effect of *Haplophylum tuberculatum* plant extract on root knot nematode. *Int. Nematol Network Newsl.* 6 : 31-32.

Suatmadji, R.W. 1969. Studies on the effect of *Tagetes* species on plant parasitic nematodes, Stiching Fonds Landbouw Export Bureau Publicate 47. Veenan und Zonen N.W., Wageningen, The Netherland.

Subramaniyan, S. 1985. Effect of *Eupatorium odoratum* extracts on *Meloidogyne incognita. Ind. J. Nematol.* 15 (2) : 247.

Sunderbabu R., Sankaranarayana C and Vadivelu S. 1990. Nematode management with plant products. *Ind. Journal of Nematology* 20 (2) : 177-178.

Swarup, G. and Sharma R.D. 1967. Effect of root extract of Asparagus racemosus and Tagetes erecta on hatching of eggs of *Meloidogyne arenaria. Indian J. Exp. Biol.* 5 : 59.

Tada, M., Hiroe, Y., Kiyohara, S. and Suzuki S. 1988. Nematicidal and antimicrobial constituents from *Allium grayi* Regel and *Allium fistulosum* L. var. caespitosum. *Agric. Biol. Chem.* 52 (9) ; 2383-2385.

Tanoa, A.S. Atwal and Y.P.S. Bajaj 1989. In Vitro inhibition of Root knot nematode *Meloidogyne incognita* by sesame root exudate and its Amino Acids. *Nematologica* 35 : 115-124.

Tarjan, A.C., and Cheo, P.A. 1956. Nematicidal value of some fatty acids. Bull. Rhode Island *Agr. Exp. Stn.* No. 112 : 41 pp.

Tasuda, Y., Goto, Y., Mikase, M. and Kondo K. 1991. Nematicidal principles in ulmi pasta, *Shoyakugaku Zasshi* 45 (3) : 270-273.

Tyagi, S.A., Mukhtar, J. and Alam, M.M. 1985. Preliminary studies on the nematicidal nature of two plants of the family compositae. *Int. Nematol, Network Newsl. Lett.* 2 (3) : 19-21.

Tyagi, S.A., Bano, M.L. and Alam M.M. 1988. Evaluation of nematicidal potential in some plant species belonging to family compositae. *Indian J. of Nematology*, 18 (2) : 228-231.

Uhlenbrock, J.H. and J.D. Bijloo 1959. Nematicidal substance in marigold, *Tagetes erecta* Linn. Recl. Trav. Chem. Pays Bas Belgium 78 : 289.

Valette, G., Cavier, R. and Debetmas, I. 1953. Action helminth ders constituents des essences vegetables. *Am. Pharm.* France 11 : 649-653.

Varma, A.C. and A. Anwar, 1995. Integrated nematode management by using leaves of medicinal plants in printed gourd. *Trichosanthes dioica* field. National Symposium in nematode problems of India - An appraisal of nematode managemetn with ecofriendly approaches and bio compounds, IARI, New Delhi 24-26th March, 1995 : 18.

Varma, M.K., Sharma H.C. and Pathak, V.N. 1978. Efficacy of *Tagetes patula* and *Sesamum orientale* against root knot of egg plant. *Plant Disease Reporter* 62 : 274-275.

Viala, De, S.S., Brodie, B.B. Rodriguez, E. and Gibson D.M. 1998. The potential of thiarubrinec as a nematicidal agent aginst plant parasitic nematodes. *J. nematol.* 30 (3) : 192-200.

Vats, R. and Nandal, S.N. 1995. Evaluation of differently prepared plant extracts as baveroot dip treatment of tomato seedlings against *Meloidogyne javanica. Crop Research* Hisar, 1995, 9 : 1, 123-128.

Verma RR 1993. Effect of different materials on the population of root knot nematode, *Meloidogyne incognita* affecting tomato crop. *Indian Journal of Nematology* 23 (1): 135-136.

Vijailakshmi, K. and B.K. Goswami, 1985. Effect of aqueous extracts of madar *Calotropis gigentea* and amerbel, *Cuscuta reflexa* on larval mortality hatch from egg masses and subsequent penetration in to tomato root. *Ind. J. nematol.* 15 (2) : 265-266.

Vijailakshmi, K. S.D. Mishra and Prasad, S.K. 1979. Nematicidal Properties of some indigenous plant materials against second stage Juveniles of *Meloidogyne incognita* (Kofoid and White) Chitwood. *Indian Journal Entomology* 41 (4) 326-331.

Winoto, S.R. 1969. Studies on the effect of *Tagetes* species on plant parasitic nematodes. Sticht Founds Landbouw Export Bur Pub. 47, Weemann & Zonen, N.V.

Wright, D.J. 1981. Nematicides : Mode of action and new approaches to chemical control. In B.M. Zukerman & Rohde R.A. (eds.) Plant Parasitic Nematodes Vol. III New York & London, Academic Press : 421-449.

Yuhara, I 1971. Effect of soil application of chopped plant material of *crotaleria* or marigold on *Meloidogyne hapla* population. Ann. Reptr. Soc. Pl. Prot. 22 : 62.

Zaki, F.A. and Bhatti, D.S. 1989. Effect of castor (*Ricinus communis*) leaves in combination with different fertilizer doses on *Meloidogyne javanica* infecting tomato. *Indian J. Nematol.* 19 (2) : 171-176.

Zavaleta-Mejia E., Castro, A.E. and Zamudio, V. 1993. Effecto del cultivo e incorporacion de *Tagetes erecta* L. sobre la problacion e infection de *Meloidogyne incognita* (Kofoid and white) Chitwood in chilli (*Capsicum annum* L.) *Nematologica* 23 : 49-56.

CHAPTER 8

ANTAGONISTIC BOTANICALS TO PLANT PARASITIC NEMATODES AND THEIR USE IN NEMATODE MANAGEMENT

Archana Mittal and B.K. Goswami

INTRODUCTION

Plant parasitic nematodes are worst enemics of mankind because of the great damage they cause to all kinds of plants growing in different ecological conditions. They cause severe losses to economically important crops e.g. – vegetables, pulses, cereals, oilseed crops, etc. Sasser and Freckman (1981) have indicated an annual crop losses due to plant - parasitic nematodes on worldwide basis to the tune of $100 billion. The ave.age loss of all crops has also been estimated to be about 10% due to nematodes (Sasser, 1989). In India, many similar reports are available, which confirms the severity of these tiny organisms. Thus, application of chemical nemticides, though effective in some cases are not always cost collective. Moreover, these nematodes cause environmental pollution, phytotoxicity, contaminate grounwater and adversely affect the land and its biotic environment. The demerits of hazardous chemicals have created interest in searching alternate methods of plant nematode management. Biopesticides of botanical organ have become the focus of attention today for facing the nematode problems especially when we think for an eco-friendly approach.

In India, botanical pesticides are available in many plants for which deep search and testing is required, as many of them are still unexplored. As they are available in bulk, they are expected to be cheaper in comparison to synthetic chemicals. Moreover, the farmers can prepare their crude extracts for their use in field. Many nematodes in India as well as in abroad have tried to reduce nematode infestation by using botanicals in different ways viz.

(A) by using parts of the botanicals directly.

(B) by using the extracts of parts of botanicals,

(C) by using the botanicals as organic manure,

(D) by using mature crop residues and products of botanicals.

1. NEMATICIDAL PLANTS

A list of angiospermic plants having nematicidal activity and part of the botanical used against plant parasitic nematodes is given in Table 1.

DIFFERENT WAYS OF USING BOTANICALS

(a) Parts of the botanicals used for Nematode Management

Many plants have been showed to possess nematicidal activity. The effective parts are stems, leaves, roots, fruits and seeds. Singh and Sitaramaiah (1967) reported that application of 5-10% (w/w) of leaves of *Azadirachta indica, Melia azederach, Cassia fistula, Sesbania aculeata* and *Crotolaria juncea* to infested soil reduced incidence of root knot nematode on tomato and okra. Lal *et al.*, (1977) studied the effect of chopped leaves of various plants on growth of brinjal plants. Chopped leaves of *Azadirachta indica* proved to be most effective in promoting growth parameters of brinjal and reducing population of *R. reniformis* on roots. Nath *et al.*, (1982) used plant materials of *Argemone mexicana* against *M. incognita* in different forms. It lowered nematode population in field and showed nematostatic properties.

KaliRam and Gupta (1982) found that neem leaves, aldicarb and potassium significantly reduced the number of gall per plant individually where as datura leaves and carbofuran increased the various growth parameters. However, combined effect of these treatments was significant in reducing the number of galls caused by *M. javanica* infecting chickpea. Maximum reduction occurred in treatments where neem leaves were added with potassium and aldicarb.

Castillo (1985) reported reduction in *R. reniformis* population in the pots where fresh leaves of four plant species viz. (*Asystacia gangetica, Polytrias amaura, Gliricidia sepium* and *Leucaena leucocephala* were added. He also observed a marked reduction in root population of *R. reniformis* on cotton. Dutt and Bhatti (1986) found chopped castor leaves effective in reducing number of galls of *M. incognita* and enhanced plant growth of tomato when applied @ 40 and 60% g/kg soil.

The dried materials of *Argemone mexicana, Eucalyptus globorus, Datura metel, Phyllanthus niruri* and oilcakes of *Calophyllum inophyllum, Madhuca indica* and *Shorea robusta* @ 1,2 and 3 per cent w/w, against *M. incognita* infecting tomato were applied by Goswami and Vijailakshmi (1986). *A. mexicana* and *P. niruri* and oilcakes of *S robusta* and *C inophyllum* showed delayed and retarded hatching of root knot nematode with less oviposition. Tiyagi *et al.*, 1988) observed that addition of chopped shoot parts of some plants of the family compositae into pot soil effectively retarded final population of *R. reniformis* and *M. incognita* thereby consequently improving growth of aubergine seedlings, Siddiqui *et al.*, (1987) reported that incorporation of chopped shoots of latex bearing plants significantly suppressed the population build up of *R. reniformis* and *Tylenchorhynchus brassicae* and reduced the root knot development caused by *M. incognita.*

Table 1. List of angiospermic plants having nematicidal effect on plant parasitic nematodes

Name of angiospermic plants		Family	Nematodes test species	Nematicidal Plant Part
Scientific	English			
Ageratum conyzoides L.		Asteraceae	*M. incognita*	Leaf
A. sativum L.	Garlic	Liliaceae	*M. incognita*	Bulb
Amaranthus gracilus Dest	Amaranth	Amaranthaceae	*M. javanica*	Leaf
Anacardium occidentale L	Cashew	Ancardiaceae	*M. incognita*	Leaf
Arachis hypogaea L.	Groundnut	Papilionaceae	*M. incognita*	Testa
Argemone mexicana L.	Mexican Prickly	Papaveraceae	*M. incognita*	Leaf, Seed
Artocarpus heterophyllus lank	Jackfruit	Moraceae	M. incognita	Shoot
Asparagus sp.	Asparagus	Liliaceae	*Trichodorus christie*	Root
Atropa belladonna L.	Belladonna	Solanaceae	*M. incognita*	Berr
Azadirachta indica A. Juss	Neem	Meliaceae	*M. incognita*	Leaf, Shoot
Bougainvillea spectabilis willid	Bougainvillea	Nyctaginaceae	*Aphelenchoides composicola*	Leaf
Brassica sp.	Mustard	Brassicaceae	*M. incognita*	Seeds
B. nigra Koch	Black mustard	Brassicaceae	—	As cover crops
Calendula officinalis L.	Pot marigold	Asteraceae	*M. incognita*	—
Calotropis gigantea Ait	Madar	Asclepidiaceae	*M. incognita*	Leaf
Cannabis sativa L.	Hemp	Cannabinaceae	*M. incognita*	Whole plant
Capsicum annum L.	Chilli	Solanaceae	*M. incognita*	Pod
Carthamus tinctorius L.	Safflower	Asteraceae	*Meloidogyne*	Flower, Root
Carthamus roseus G. don	Periwinkle	Apocyaceae	*M. incognita*	Leaf
Cedrela toona Rob. ex. Rottl.	Red Cedar	Meliaceae	*Aphelenchoides composticola*	Leaf
Chrysanthemum coronarium L.	American Chrysanthemum	Asteraceae	*M. incognita*	Leaf
Citrus aurantium L.	Sourorange	Rutaceae	*Ditylenchus destructor*	Fruit skin
Coffee spp.	Coffee	Rubiaceae	*M. exigua*	Beancover, Straw
Coriandrum sativum L.	Coriander	Apiaceae	*M. incognita*	Seeds
Crotalaria spp.		Papilionaceae	*M. hapla*	Whole plant
Curcuma domestica val.	Turmeric	Zingiberaceae	*M. incognita*	Rhizome
Cuscuta reflexa Roxb.	Dodder	Convolvulaceae	*M. incognita*	Flower, whole plant
Datura metel L.	Datura	Solanaceae	*M. incognita*	Leaf
Eichhornia speciosa kunth	Water hyacinth	Pontederiaceae	—	Whole plnat
Eragrostis curvula Nees	Weeping lovegrass	Poaceae	*Meloidogyne* spp.	Root
Festuca rubra L.	Red fescue	Poaceae	*Pratylenchus penetrans*	Root
Ficus carica L.	Fig	Moraceae	*T. brassicae*	Shoot
F. elastica Roxb.	Rubber	Moraceae	*Meloidogyne* sp.	Latex
Gaillardia spp.		Asteraceae	*Meloidogyne* sp.	—
Glycine max Merr	Soyabean	Papilionaceae	*Pratylenchus*	Seed meal
Gossypium Spp.	Cotton	Malvaceae	*M. exigua*	Seed
Helenium sp.	Bitter Sneeze Weed	Asteraceae	—	—

Table 1 contd...

Table 1 contd...

Name of angiospermic plants		Family	Nematodes test species	Nematicidal Plant Part
Scientific	English			
Jasminum arborescens Roxb.	Tree Jasmine	Oleaceae	*M. incognita*	Leaf
Leucaena leucocephala de Wit	Subabul	Mimosaceae	*M. incognita, Helicotylenchus dihystera*	Leaf
Linum usitatissimum L.	Flax, Linseed	Linaceae	*M. incognita, Rotylenchulus*	Seed
Manihot esculenta crantz	Tapioca	Euphorbiaceae	Meloidogyne sp.	—
Mangifera indica L.	Mango	Anacardiaceae	*M. incognita*	Leaf
Melia azadirach L.	Persian lilac	Meliaceae	*M. incognita*	Leaf
Nicotiana tabacum L.	Tobacco	Solanaceae	*Tylenchorhynchus* sp.	Leaf, Stem
Ocimum sanctum L.	Holy basil	Lamiaceae	*M. incognita*	Leaf
Oenothera sp.	Primorse	Onagraceae	*M. incognita*	—
Parthenium hysterophorus L.	Parthenium	Asteraceae	*M. incognita Heliocotylenchus dihystera*	Leaf
Pennisetum purpurcum K. Schum	Elephant grass	Poaceae	*Pratylenchus branchyurus*	Straw
Phaseolus lunatus L.	Lima bean	Papilionaceae	*M. incognita*	Seed
P. Vulgaris L.	Kidney bean	Papilionaceae	*M. incognita*	Leaf, Stem
Phyllanthus fraternus Webster		Euphorbiaceae	*M. incognita*	
Polygonum hydropiper L.	Water pepper	Polygonaceae	*M. incognita, Rhabditis*	Leaf
Raphanus sativus L.	Radish	Brassicaceae	*Pratylenchus, Tylenchorhynchus*	Cover crop
Ricinus communis L.	Castor	Euphorbiaceae	*Tylenchorhynchus brassicae, M. javanica*	Leaf
Sesbania sp.	Sesbania	Papilionaceae		Whole plant
Solanum hispidum Pers.		Solanaceae	*M. incognita*	
Tagetes erecta L.	Marigold	Asteraceae	*Pratylenchus* sp.	Root
T. patula L.	French marigold	Asteraceae	*M. incognita*	Root
Trichosanthes anguina L.	Snake gourd	Cucurbitaceae	*M. incognita*	Seed
Tridax procumbens L.		Asteraceae	*M. incognita*	Leaf
Zingiber Officinale Rosc.	Ginger	Zingiberaceae	*M. incognita*	Rhizome

Alsayed *et al.*, (1992) found that when leaf powders of *Canna indica, Ricinus communis, Lantana camara, Clerodendrum incrme* and *Tagetes erecta* were added @ 5, 10 or 15 g/kg soil, all treatments suppressed reproduction of *R. reniformis* and improved soybean shoot and root weights. El-Nagar *et al.*, (1993) reported that addition of dry ground leaves of the Egyptian neem, *Melia azedirach Asparagus, Sprengeri, Eucalyptus* sp. and *Psidium guajava* at the rates of 2.5 and 5.0 g per pot significantly reduced the number of eggmasses of *R. reniformis* per root on sunflower.

(b) Extracts of the parts of botanicals used for Nematode Management

Methods for the screening of plants for their nematicidal activity have been tried. These include preparation of plant extracts and testing these extracts both *in vitro* and *in vivo*, on nematodes. Desai *et al.*, (1973) tested the extracts of different parts of 26 plant species against a mixed population of larvae of *M. javanica* and *M. incognita acrita*. Thirteen plant species which showed strong nematicidal properties were *Ageratum conzoides* L. (leaf), *Allium sativum*, L. (bulb), *Anacardium occidentale* L. (leaf) *Argemone mexicana* L. (leaf), *Calophyllum inophyllum* L. (leaf), (deoiled cake), *Citrus reticulata* Blanco (Fruit Skin), *Datura stramonium* L. (leaf), *Hyduocarpus laurifolia* (Deoiled cake), *Holarrhena antidysenterica* Wall (seed) and *Uinca rosea* L. (leaf). Vijailakshmi *et al.*, (1979) screened 50 indigenous plant, products for their nematicidal value against the second stage juveniles of *M. incognita* under laboratory conditions. Kali Ram and Gupta (1980) applied aqueous extracts of fresh neem leaves @ 10, 20, 30 and 40 g leaves per kg treated soil to chickpea plants grown in *M. javanica* infested soil and found good plant growth and reduction in root galling with the highest application rate.

Majumdar and Mishra (1991) reported that soaking of chickpea seed in the aqueous extracts of leaves of bhang, bhangra, dhatura, Kateli (*Argemone mexicana*) and neem reduced the penetration of *M. incognita* in chickpea seedlings, Siddiqui and Alam (1987a) found that when seeds of tomato, aubergine and okra were coated with water soluble extracts of *Azadirachta indica* and *Melia azedarach* and the resultant seedlings were inoculated with *M. incognita* or *R. reniformis*, there was significant reduction in root galling and *R. reniformis* population builp up.

The presence of an alkaline water soluble nematostatic chemical constituent in leaf extract of *Tagetes patula* was reported by Indra Rajavanshi *et al.*, (1985) and tested against *Xiphinema basiri*. The aged extract showed better nematostatic property at longer exposure periods in comparison to fresh and autoclaved leaf extracts. Siddiqui and Alam (1988) observed water extracts of different parts of marigold (*Tagetes lucida*) highly deleterious to *Meloidogyne incognita*, *Rotylenchulus reniformis*, *Hoplolaimus indicus*, *Tylenchorhynchus brassicae*, *Helicotylenchus indicus* and *Tylenchus filiformis* however to varying extent. Flower extracts caused greatest nematode mortality and inhibition in juvenile hatching followed by seed, leaf and root extracts. They also reported that the root exudates of all the three species of Tagetes i.e. *T. lucida*, *T. minuta* and *T. teemifolia* have nematotoxic principles to varying degree.

Greval and Sohi (1988) tried toxicity of some plant extracts viz., *Eucalyptus hybrid*, *Leucaena leucocephala*, *Populus dettoides* *Tectona grandis* against *Aphelenchoides compesticola*. The extracts of *L. leucocephala* were most effective followed by *P. deltoides*, *E. hybrid* and *T. grandis*. Chattopadhyay and Mukhopadhyay (1989) tried water extracts of leaves of 'Kathal Campa' (*Artobotrys odoratissimes*) against *Meloidogyne incognita*. Significant reduction in larval hatch from the eggmasses was observed after 24, 48 and 72 hours exposure. Maximum reduction in hatching was observed in 72 hours.

Subramanian and Selvaraj (1988) observed slow movement of *Radopholus similis* by applying leaf extracts of *Tagetes patula* within 2 hours of the treatment. In 4 hours all the nematodes were inactivated or killed in 1 : 1 x 1 : 5 dilution extracts. Nematodes failed to regain activity when transferred to fresh distilled water. Jasy and Koshy (1992) reported that extracts of leaves of *Glyricidia machlata*, *Ricinus communis* and *Crotolaria juncea* were found lethal to *Radopholus similis* at dilutions of 1.5 withing 24 hours. The nematode mortality decreased with increase in dilutions of the extract.

(c) Products of the botanicals used for Nematode Management

The plants whose nematicidal properties have been reported by many workers over the years are *Tagetes erecta*, *T. patula*, *Azadirachta indica*, *Brassica campestris, Ricinus communis, Arachis hypogaea* and *Madhuca indica.* Of these 7 plants species of *Tagetes* were used mostly in inter- culture with other crops and the rest yielded oilcakes which served as nematicides. Oilcakes from the seeds of *Callophyllum inophyllum, Tectona grandis, Pongamia glabra, Carthamus tinctorius* and *Shorea robusta* have also been used with success against plant parasitic nematodes. The oilcakes undergo decomposition in soil by microorganisms and serve as fertilizers there by helping in plant growth. The plant products like the leaves of *Leucaena leucocephala, Azadirachta indica* and *Pongamia,* Cassava root pulling, Locust bean husks manipueira and citrus fruit canning factory waste, when used as soil amendments reduced nematode disease and increased crop yield.

The effect of four fractions of neem kernals viz., pure neem oil, Linonoids-1, Linonoids-2 and Neemmark on *M. incognita* was observed by Dev Kumar *et al.*, (1985). The pure oil was inactive while Limonoids were highly active in respect with larval mortality and larval emergence. The neem mark was more active in exhibiting larval emergence than causing mortality. Prasad *et al.*, (1987) tested biotoxicity of groundnut testa's extract and various concentration viz., 1000, 500, 250, 100 and 50 ppm against *M. incognita.* Nematode immobility was essentially proportional to concentration and time exposures with an increasing concentration from 5 to 1000 ppm. There was a progressive increase in per cent immobility of root knot larvae, 100 ppm was sufficient to immobilise more than 70 per cent nematode larvae after 72 hours. The toxicity also increased with duration of exposure from 24 to 72 hours.

Gokte and Swarup (1988) applied dried neem seed kernel to wheat (cv. Sonalika) seed with the help of gum at 3 concentrations i.e @ 400, 600 and 800 mg powder/100 g seed and observed its effect on the development of *Anguina tritici.* The results indicated the inhibitory property of powdered neem seed kernel on larval penetration to the growing point of wheat seedlings.

Some organic oils such as neem oil (*Azadirachta indica*), Karanj oil (*Pongamia glabra*), *Chalmogra oil (Hydnocarpus kurzii)* and polanga oil (*Calophyllum inophyllum* as seed treatment seedling root dip and soil treatment, to determine their efficacies on the infectivity of *M. incognita* affecting tomato were tested by Pradhan *et al.*, (1989), oils of chalmogra, neem and Karanj in higher concentrations proved effective in preventing larval penetration and gall production in the roots. Among the three methods of treatment, seedling root dip and wet seed dressing were relatively more effective than soil treatment.

Chattopadhyay and Mukhopadhyay (1989) tested the nematicidal activity of corn powder of *Typhonium trilobatum* and seed dust of *Melia azedarach* against *Meloidogyne incognita, T. trihobatum* not only exhibited potentiality as nematicide than *M. azedarach* but acted as a better growth promotor also. In preplant application, *T. trilobatum* at higher dose controlled 18 per cent more nematode than that of *M. azedarach.* The application of numbin (a triterpenoid from *Azadirachta indica*) as seed dressing significantly reduced the root knot of tomato and nematode population of *M. incognita, R. reniformis* and *Tylenchorchynchus brassicae* with as consequent improvement in plant growth (Siddiqui and Alam (1990).

Majumdar and Mishra (1992) found neem seed kernel and neem seed coat as nematicidal when used as seed against *M. incognita* infecting mungbean. There was significant reduction in the penetration of second stage juveniles in roots. Neem seed coat could also be used for the management of *M. incognita* infecting chickpea.

Majumdar and Mishra (1992b) found that soil treatment @ 20% w/w when combined with seed coating @ 20% w/w provided 66.5% reduction in root knot galls and was also found to be effective in respect of plant growth characters.

(d) Effect of Biopesticidal or Nematicidal compounds on Nematodes

Nematicidal compounds from plants would be economical only when such compounds are available in large quantities either through their isolation from plants or through their synthesis on a commercial scale. If none of the two processes is cost effective the compounds even very effective against nematodes, will have limited used only in high value crops. At present nematicidal principles of only a few plants are known. There exists a need for research in this area so that a comprehensive knowledge of the range of nematicidal compounds present in plants is available. These compounds are mainly glycosides, quinones, unsaturated and saturated hydrocarbons, heterocyclic compounds, organic acids, aromatic compounds, ester, sulphur compounds and terpenes. Among the terpenes, monoterpenes showed the strogest nematicidal activity followed by sesqueterpenes. Essential oils from the compositae and citrus plants proved toxic to nematides. The pure compounds as for example eugenol are sometimes systemic in action and effective at extremely low concentrations. A vast body of literature (Table 2) has been accumulated on biopesticides occurring in plants during the last two- three decades in India and abroad. Many well known biopesticides developed from biochemicals from trees (viz., neem, Karanj, Mahua) and gain popularity.

The leaf extract of *Polygonum hydropiper* contain formic and acetic acid and showed strong nematicidal action. Formic acid proved stronger showing its effect at 0.003% concentration. (Sukul, 1970). The n-pentyl, n-propyl, n-butyl, n-heptyl and z- methyl substituted forms of primin derived from *Primula obconica*

Table 2. Nematicidal principles found in botanicals.

Name of botanicals		Family	Nematode species tested	Active Principle
Scientific	English			
Asparagus sp.	Asparagus	Liliaceae	*Trichodorus christiei*	Glucoside ($C_6H_{20}4$) Asparagusic acid
Azadirachta indica A. Juss	Neem	Meliaceae	*M. incognita*	Nimbidine, Thionemone
Carthamus tinctorius L.	Sunflower	Asteraceae	*Aphelenchoides besseyi*	3-cis, 11-Transtrideca-1, 3, triene, 5,7,9-triyne and 3 transisomer
Cymbopogon flexuosus wats	Malabar	Poaceae	*Meloidoggyne* sp.	Citral, geraniol, citron ellol
Nicotiana tabacum L.	Tobacco	Solanaceae	*M. incognita*	Nicotine
Ocimum sanctum L.	Tulsi	Lamiaceae	*M. incognita*	Eugenol
Physostigma venenosum Balf.	—	Fabaceae	—	Physostigmine
Polygonum hydropiper L.	Water pepper	Polygonaceae	*M. incognita, M. javanica, Mononchus, Dorylaimus*	Formic acid, acetic acid, essential oil
Solanum viarum Dunal	—	Solanaceae	*M. incognita*	Solamrgine
Tagetes sp.	Marigold	Asteraceae	*Meloidogyne*	Polythienyes, X-terthienye
T. patula	French Marigold	Asteraceae	*Meloidogyne* sp.	Hydroxy trematone bithienyl derivative

showed some nematicidal activity against *Ditylenchus* at strengths of 10-100 ppm (Schildenecht and Lubosch, 1987). Eugenol (C_{10} H_2 C_2), an active nematicidal principle from *Ocimum sanctum*, also showed strong systemic effect reducing *M. incognita* infestation of okra at a concentration of 0.2 ml/1. Foliar spray was effective than soil drench. (Bala and Sukul, 1987).

Davis *et al.*, (1989) reported that Phytolectins, particularly concanavalin A, from jack bean *canavalia ensiformis* reduced *M. incognita* infestation of host plants *in vivo* expts. Das *et al.*, (1989) found that lectins may play some role in the natural defense mechanism of plants and are accumulated in galled regions of the roots of okra infested with *M. incognita* while in uninfested roots lectins are randomly distributed.

Allium grayi and *A. fistulosum* var. *Caespitosum* contain thiosulfinate compounds, disulfides and thiosulfonates and showed nematicidal properties. (Tada *et al.*, 1988). Davis and Rich (1987) found that nicotine at a concentration of 1-100 mg/ml reduced mobility of *M. incognita* larvae and their ability to induce root galls in tomato. Nicotine content in tobacco roots increased greatly in resistant than in susceptible varieties of tobacco upon exposure to *M. incognita*. Ascorbic acid and some amino acids such as glutamic acid, L-glutamic acid and L-arginine applied by foliar spray before inoculation reduced *M. incognita* infestation of tomato. (Al-Sayed and Montasser, 1986). Avermectins, the macrocylic lactones derived from soil actinomycete, *Streptomyces avermitilis* proved effective against *M. incognita* (Wright *et al.* 1983; Garabedian and Van Gundy, 1983). Jietacins A and B, isolated from the culture broth of a strain of Streptomyces exhibited 10 times higher nematicidal activity against (*Bursaphelenchus lignicolus*, than did avermectin B (Imamura *et al.*, 1989).

Mahajan *et al.*, (1986) screened 18 naturally occurring terpenoid lactones for their nematicidal activity against *M. incognita* larvae. They found that maximum nematicidal activity was associated with a methylene lactone moiety present in 6 compounds.

2. CONCLUSION

Botanicals for the management of plant parasitic nematodes are observed to be easy in application, free from environmental pollution and improve soil health structurally and nutritionally. Scientists all over the world are now looking towards eco-friendly selective, biodegradable, safe and economical biochemicals for use in programmes of integrated nematode management. The bio-pesticides appear to have a promising role for the development of future commercial pesticides not only for agricultural crop productivity but also for the safety of the environment and public health. The plant world is a rich store-house of biochemicals that could be tapped for use as pesticides. Botanicals being the only alternative to chemicals which will fulfil all requirements of farmers and will also bring sustainability to agriculture as it involves integration of cultural, biological and natural inputs for the management of nematodes.

REFERENCES

Alsayed, A.A. and Mnotasser, S.A. 1986. The role of ascorbic and glutamic acids in controlling the root knot nematode, *Meloidogyne javanica*. *Egyptian J. Phylopathol* 18 : 143- 148.

Alsayed, A.A., Ahmad, S.S. and Abdet Hameed, S.H. 1992. Influence of decomposing dry leaf powders on *Rotylenchulus reniformis* and growth response of soybean. *Annals of Agric. Sci.*, Moshtohor, 30 : 615-620.

Bala, S.K. and Sukul, N.C. 1987. Systematic nematicidal effect of eugenol. *Nematropica* 17 : 219-222.

Castillo, M.B. 1985. Some studies on the use of organic soil amendments for nematode control *Philippine Agriculturist* 68 (1) : 76-93.

Chattopadhyay, P.R. and Mukhopadhyay, M.C. 1989a. Effect of leaf extracts of *Artotrys Odoratissimus* R. (Fam. Anonaceae) on the hatching of the eggs of *Meloidogyne incognita* (Kofoid and white', 1919) Chitwood, 1949. *Indian J. Nematol* 19 (1) : 29-31.

Chattopadhyay, P.R. and Mukhopadhyay, M.C. 1989b. Comparative studies on the nemiticidal properties of *Typhonium trilobatum* and *Melia azedarach. Indian J. Nematol.*, 19 (1) : 5-9.

Das, S., Sukul, N.C., Mitra, D. and Sarkar, M. 1989. Distribution of lecting in nematode infected and uninfested roots of *Hibiscus esculentus. Nematol Medit.* 17 : 123-125.

Davis, E. and Rich, J.R. 1987. Nicotine content of tobacco roots and toxicity to *Meloidogyne incognita J. Nematol.* 19 : 23- 29.

Davis, E.L. Kaplen, D.T., Dickson, D.W. and Mitchelss, D.J. 1989. Reproduction of lectin treated *Meloidogyne* spp. in two treated soybean cultivars. *Rev. Nematol* 12 : 257-263.

Desai, M.V., Shah, H.M. and Pillai, S.N. 1973. Nematicidal property of some plant species. *Indian J. Nematol* 3 : 77- 78.

Devakumar, C., Goswami, B.K. and Mukerjee, S.K. 1985. Nematicidal principles from neem (*Azadirachta indica* A. Juss.) Part I. Screening of neem kernel fractions against *Meloidogyne incognita. Indian J. Nematol* 15 : 121-124.

Dutt, R. and Bhatti, D.S. 1986. Determination of effective doses and time of application of nematicides and castor leaves for controlling *Meloidogyne javanica* in tomato. *Indian J. Nematol* 16 (1) : 8-11.

El-Nagar, H.I., Hendy, H.H., Abdul-Hameed, S.H. Farhat, A.A. and Osman, A. A. 1993. The role of dry ground leaves of some plants in controlling the reniform nematode *Rotylenchulus reniformis* infecting sunflower. Bull. Faculty Agric., Univ. Cairo. 44 : 205-216.

Garabedian, S. and Van Gundy, S.D. 1983. Use of avermectins for the control of *Meloidogyne incognita* on tomatoes. *J. nematol* 15 : 503-510.

Gokte, N. and Swarup, G. 1988. Effect of neem (*Azadirachta indica*) seed kernel powder treatment on penetration and development of *Anguina tritici* on wheat. *Indian J. Nematol.* 18 : 149.

Goswami, B.K. and Vijaylakshmi, K. 1986. Effect of some indigenous plant materials and oilcake amended soil on the growth of tomato and root knot nematode population. *Ann. Agric. Res.* 7 : 363-366.

Grewal, P.S. and Sohi, H.S. 1988. Toxicity of some plant extracts to *Aphelenchoids composticola, Indian J. Nematol* 18 : 354- 355.

Imamnnra, N., Kuga, H., Otoguro, K., Tanaka, H. and Omura, S. 1989. Structures of jietacines unique B-unsaturated azoxy antibiotics. *J. Antibiotics* 42 : 156-158.

Indra Rajvanshi, Verma, M.K. and Yadav, B.S. 1985. Nematostatic properties of *Tagetes patula* L. Aqueous leaf extract on *Xiphinema Basiri* Siddiqui, 1959. *Indian J. Nematol.* 15 (2) : 195-196.

Jasy, T. and Koshy, P.K. 1972. Effect of certain leaf extracts and leaves of *Glyricidia maculata* (H.B. and K.) stend as green manure on *Radopholus similis. Indian J. Nematol* 22 (2) : 117-121.

Kali Ram and Gupta, D.C. 1980. A note on the efficacy of neem leaf extract in the control of *Meloidogyne javanica* infecting chickpea (*Cicer arietinum*). *Indian J. Nematol.* 10 : 96-98.

Kali Ram and Gupta, D.C. 1982. Efficacy of plant leaves, nematicides and fertilizers alone and in combination against *Meloidogyne javanica* infecting chickpea. (*Cicer arietinum*) *Indian J. Nematol.* 12 : 221-225.

Lal, A., Yadav, B.S. and Nandwana, R.P. 1977. Effect of chopped leaves of various plants and sewage on the plant growth and reniform nematode, *Rotylenchulus reniformis Indian J. Mycol Pl. Path.* 7 : 68-69.

Mahajan, R., Singh P., Bajaj, K.L. and Kalsi, P.S. 1986. Nematicidal activity of some sesqneterpenoides against root knot nematode (*Meloidogyne incognita*). *Nematologica* 32 : 119- 123.

Majumdar, V. and Mishra, S.D. 1991. Nematicidal efficacy of some plant products and management of *Meloidogyne incognita* in pulse crops by soaking seeds in their aqueous extracts. *Current Nematology* 2 (1) : 27-32.

Majumdar, V. and Mishra, S.D. 1992a. Effect of seed soaking in aqueous extracts of neem seed on germination of mungbean and penetration of second stage juveniles of *Meloidogyne incognita Ann. Agric. Pes.* 13 (3) : 297-299.

Nath, A., Sharma, N.K., Bhardwaj, S. and Thapa, C.D. 1982. Nematicidal properties of garlic. *Nematologica* 28 : 253- 255.

Pradhan, P., Das, S.N. and Dora, D.K. 1989. Effect of some organic oils on the infectivity of root knot nematode (*Meloidogyne incognita*) affecting tomato. *Indian J. Nematol.* 19 (2) : 162-165.

Prasad, D., Agarwal, A.K., Bhuchare, V.M. and Sethi, C.L. 1987. Biotoxicity of groundnut testa's extract against *Meloidogyne incognita. Indian J. Nematol.* 17 (1) : 121.

Sasser, J.N. and Freckman, D.W. 1987. A world perspective on Nematology. The role of the society. In vistas on Nematology a commemoration of twenty fifth anniversary of the society of Nematologists (Eds. J.A. Veech and D.W. Dickson). Society of Nematologists. Inc. Hyattsville, M.O., 7-14 pp.

Sasser, J.N. 1989. Plant parasitic nematodes : The farmers hiddern enemy. Dept. of plant Pathology, North carolina state university, Raleigh, 115 pp.

Schildknecht, H. and Lubosch J. 1987. Plant derived biocides XXI. Molluscidal properties of quinones. Pflenzenalwchrstoffe XXXI. Mulluskizide Eigenschaftern Von Chinonen. *Zeitschrift fur Naturforschung* C (Biosciences) 42 : 31-31.

Siddiqui, M.A. and Alam, M.M. 1987a. Efficacy of seed dressing with extracts of neem and Persian lilac against *Meloidogyne incognita* and *Rotylenchulus reniformis. Nematol Mediterr* 15 : 99-403.

Siddiqui, M.A., Haseeb, A. and Alam, M.M. 1987b. Evaluation of nematicidal properties in some latex bearing plants. *Indian J. Nematol.* 17 : 99-102.

Siddiqui, M.A. and Alam M.M. 1988. Toxicity of different plant parts of *Tagetes lucida* to plant parasitic nematodes. *Indian J. Nematol.* 18 : 181-185.

Siddiqui M.A. and Alam, M.M. 1990. Control of root knot and reniform nematodes by bare root dip in leaf extracts of margosa/neem and Persian lilac. *Zeitchrift fur Pflanzenkrankheiten and Pflenzenschutz* 95 : 138-142.

Singh, R.S. and Sitaramaiah, K. 1967. Effect of decomposing green leaves, sawdust and urea on the incidence of root knot of okra and tomato. *Indian phytopath* 20 : 349-355.

Subramanian, S. and Sellvaraj, P. 1988. Effect of *Tagetes patula* L. leaf extract on *Radopholus similis* (cobb : 1983). Thorne, 1949. *Indian J. Nematol* 18 : 337-338.

Sukul, N.C. 1970. Inhibition of nematode infestation of wheat seedlings by *Polygonum hydropiper. Nature* 226 : 771-772.

Tada, M., Throl, Y., Kiyohara, S. and Suzuki, S. 1988. Nematicidal and microbial constituents from *Allium gravi* Regel and *Allium fistulosum* L. var. Caespitosum. *Agric. Bial. Chem.* 52 : 2383-2385.

Tiyagi, S.A., Bano, M. and Alam, M.M. 1988. Evaluation of nematicidal potential in some plant species belonging to family compositae. *Indian J. Nematol* 18 : 228-231.

Wright, D.T., Birtle, A.J. Crops, A.E. Dybas, R.A. 1983. Efficacy of avermectins against a plant parasitic nematode. *Ann. Appl. Biol.* 103 : 465-470.

CHAPTER 9

NON-PESTICIDAL MANAGEMENT OF NEMATODES IN FORAGE CROPS

N. Hasan

INTRODUCTION

Forage crop production in all of its ramification is of major importance to agriculture. This is another way of saying that forage crops are of vital concern to man. It is not only the cheapest source of nitrogen and palatable nutrients for livestock, but is also valuable in restoring and maintaining soil fertility and structure. Moreover, in the milkshed areas where the major enterprise of the farmers centres around the milk production, continuous supply of green fodder round the year has become the basis for success of such industry. Therefore, cultivation of nutritive fodder crops as such or as part of the normal crop rotation practices has gained importance in India and elsewhere in the world.

Forage crops virtually in every part of the world are also impaired to a greater or lesser extent by nematode pests in the same way as in other crops. However, the most evident damage may occur in tropical and subtropical regions in which besides other favorable conditions, crops are grown all round the year without a vegetative rest period. In such a case the nematodes, particularly the polyphagus ones get the opportunity of uninterrupted food supply and thus multiply to high population resulting in severe crop losses. The plant damage is sustained in several ways which together lead to poor growth, low yield and impaired quality.

Few reviews are available on nematode diseases of forage crops which have already been recognized as major agricultural problem especially in the temperate countries of the world (Binefors, 1969; Cook and Yeates, 1993; Erickson, 1972). However, in India and other tropical countries of the world, importance of the nematode problems in forage crops has been realized only during a couple of decades back (Bhaskar *et al.*, 1999, Hasan, 1990; Jain & Hasan, 1995, 1998a; Pandey *et al.*, 1996). The objective of this review is to emphasize the importance of nematode pests in relation to forage crop production and also to focus on the progress made in various facets of their management especially in the area of non pesticidal approaches.

1. NEMATODE PROBLEMS

A. Cultivated fodder

A number of plant parasitic nematode genera belonging to various species were found to be associated with a variety of fodder crops grown in different parts of the country (Hasan, *et al.*, 1998; Pandey *et al.*, 1995). It is difficult to include here every nematode species which might injure a crop. Therefore, only major nematodes are considered which have been identified on the basis of their actual or potential economic importance or on their frequency of occurrence/population density (Table 1).

Nematode can be a real problem in forage production when they cause a reduction in the growth and vitality of the crop plant leading to an economic loss. Moreover, certain nematodes have the

Table 1. List of plant parasitic nematodes associated with major forage crops in India.

* *Meloidogyne incognita* (Kofoid & White) Chitwood
* *Meloidogyne javanica* (Treub) Chitwood
Meloidogyne arenaria Chitwood
* *Rotylenchulus reniformis* Linford & Oliveira
* *Tylenchorhynchus vulgaris* (Upadhyay) Swarup & Sethi
* *Tylenchorhynchus mashoodi* (Siddiqui & Basir)
* *Pratylenchus zeae* Grahm
Pratylenchus spp.
Hoplolaimus indicus Sher
Hoplolaimus columbus Sher
* *Helicotylenchus dihystera* (Cobb) Sher
Heliocotylenchus indicus Siddiqui
Paratylenchus sp.
Aphelenchus avenae Bastian
* *Heterodera cajani* Koshi
* *Heterodera zeae*
Tylenchus filiformis Butschii
Longidorus elongatus (De Man) Thorne & Swander
Trichodorus sp.
Xiphinema sp.
Criconemoides spp.
Basirolaimus spp.
Radopholus similis
Rotylenchus sp.
Psilenchus spp.
Ditylenchus spp.
Telotylenchus indicus.

* Key nematodes of forage crops.

additional ability to transmit viruses. Other alter the physiology of their hosts so that it becomes more vulnerable to fungal diseases, or they provide avenues of entry for pathogenic bacteria and these are often of greater economic importance than the direct feeding activity of the nematode. Examples of certain disease complexes involving nematodes and other pathogens in forage crops are given in table 2. In addition, certain nematode species have also been reported to cause decline in *Rhizobium* nodules resulting into reduced nitrogen fixation capacity of forage legumes (Hasan & Jain, 1984a; Hasan & Jain, 1988; Kanwar & Gupta, 1988; Yeates *et al.*, 1977). It is difficult to assess with any accuracy the exact losses caused by nematodes in forage crops. However, in some temperate countries losses of 3-11% equivalent to several million dollars per year have been reported to occur in some important forage legumes (Table 3). Presumably the losses in most of the tropical and subtropical countries with less advanced agricultural techniques may even be higher.

B. Grasslands

Plant parasitic nematodes are widely distributed in both tropical and temperate grasslands of the world (Curry, 1987;

Table 2. Examples of disease complexes involving nematodes and other pathogens

Crops	Nematodes	Other organisms (Fungi/bacteria)	Disease syndrome	References
Cowpea	*Meloidogyne incognita*	*Rhizoctonia solani*	Root-rot	Khan & Hussain, 1988
	M. javanica	*Fusarium* sp.	Wilt	Thomson et al., 1959
		R. solani	Root-rot	Kanwar et al., 1988
	Rotylenchulus reniformis	*R. solani*	Root-rot	Khan & Ḥussain 1988
Lucene	*Ditylenchus dipsaci*	*Corynebacterium insidiosum*	Wilt	Dumbier et al., 1979
	M. incognita	*C.insidiosum*	Wilt	Erickson, 1972
	M. hapla	*Phytophthora megasperma* var. *medicagnis*	Root-rot	Welty et al., 1980
	M. incognita acrita	*F. oxysporum* var. *batatas*	Wilt	Kushner & Crittenden, 1967
	Pratylenchus penetrans	*F. oxysporum*	Wilt	Edmunds & Mai, 1967
		Trichoderma viridae	Root-rot	As above
		F. solani	Wilt	Manza, 1974
White clover	*D. dipsaci* *Meloidogyne* spp.	*Pythium middletoni*	Death of stolon	Irwin & Jones, 1977
Berseem	*Tylenchorhynchus vulgaris*	*F. semitectum*, *R. solani*	Root-rot	Hasan & Bhaskar, 1992
Soybean	*M. arenaria*	*Sclerotium rolfsii*	Southern blight	Minton et al., 1975
Maize	*T. claytoni*, *Tylenchus agricola*	*F. roseum*, *P. ultimum*	Root-rot	Kissiel et al., 1969
	T. vulgaris, *P. Zeae*	*F. moniliforme*	Root-rot	Upadhyay & Swarup 1981
	P. brachyurus	*F. moniliforme*	Seedling rot	Jordan et al., 1987
	Hoplolaimus indicus	*F. moniliforme*	Root-rot	Nath et al., 1974
Pearlmillet	*M. incognita*	*Sclerospora graminicola*	Downy mildew	Vaishnav & Sethi, 1978a
Kentucky blue grass	*T. dubius*	*F. roseum*	Blight	Vargas & Laughlin, 1972

Table 3. Estimated annual crop losses due to plant parasitic nematodes (mainly after Sasser & Freckman, 1987)

Crops	% Yield losses	Monetary losses (mil. US $)
Maize	10.2	6,736
Field bean	10.9	1,156
Soybean	10.6	2,687
Cowpea	55.0	—
Lucerne	3.0	48.6
Lespedeza	3.0	0.457
White clover	—	1.00
Grasslands (temperate)	12-28%	—
Grasslands (pairie)	28-59%	—
Hay	5-10	—
Forages (world wide basis)	8.2	—

Erickson, 1972; Hasan, 1990a; Yeates, 1979). These plant parasitic group in grassland include ecto-parasitic migratory, *Longidorus*, *Xiphinema*, *Tylenchorhynchus* and *Heliocotylenchus*; endoparasitic migratory, *Pratylenchus* and *Ditylenchus* spp. and non migratory cyst forming *Heterodera* spp., root knot *Meloidogyne* spp. and seed gall forming *Anguina* spp. (Jones & Jones, 1984; Labruyere, 1979; Spaul & Cleament, 1982; Hasan *et al.*, 1986). Although grassland normally supports and can tolerate large numbers of nematodes, recent evidence suggests that there are some situations where nematode damage is more likely to occur i.e. on light sandy soil especially during the establishing period of young seedlings (Hasan *et al.*, 1986; Spaull & Clement, 1982; Williams, 1984). Similarly, high populations of certain important nematode species of the grassland soil have also been implicated in the resowing problem of old pastures (Labruyere, 1979).

Precise losses in grasslands due to nematodes are poorly documented. However, ample data showing percentage yield reduction has appeared in the literature and though mainly derived from pot experiments, illustrate the detrimental effect on plant growth caused by nematodes (Erickson, 1972; Jain & Hasan, 1986b). In addition herbage losses of 28-59% have been attributed to root feeding nematodes in prairie grassland (Smolik, 1974), as well as 5-10% in hay in USA (Hodges, 1973) and significant depression in production in new Lolium plots in England (Spaull *et al.*, 1983).

2. NON-PESTICIDAL MANAGEMENT (NPM) STRATEGIES

The use of synthetic chemicals to control nematodes in forage crops holds little promise of ever being an economic possibility. Moreover, it has also other problems like toxic residues, development of resistance in pests, killing of beneficial organisms and many environmental and health hazards. Over the year a number of NPM methods have been experimented to suppress nematode population in forage crops with varying degree of success. These method of nematode management although offer far less certainty of killing or stalling the action of nematodes as compared to the nematicides. Nevertheless, NPM approaches utilized for bringing down the nematode populations below damaging threshold level are ecologically safe especially when considered in relation to biodiversity conservation of beneficial micro-organisms and sustainability of agro ecosystem (Bingefors, 1969; Cook & Yeates, 1993; Erickson, 1972; Jain & Hasan, 1995, 1998a). In this article an effort has been made to deal in detail, the principles, problems and progress made in the area of NPM of phytophagus nematode in forage crops with special reference to the work done in Indian and other countries of the world.

(a) Crop rotation and inter-cropping

Too frequent cropping of land with same crop can lead to a serious buildup of nematodes in the soil. Among various cultural practices, crop rotation as nematode management technique particularly in annual crops continues to be one of the important recommendation all over the world. Sequential alternation of crops with varying degree of susceptibility to the nematode pests can produce striking results in the reduction of crop losses especially in tropical developing countries. The basic principle underlying crop rotation for nematode management are, 1-the differential host preference of nematodes or host status of the crops classified as good, poor, non-host (immune) or antagonistic and 2- the rate of population decline in the absence of the host. In addition to these the non host crops (alternative crops) to be grown between the host crops must satisfy certain agronomical and economical requirements of the grower. However, incorporation of resistant cultivars of the susceptible crops normally grown in the cropping system is considered as the first alternative.

A number of reviews dealt with the question whether cropping systems or crop rotations are sufficient as mean to manage nematodes. With very few exceptions the importance of crop rotation in regulating nematode population has been universally admitted (Norton, 1978; Nusbaum & Ferris, 1973; Sethi & Gaur, 1990). In consequence, research efforts in evolving suitable cropping sequences consisting forage as main or alternative crop for nematode population regulation in different parts of the world has been studied (Erickson, 1972; Sethi & Gaur, 1990).

Red clover and lucerne were found to be highly susceptible to *D. dipsaci* and *P. penetrans* in most of the temperate countries and advised to be grown only after (8-10 yrs.) beet which was found to suppress the population of these nematodes (Decker, 1969; Oostenbrink 1961). Similarly, *H. trifolii* infecting red clover may be eliminated kif rotated with corn or soybean (Norton, 1967). In USA lucerne is usually rotated with cotton for two years because of the damage caused by *M. arenaria* (Reynold & O'Banon, 1960). Moreover, this crop was found to be consistently reduced the population of pin nematode that has been increased by pre-sowing with birds foot trefoil (Hawan & Kuzub, 1978). Shagalina (1973) reported that two Mediterranean lucerne (*M. polia*) varieties grown after cotton together with elimination of weeds considerably reduced *M. incognita* and *M. javanica* population. Population of *Tylenchorhynchus* and *Pratylenchus* species were reported to be the lowest following berseem in the rotation (Shafshak *et al.*, 1974). *Kobe lespedeza* has been advised not to be grown in rotation with soybean because of the susceptibility to *H. glycine* (Erickson, 1972). Population of *M. naasi* was found to be thirty times greater after Italian rye grass but reduced after potato and wheat (Gooris & Herde, 1967).

In most of the tropical and sub-tropical countries including India, work on this approach of nematode, control particularly in forage crops are limited. This is perhaps due to 1-the multiple growing season in which 3-4 or even more crops are grown in succession on the same piece of land during a year. In such a case not only the crop replacing the main crop but also the intervening crops have to meet the requirements of a rotation crop, 2-the nematodes particularly the polyphagus and continuously reproducing ones get almonst uninterrupted food supply and thus multiply to high population making selection of suitable non-host crop a difficult task and 3- the poor recognition of the nematode problem. Inspite of the above fact, recently in Nigeria intercropping of maize with leucaena or growing *Stylosanthes gracilis* prior to tomato, reduced root knot nematode (Egunjobi, 1985. However, in India all the three species of stylo i.e. *S. gracilis*, *S. similis* and *S. hamata* grown after banana were heavily infested with *R. similis* (Subramaniyan, 1985). Chawla and Prasad (1973) observed little or no difference in the nematode population following either sorghum wheat cowpea or bajra wheat cowpea cropping sequences. As a consequence, Prasad & Gaur (1974) suggested that often the combined damage caused by polyspecific communities, present is more important than that due to any one species. Moreover, Oostenbrink (1964) observed that the composition of the population is markedly influenced by the last grown crop or crops and thus it is this principle which probably contributes more towards the utility of crop rotation in nematode control. Population of *Tylenchorhynchus* following berseem was observed to be relatively low than after maize, wheat or sugarcane in a cropping sequence (Mukhopadhyay & Prasad, 1968, Similarly, Srivastava and Sethi (1986) reported that stunt and reniform nematodes maintained their population without major fluctuation while *H. zeae* showed host specificity to maize cowpea wheat rotation. Sethi and Gaur (1986) suggested that employing crop rotation especially against the above nematode species with limited or delineated host range prove to be more effective in population reduction. In India some of the major nematode species like *H. avenae*, *H. zeae*, *H. sorghi* are found to be mainly parasitic on graminaceous forages while *H. cajani*, *M. incognita* and *M. javanica* (Table 4) generally prefere leguminous crops (Sethi & Gaur, 1986; Hasan, 1990c). Therefore, it would not be impossible to find a suitable and acceptable alternative crops to replace main host crop in most part of the country for short or even long rotations.

Research efforts during past few years on the role of certain important tropical pasture grasses in effective nematode population suppression when introduced as alternative crop in the crop rotation system or intercropped with susceptible crops has also worked out. Among these grass species or varieties, Pangola grass *Digitaria decumbans* love grass *Eragrostis curvula* cv., Ermalo, Guinea grass *Pannicum maximum* cv. Umtali and Sabi, Rhode grass *Chloris gayana* cv. Katombora and coastal bermuda grass have been successfully utilized in tobacco or cotton rotations for effective control of dominant root knot nematode species in different countries of the world (Ayala, 1967; Burton, 1987; Daulton, 1962; Koen & Grobbelaar, 1968; Shephard, 1978, 1979; Winchester, 1960). Similarly, many legume cover crops or green manure crops which also gives valuable forages for livestock were found to reduce a wide range of nematode species. Of these *Crotalaria spectabilis*, *C. brevidence*, *Mucuna deeringiana*, *Indigofera hirsuta* were reported to be the best alternative crops for efficient control of the *Meloidogyne* spp., *Belonolaimus longicaudatus*, *Pratylenchus brachyurus*, *Criconemoides* and *Trichodorus* spp. (Brodie *et al.* 1970; Ferraz, 1985; Reddy, 1983; Rhoades, 1976; Soffes *et al.*, 1980). These crops can reduce or

Table 4. Host status of forage crops to major nematodes species

Crops	Nematode species*\Host suitability**								
	Mi	Mi	Rr	He	Hs	Ha	Pz	Tv	Hd
Legumes									
Berseem	+	+	0	–	–	–	+	+++	0
Lucerne	+++	+++	0	–	–	–	+++	–	0
Cowpea	+++	+++	+++	+++	–	–	0	0	+
Clusterbean	+++	+++	+	+++	–	–	0	0	0
Mothbean	+++	+++	+++	+++	–	–	0	0	0
Fieldbean	+++	+++	+++	+++	–	–	0	0	0
Stylo	+	+	0	+	–	–	0	0	+
Siratro	–	+	0	–	–	–	0	0	0
Clitoria	+++	+++	+	–	–	–	0	0	0
Grasses									
Sorghum	+	+	+	–	+++	–	+++	+++	+++
Bajra	+	+	+++	–	+	–	+++	+++	+
Maize	+	+	+	–	+	+	+++	+++	+
Oat	+	+	+	–	+	+++	+++	+++	+
Buffel Grass	–	–	0	0	0	–	+	+++	+++
Marvel Grass	–	–	0	0	0	–	+	+++	+++
Setaria Grass	–	–	0	0	0	–	+	+++	+++

*Mi = *M. incognita*; Mj = *M. javanica*, Rr = *R. reniformis*; Hc = *H. cajani*; Hs = *H. sorghii*; Ha = *H. avenae*; Pz = *P. zeae*; Tv = *T. vulgaris*; Hd = *H. dihystera*
** (o) = Not known; (-) = Non host; (+) = Poor host; (++) = Fair; (+++) = Good host.

eliminate the use of chemical nematicides and nitrogen fertilizers in cropping systems (Soffes *et al.*, 1980). In addition certain tropical pasture legumes has also been identified as non hosts or resistant to some important nematode species which may be utilized in crop rotation systems (Hasan, 1990a; Kretschmer *et al.* 1980). Further it needs to be emphasized that incorporation of these legumes and grass species or varieties would be economically attractive in localities where livestock form a part of farming system. Noe (1986) also suggested that the most profitable management alternative for particular plant parasitic infestation would be to use the land as pasture for a number of yerars, thus rotating to a profitable non-host.

Inter-crops are said to be better than mono-cultures because they, 1-protect against pests, 2-provide a more balanced human and animal diet, besides other many more advantages. However, of variety of factors that might be involved in the facilitative production potential the one most cited and perhaps best documented is the reduction in pest attack frequency found in inter crops. There are also evidence that inter crops reduce nematode attack as already been described in the previous paragrah. Of the different mechanisms, the disruptive crop hypothesis, in which a second species disrupt the ability of the pest to attack its proper host efficiently seems to be the most accepted mechanism in case of nematode control The disruptive plant species may serve as repellant or killed the nematode by releasing toxic chemicals in the form of root exudates in the rhizosphere of the same environment (Vandermeer, 1990).

Weeds as foci of infestation within crops and as carry over hosts usually negate the effect of crop rotation for nematode cotnrol (Hasan 1985a, Hooper and Stone, 1981). Moreover, nematode control via weed control presents greater problem in tropical and sub tropical agriculture because of the greater abundance of the weeds together with the wide host range of some important nematode species. Therefore, thorough weed control either with the weedicides or by mechanical means must be essentially combined with crop rotation.

(b) Organic amendments

The agro ecosystem can be modified in many ways to improve the benefits obtained from biological control of nematodes. Natural enemies population can be enhanced by increasing the necessary habitat and/or by altering habitats to be more suitable for the colonization or subsequent population. The addition of organic matter in the soil appears to be the only factor that can rapidly alter the physical, chemical and biological properties of the soil and may be the reason why the additional organic matter succeeds in altering nematode population. Therefore, until recently the principle applied approach to manage nematode by organic soil amendments has been to study the possible mechanisms involved in the resulting control. The two most prevalent hypotheses are, 1-that decomposition products released from amendements into the soil are directly toxic (Alam, 1976; Hasan, 1977; Sitaramaiah, 1990) or 2- that increase of the soil microbial and faunal population by addition of amendments stimulates parasites and predators activity (Goday *et al.*, 1983; Khan, 1990; Van Gundy, 1985). Both are the important and functional hypotheses, largely attributed to the biological control of plant parasitic nematodes.

Unfortunately most organic additives be used in such a large quantities to make a significant change in the ecosystem that are neither practical nor economical unless there is an abundant supply available locally. Nevertheless, Mullar & Gooch (1982) has emphasized the advantages of using organic amendments for nematode control particularly in tropical countries and may not be ignored altogether. Recently in India of the various oil cakes used application of need cake @ 10-15 q/ha once in a year, considerably reduced the nematode population in bajra-berseem and cowpea oat in two successive cropping seasons. Moreover, this treatment also resulted in the increase of the photosynthetic pigments of the crops (Hasan & Jain, 1984b; Jain & Hasan, 1986a). Similarly, in another field trial incorporation of *Parthenium* chopped plants @ 50 q/ha, sawdust 30 q/ha, neem cake 15 q/ha, 7-10 days before sowing not only reduced the nematode population but also increased both forage yield and nutritional quality in two successive cropping seasons (Hasan & Jain 1993). Further in a pot experiment incorporation of *Leucaena* leaves into the root knot infested soil resulted into significant reduction in the nematode incidence of cowpea (Jain & Hasan, 1984). Neem cake @ 500 kg/ha was also found to be effective against *R. reniformis* and *H. dihystera* infecting field bean (Sundaram & Velayutham, 1988).

(c) Other cultural methods

Deep summer ploughing, burning of grasslands after harvest, weed control, crop residue destruction, use of clean seeds, bush fallow, flooding and disinfection of farm equipments before moving it from heavily infested to uninfested fields are some of the cultural practices reported and to reduce the nematode population upto some extent (Thurston, 1992). Of these deep summer ploughing especially in climates where there is no rain fall for 2-3 months during hot and dry season, some types particularly the root-knot and reniform nematodes populations can be reduced. In this case nematode juveniles and eggs die quickly when exposed to sun light and drying. However, these practices are employed only under special conditions and may be more effective when integrated with other more effective measures like crop rotation resistant varieties and chemicals (Alam, 1991).

(d) Resistant varieties

Potentially the most economic and effective method of nematode management in forage crops is to use nematode resistant plants or varieties. Host plant resistance represents the inherent ability of crop plants to restrict, retard or overcome the pest/pathogen infestations and thereby to improve the yield and or quality of the harvestable crop products. To achieve this the primary objective, therefore, is to develop crop varieties resistant to one or more nematode species along with other desirable agronomical characters. Considerable progress in resistant breeding in forage crops has already been made resulting into a number of resistant varieties (Table 5) throughout the world particularly in temperate countries (Bingefors, 1969, 1971; Hasan, 1990a; Jain & Hasan 1995, 1998).

Genetic variability for susceptibility to root knot in cowpea was demonstrated as early as in 1902 by Webber and Orton and genetic resistance could be transferred to a susceptible cultivar. As a result the first resistant to root knot variety was developed in cowpea. Similar genetic variation for nematode susceptibility/resistance of various other forage legumes e.g. alfalfa, white clover, red clover, lespedeza to major nematodes i.e. *Meloidogyne, Heterodera* spp. and *D. dipsaci* were also demonstrated from different parts of the world particularly from America and European countries (Bingefors, 1969, 1971, Cook & Yeates, 1993, Erickson 1972). Genetic variation for susceptiblity to certain nematode species were also reported in some grasses but resistance breeding in these crops is an almost virgin area of research (Bingefors, 1969). Availability of genetic resistance appears to be more common against specialized endo and or semi endo parasites particularly in forage legumes but for many ecto-parasitic nematodes very little is known. During last few years considerable research has been made in understanding the genetic basis of plant nematode interaction (Sidhu and Webster, 1981a) and also current methods of developing nematode resistant cultivars (Bingefors, 1971; Kaplan, 1982; Sidhu & Webster, 1981B). Although some nematode species and other organisms such as insects, fungi bacteria and viruses (Table 6), but further progress on the aspects of multiple pests and disease resistance breeding is still meager.

Traditional approaches to the development of nematode resistant varieties are time consuming and often limited by crossing barriers. Biotechnology or genetic engineering can address these bottlenecks of time and gene source (Keen, 1986). Though limited, but significant achievement have been made in recent years on the possible application of these techniques in breeding for nematode resistance. Regenerating plant from organs, tissues or cells are appear to be an effective way of obtaining somaclonal variants with enhanced disease resistance (Litz, 1986). Cytoplasmic techniques for transferring resistant genes from wild host species to crops, the use of tissue culture and protoplast fusion method and the possible implication of recombinant DNA are some of the effective ways beginning to play a significant role in breeding for nematode resistance (Fessuliotis, 1985, 1987).

Table 5. Examples of nematode resistant sources/varieties in forage legumes.

Crops and nematode species	Resistant sources/varieties	Main references
Alfalfa		
M. incognita	Moapa, Sonora	Bingefors, 1969, 1971
M. javanica	Moapa	
M. hapla	Plants in cv. Vernal and cv. Hilmar	
Ditylenchus dipsaci	Alfa II, Vertus, Nematol (Europe); Frontier, Pergamino, San Martin, Nemasint (S. America); Nemastan, Lahontan, Talent, Washoe (USA) etc.	
Red clover		
Meloidogyne spp.	Plants in different varieties	
D. dipsaci	Merkur, Resistant, Hermes, Hermes II, Brita Disa, Reko (Sweden), Corona, Resident, Otofte, Hera (Denmark); Dorset, Marl, Cotswold, Essex, S123, Montogomery (Great Britain) etc.	
White Clover		
M. incognita	Plants in Swedish and Dutch varieties	
Heterodera trifolii	Plants in cultivated varieties	
D. dipsaci	Wild english and other local varieties, Milka, Milkanova, Lodi, Mira, Trifo, Zero, Mega, Adefa (Denmark) etc.	
Lespedeza		
M. incognita	Rowan	
M. hapla	Lines in L. cuneata	
Soybean		
Meloidogyne spp.	Jackson, Hill, Delmar, Palmetto, Lardeo, Dyer, Forest, JS-2, Selec. DS 74-20-1	Bingefors, 1971; Dasgupta and Ganguly, 1986
H. glycines	Dyer, Pickett, Custer, Nemasherazu etc.	
R. reniformis	Dyer, Pickett, Peking, Centannial, Cluster etc.	
Cowpea		
M. incognita	Iron, Clay, Chinese Red, Groit, New era, Mississipi purple, Mississipisilver, Viktor K 798, Broweye 7, Brabham Monetta, Catjang, Columbia etc.	Dasgupta and Ganguly, 1986; Hartman, 1976; Singh *et al.*, 1974.
R. reniformis	V-16	
H. cajani	Sel A, Sel 1-2, Sel 4-1, 24-8-2, P 33-2-1, P 85-2, etc.	
Cluster bean		
M. incognita	ILO 1539-1, IL 79-1028-1, ILO 77-4-1 (germplasm lines)	Hasan and Jain, 1988
Moth bean		
M. incognita	IL 1134, IL 1202 (germplasm linesw from indigenous collection)	Hasan and Jain, 1984c
Clitoria		
M. incognita	ILO 160-1, ILO 1442 (germplasm lines from indigneous collection)	Hasan and Jain, 1985b
Sesbania		
M. incognita	S. tetraptera, S. macrocarpa 1535	Hasan and Jain, 1985a

Table 6. Examples of multiple pests and disease resistant sources/varieties in forage legumes

Crops and nematode species	Other pest and diseases (insects, fungi, bacteria and viruses)	Resistant source/variety	References
Alfalfa			
D. dipsaci *M. hapla*	*Achyrthosiphon pisum, Therioaphis maculata, Colletoricum falcatum, Glomerella tucumensis, P. megasperma, C. insidiosum*	Arc x Agate x Apalachee W 10	Elgin, 1982
M. hapla	*P megasperma, F. oxysporum* and two insect spp.	Plants in *M. sativa* lines, U.C. Cibola	Lehman *et al.*, 1983
D. dipsaci	Verticillium albo-atrum	Plants in M. sativa lines, Vernema	Peaden, *et al.*, 1983
D. dipsaci	*C. insidiosum*	Plants in cv. Beaver, Trek	Hanna & Hawn, 1975
Red Clover			
D. dipsaci	*Sclerotinia trifoliorum*	Plants in M. sativa lines, Britta	Johnson, 1976
Cowpea			
M. incognita	*Fusarium sp.*	Iron	Orton & Webber, 1902
Lespedeza			
M. incognita acrita	*Rhizoctonia spp.*	Belsville 23-864 x AL 2193, AL-Lotan	Donnelly, 1981
Soybean			
M. incognita Race 3, H. glycine	*P. sajae*	Plants in M. sativa lines, Mack, Centennial	Caveness, *et al.*, 1977
M. incognita Race, 1,3, R. reniformis, H. glycines Race 1,2,3	*P. megasperma, X. phaseoli, Corynespora cassicola*	D 64-4636 (Hill 49-2491 (4) x Jackson) x Pickett 71. Centennial	Hartwig & Epps. 1977
Snap bean			
M. ingonita	*Uromuces phaseoli*, NY 15 strain of mosaic	PI 165426, Nema snap	Wyatt *et al.*, 1983

A variety of genetic engineering strategies are being developed recently that will result in a significant extension of the range of crops genetically resistant to nematodes. Several genes with regulatory functions and other genes sustaining the biochemical makeup especially in root knot and cyst nematodes have been characterized recently. These genes expressed mainly in the form of -proteinase inhibitor e.g. CpTI (trypsin inhibitor) from cowpea, OCI (oryzacystatin) and OCII (cystatin) from rice have been reported to interfere with nematode feeding, digestion and feeding cell formation and conferring resistance against nematodes (Vrain, 1998). The expectation is that these genes will transfer nematode resistance when expressed in transgenic plants. As early as in 1975, Dasgupta and Ganguly showed a trypsin like activity in a root knot nematode *M. incognita* and indeed, when the CpTI gene is expressed in transgenic potato, it impedes reproduction of this nematode, and it also interfere with the development of a potato cyst nematode, *G. pallida* (Hepher & Atkinson, 1992). Serine or cystatin inhibitors from soybean and rice disrupt digestive processes, and prevent or delay nematode development and reproduction. Many proteinase inhibitors and lactins are available in this strategy can be tailored to control speciefic nematodes. The physiological effects of certain lectins on nematodes are effective but not as well understood as the effects of proteinase inhibitors. Research is necessary to express proteinase inhibitors and lectins together in transgenic plants. The expression of two or more genes for resistance will prevent nematodes pests from adapting and prolong the durability of genetic resistance (Vrain, 1998). Further genetic engineering in conjunction with cell biology allows the introduction of genes from any source into host plants. This means that genes become available from bacteria, which produce nematicidal/insecticidal proteins from bacteria or fungi, which produce antibiotic or other biologically active chemicals against plant pathogens. The promise of these technology is in its ability to teach plants to produce their own nematicides/insectisides or fungicides. Even more sophisticated improvement is possible. For examples, if a microbe produce a molecule which is nematicidal and the plant has the biosynthetic machinery to make a key intermediate of this molecule, then perhaps the gene coding for the enzyme necessary to complete the biosynthetic pathway, could be introduced into the plants through available vector/techniques, causing the plants to produce its own nematicide. Development of such "super varieties" through these novel approaches is a dream yet to be fulfilled.

Foods derived from engineered crops must be safe to eat, and for that reason many genes for resistance found in certain plants, even in commercial crops, can not be expressed in other crops. For example while several lactins from beans are very effective against many pests, these lactins are quite toxic and they should not be expressed in other crops, which are consumed either by human beings or animals. Trypsin proteinase inhibitors are effective against many insect and nematode pests but some of them also affect mammalian digestion, making it doubtful that health regulatory bodies will allow their expression in transgenic crops. However, these toxic proteins will undoubtedly be used to protect nonfood crops, such as in floriculture or in forestry (Vrain, 1998).

The new biotechnology is advancing rapidly. While many of the procedures may not be applicable to nematology today, nematologists and plant breeders may one day be able to utilise this technology to develop nematode resistant varieties that have been impossible for scientists to produce with conventional breeding procedures. However, there is still much work to be done in plant breeding by conventional methods with the germplasm in hand. According to Vrain 1998 genetic engineering like most emerging technologies has polarized opinion, and its potential for nematode management probably lies somewhere midway between extremes that have been proclaimed. There is no doubt that engineering resistance against nematodes will improve agricultural crops. However, it is too early to predict with strategy will actually develop into successfull cultivators with durable resistance.

(e) Biological control

Since long scientists have adopted various definitions for biological control. However, the most preferred (49%) Baker & Cook's (1974) definition of which a shortened version is found in their recently published book (Cook & Baker, 1983) : "Biological control is the reduction of the amount of inoculum or disease producing activity of a pathogen accomplished by or through one or more organisms other than man." Biological nematode control in relation to crop production system is a subject of considerable current interest because of a perceived urgency to develop and adopt safe, economic and efficient method for managing nematode pests. This can be achieved through a number of pathogens, parasites and predators like viruses, fungi, bacteria, turbellarians, enchytreids, collembola and mites etc. and some of them have been extensively studied and reviewed during recent past (Hasan & Jain, 1992; Jatala, 1986; Kerry, 1984; Khan, 1990; Swarup & Gokte, 1986; Van Gundy, 1985).

(i) Soil antagonists : Of a large list of over 200 soil antagonists identified as so called biocontrol agents at least 75% are soil fungi that parasitize or prey on nematode juveniles or adults and eggs or cysts and on which main emphasis appears to be concentrated (Hasan & Jain, 1992). Reports on the use and efficiency of some potential biocontrol agents of nematodes infecting forage crops either in the green house or field experiments are few as compared to other economically important crops. However, recently in India and elsewhere *Paecilomyces lilacinus* has been successfully cultured at mass scale on maize, bajra, sorghum and water hyacinth leaves (Devide, 1985; Sharma & Trivedi, 1986). Cowpea seed treatment with *P lilacinus* spores were reported to be more effective in reducing the root knot infestation level when egg masses or the eggs were used as source of inoculum (Midha, 1985). In a field experiment application of *P. lilacinus* cultures grown on sorghum seeds @ 20 kg/ha considerably reduced the root knot nematode intensity in cowpea. However, combined application of this fungus with certain organic matters was found to be more effective than when used alone. Moreover, their residual effect in reducing the nematode incidence persisted even in the subsequent lucerne crop. Further, carbofuran as seed treatment in combination with the fungus was also effective in nematode management showing their compatibility (Hasan & Jain, 1993). Few products were also marketed commercially, like Royal 300 and 350 which contains nematode trapping fungi of the genus Arthrobotrys in France (Mankau, 1981) and Biocon of *Paecilomyces* in Philippines (Jatala, 1986). Of the several antagonists of nematode pests, Khan (1990) attributed most of the success to *Arthrobotrys, Nematophthora gynophila, Dactylella oviparasitica, P. lilacinus* and a gram positive bacteria *Bacillus penetrans*, since these organisms have not only shown promising results in the initial studies, but fortunately also parasitises the most destructive group of nematodes i.e. *Meloidogyne* and *Heterodera* spp. and therefore, deserve to be called as promising biocontrol agents.

There has been very little effort and incentive to research for practical biocontrol alternatives in nematology. The recent information of CSRS, science and education, USDA, on biological control research suggest that approximately only 16% of the effort in 1982 was in plant pathology and nematology of which less than 2% is in nematology (Van Gundy, 1985). Another deterrent to biological control of nematodes has its foundation in nematode control philosophy based on the use of synthetic chemicals designed to reduced nematode population by more than 90%. Also biological control of nematodes was considered only a curiosity instead of a serious practical approach to nematode control. Fortunately, the expansion of IPM strategy in the 1970s has encouraged management approach with integrated multiple control practices, one of which may be the biological control. In addition there are a number of scientific barriers to the commercial development of biocontrol agents of plant parasitic nematodes. These include, 1-stability of the natural biological balance of the ecosystem, 2-antagonism of soil organism to new introductions through the soil profile,

3-establishment of the introductions in agro-ecosystems, 4-the zone of activity of the biological control agents and 5- the assessment of biological activity and synchronization with nematode biology.

(ii) Vesicular-Arbuscular Mycorrhizal (VAM) Fungi : Plant parasitic nematodes and VAM fungi commonly occur together in the roots or rhizosphere of the same plant, each having a characteristic but opposite effect on plant vigor. VAM increases the supply of the nutrients particularly phosphorus to crops by better exploration of the soil beyond the depletion zone and receiving the fixed carbon from the plants, thus acts as symbionts. Whereas the nematodes as pathogen or parasites take out vital nutrients from the roots making reduced plant growth. The single most common effect of VAM fungi on nematode susceptible plants is promoting tolerance to nematodes. Thus plants heavily colonized by mycorrhizal fungi are able to grow well inspite of the presence of damaging levels of plant parasitic nematodes. The interaction between VAM fungi and soil borne pathogens particularly plant parasitic nematodes has been extensively studied and reviewed by several workers (Hasan & Jain, 1992; Husey & Roncardori, 1982;p Jain & Hasan, 1998b; Schenck *et al.*, 1975; Shonbeck 1979; Smith, 1987), but their effect on nematode incidence varied. Many workers have demonstrated a reduction in neamtode population densities (Cooper & Gradison, 1986; Grandison & Cooper, 1986; Jain & Hasan, 1988; Sikora & Schonbeck, 1975), but cases where nematode population remain unaffected (Hasan & Jain, 1987b; Jain & Hasan, 1988) or even increase (Atilano *et al.*, 1981) suggested that these differences in the interaction studies may therefore, be the result of variable root colonization by VAM fungi and not necessarily due to differences in nematode - VAM host combination. He has further emphasized that each nematode - VAM host combination is unique and therefore, not subject to generalization and thus require study of each combination.

On the basis of the findings of a number of workers it has been observed that regardless of the species, less nematodes were found in most of the mycorrhizal plants (Table 7). However, the mechanism involved in the reduction in nematode numbers in the mycorrhizal plant may vary in the following way (Shonbeck, 1979), 1-penetration of the larvae in the root was not inhibited but their development within the mycorrhizal root was restricted 2-number of female s within the egg sacs of mycorrhizal plants was half than non mycorrhizal plants, 3-condition for nematode development in the roots appear to be detrimental due to mycorrhiza. Later, Smith (1987) while supporting the above hypotheses and considering them as base, also discussed regarding the mechanism of mycorrhizal induced host resistance or tolerance to nematodes which may be due to, 1-increased root growth and function, 2-nutritional effect other than P, 3-alteration in root exudation, 4-competition for host photosynthates, 5-production of nematistic compounds, 6-competition for space and 7-parasitism of eggs. He has also suggested that probably VAM fungi do not directly interact with plant parasitic nematodes inspite of their proximity in root tissues but more likely, they alter the host either physically or physiologically and thus indirectly affect the host nematode relationship.

Though, largely based on green house tests considerable success have been achieved towards the possibility of utilizing VAM fungi to mitigate or reduce the adverse effect of nematodes through out the world. However, the future use of mycorrhizal inoculation as biological means of plant protection at field levels lies mainly, 1-in the mass production of fungal inocula

Table 7. Effect of VAM fungi on nematodes infecting forage crops

Host and nematode species	Effect in mycorrhizal plants	References
Alfalfa		
M. hapla	Improved plant growth, reduced nematode numbers and adult development	Grandison & Cooper, 1986
White Clover		
M. hapla	Improved plant growth, fewer nematode numbers	Cooper & Grandison, 1986
Berseem		
T. vulgaris	Improved plant growth, nematode population unaffected	Hasan & Jain., 1987b
Soybean		
M. incognita	Increased yield, fewer galls	Schenck et al., 1975
Cowpea		
M. incognita	Increased plant growth, fewer nematodes and galls	Jain & Hasan, 1988; Jain & Sethi, 1988a
H. cajani	Improved plant growth, reduced larval penetration and cyst production	Jain & Sethi, 1987; 1988b
Oat		
M. incognita, M. hapla	Increased plant growth, fewer nematodes	Sikora & Schonbeck, 1975
Sorghum		
P. zeae, T. vulgaris, H. dihystera	Increased plant growth, nematode population unaffected	Jain & Hasan, 1987; 1988

whose properties are known and predictable, 2-they must be genetically stable and competitive as symbionts so that they persist in the crop. At the moment the greatest draw back is perhaps its inability to culture on synthetic media. Also we know too little of the genetic variation and variability of mycorrhizal fungi and almost nothing of their breeding system and genetics. Nor do we know clearly what properties are required for efficient symbiotic relationship between any host and its fungus (Shonbeck, 1984).

Several methods for commercial production of mycorrhizal inoculum has been attempted/described in different parts of the world. However, their use are currently restricted to situations where the natural population of VAM fungi have been destroyed, as in fumigated, chemically treated areas, green house or disturbed areas (Reeves, *et al.*, 1979). Daniels and Menge (1981) presented results suggesting that VAM fungus (*Glomus epigaeus* shows great potential for commerical application. It is capable of producing extremely large numbers of spores in pot cultures. The spores can be stored for atleast three months with little reduction in germinating capacity. Recently in India considerable efforts has been made to identify the efficient native strains of VAM fungi in relation to both host and nematode species, alongwith the producing inoculum. From the foregoing it may be concluded that such interactions may some day lead to the manipulation of fungi forming VA mycorrizae by altering specific crop management system for the purpose of minimizing crop losses. Learning to deal with VAM might permit some form of nematode control requiring less reliance on pesticides and might even allow reduction in the high rate of phosphorus fertilizer currently used for certain crops (Hasan & Jain, 1998; Hussey & Roncardori, 1982).

(f) Natural plant products or Bio-pesticides

In view of various problems explicit to chemical control of plant parasitic nematodes, interest on the possibility of evolving nematicides of plant origin has been revived during recent past. As a consequence nematotoxic potential of different kinds of natural plant products especially in the form of crude extracts and certain chemical constituents of many plant species have been extensively investigated (Gommers, 1981; Suryanarayan, 1990). Besides a number of bioassay methods were developed for these bio pesticides, though, largely based on laboratory or green house tests. However, application in the main field at larger scale is still seems to be the biggest limiting factor for their success in nematode control. In the following paragraph an attempt has been made to mention briefly some of the representative references related to this aspect of nematode control.

The water extract of neem (*Azadirachta indica*), leucaena (*Leucaena leucocephala*), parthenium (*Parthenium hysterophorus*). alongwith several other plants were found to be toxic to some important nematode species especially *Meloidgyne* spp. in vitro, tests. Moreover, incorporation of either the extracts or leaves of these plants into the nematode infested soil has resulted into reduced nematode incidence (Hasan & Jain, 1984c; Jain & Hasan, 1984b; Kaliram & Gupta, 1980). The water soluble fractions of the bitter principles i.e. nimbidin, thionimone, azadiractin isolated from neem and parthinin from parthenium were reported be effective both,as inhibitor of egg hatching and larval mortality of nematodes (Hasan & Jain 1984; Khan *et al.*, 1974; Warthen, 1989).

Certain plants are known to have allelochemicals such as terthienyls, acetylenes, terpenoid aldehydes, phenoxypropionic acid derivatives etc. which have also been reported to exhibit high nematicidal activity (Hayashi *et al.*, 1983; Kimura *et al.*, 1981; Morki *et al.*, 1982; Veech, 1979). Further, terpenoids in general and terpenoid lactones especially their synthetic derivatives and certain sesquiterpenoid phytoalexins e.g. rishitin, coumestrol, glyceollin, cucurbitacin in particular are now emerging as a new group of nematotoxic compounds. These compounds have been shown to either repell or kill the nematodes at a very low concentration. Moreover, their presence in the plants have been attributed to both pre-infectional and post infectional resistance to nematodes (Alphey & Robertson, 1986; Mahajan *et al.*, 1986; Malik *et al.*, 1987, Veech, 1981). Unfortunately, the biochemical mode of action of most of the compounds are not yet known. Alphey & Robertson (1986) has emphasized that the potential nematotoxic properties shown by them may be sufficient to protect a crop from nematode attack.

Sangwan *et al.*, (1985) and Hasan (1992) have reported nematicidal properties of essential oils isolated from certain plant species. Of these *Trachyspermum ammi* and *Coleus aromaticus* oil exhibited maximium toxicity to root knot nematode in a very low concentration. In addition, the infectivity of the egg masses was found to be greatly reduced when exposed in the above oils resulting into a very low gall intensity in cowpea. Since Sesquiterpenes are the major chemical component of all essential oil, therefore, the potential nematotoxic properties shown has been attributed to the presence of some specific biologically active sesquiterpenes in it (Hasan, 1992).

Some of the major problems in the successful use of natural plant products are 1 -their large scale extraction, synthesis and availability 2-their storage and field stability because of rapid degradation and destruction of the extract or chemical constituents (Larson, 1987). Perhaps due to these reasons none of the product have yet attained commercial status except neem. Nonetheless, such biodegradable and environmentally acceptable natural products may be of potential value in nematode control in several ways, 1-they may be a source of chemical structures for screening, 2 -they may possess activity that is applicable to nematode control directly or after structural modification of the original structure and finally, recognition of their function in nature may suggest new approaches to nematode control.

(g) Integrated Nematode Management (INM)

As obvious from the preceding discussions, in many, if not most cases, various control measures by themselves are not free from one or the other limitations to provide safe, economically acceptable and efficient nematode population suppression in agricultural cropping systems. Further it needs to be emphasized that nematode diseases rarely reach the epidemic level. However, severe nematode infestations of the crops are likely to result if, 1-susceptible host is repeatedly sown 2-a high population of polyspecific/particular species or a race of nematode occur. 3-environment is favorable for nematode multiplication, development, attack and spread. If any of these three factors is limiting in any way, crop damage is not likely to be severe. Therefore, there is a great scope of developing and implementing integrated pest management (IPM), A subject of considerable current interest.

The definition of IPM that has been most used was coined by Smith & Reynold (1966) in an FAO symposium on integrated pest management as "Integrated pest control is a pest population management system that utilizes all suitable techniques in a compatible manner to reduce pest population and maintained them at levels below those causing economic injury". Integrated control achieves this idea by harmonizing techniques in an organized way, by making control practices compatible, and by blending them into a multifaceted flexible evolving system. Subsequently, Bird (1981) emphasized that integrated nematode management (INM) must also be implemented in conjunction with the broader objective of IPM. Later sethi & Gaur (1986) suggested that the aims of INM approach be, 1-to utilize several compatible control techniques in combination, 2 -to maximize natural environmental resistance to plant parasitic nematodes, 3-to apply specific and drastic control measures on as and when necessary and, 4-t0 maximize profit to the particular grower with location and resource specific recommendation. Of these economic considerations, in INM system has gained more importance in recent years perhaps because of the involvement of farmers interst. For example, Raeis *et al.*, 1983 considered 24 different methods of controlling soybean cyst nematode with combination of varieties, fertilizers, nematicides and crop rotation to select the control method that maximize the income over variable costs from the farm at Illinois, USA, A sample of the solution indicated that part of the farm would be in continuous soybean with a resistant cultivar and the balance was to be in corn-soybean rotation using a susceptible cultivar and nematicide, thus maxmizing the net return. Bird (1981) has described in more detail the component of INM for plant protection as, 1-biological and environmental monitoring, 2- agricultural production system (ASP), 3-pest crop eco-system model and system design and implementation. Further, he has also emphasized that the INM system consist of the principles of exclusion, population reduction and tolerance in relation to threshold levels and not eradication of nematode which does not appear to be a practical proposition.

In India substantial efforts has been made to combine several methods to manage nematode in forage crops, though on limited experimental stage. For e.g. certain intensive rotations combined with fertility management (sorghum/bajra-wheat cowpea x 50 kg N (urea) was reported be mostly associated with lower nematode population (Chawla & Prasad, 1973). Similarly, Hasan & Jain 1986,1987) also observed considerable population reduction following a short term rotation (bajra+ Cowpea/sorghum + cowpea-berseem+ oat/lucerne + oat) combined with certain organic matters applied only once before the first cropping season. In addition, Lal *et al.* (1983) has observed that ploughing + levelling followed by summer fallowing and green manuring with *Crotalaria juncea* resulted into complete elimination of the nematode. Combination of deep summer ploughing either with organic amendments or nematicides in lower doses sufficiently reduced the nematode population besides an increase in forage yield of cowpea and field bean (Sundaram & Velaythum, 1988). Although, the above examples are not based on ideal cost benefit analysis, but rather representation of the kind of study which project the possibilities of integration of several methods.

3. CONCLUSION AND FUTURISTIC APPROACHES

Nematode diseases of forage crops have already been recognized as major agricultural problem especially in temperate countries of the world. However, in India research on the importance of nematode pests in forage crops has lagged behind other crops with higher cash return. Presently changes are fast occurring in Indian agriculture e.g. the use of intensive fodder cropping system per unit area per unit time because of the limited land availability for forage crops. Therefore, same piece of land is intensively cultivated to maintain continuous supply of green fodder round the year especially in irrigated and milkshed areas. This provides more favorable conditions for nematode to multiply. Moreover, climatic conditions in India are ideal for nematode reproduction and development. As discussed above it may be concluded that although significant progress has been made on various aspects of non pesticidal nematode management of forage crops. However, much remains to be done before existing and new integrated nematode management options can be used in an optimal manner. To achieve this the following major areas that deserve increased attention are :

1. To exploit the use of known antagonistic forage legumes and grasses in crop rotation or in inter-cropping with other susceptible cash crops especially in those areas where livestock are part of the farming system.
2. To assess and demonstrated the full benefits of certain promising biocontrol agents at larger level under field conditions even as a short term practice.
3. To conduct more research on identification of new indigenous biopesticides and better utilization of organic matters for nematode management, conservation of beneficial soil microorganisms and soil fertility.
4. To intensity breeding and screening program for crop resistant against key nematodes and also to encourage the greater use of biotechnology especially genetic engineering in developing transgenic plants by recombinant DNA technology to develop multiple pest and disease resistant varieties.
5. To identify nematode tolerant culitvars which have one advantage over resistant cultivars in that they reduce

yield losses from nematodes without providing selection pressure on the nematode for development of more aggressive races.

6. To intensify efforts into developing an economically viable, practical and easily adoptable integrated management of pest complex rather than nematode alone by using multidimensional management approaches and also to make better use of existing knowledge on pest and disease control conducted individually.

7. To create awareness among farmers of nematode problems through various technology transfer approaches.

REFERENCES

Ahmed, S.T., Faruqui, S.A. & Hasan, N. 1988. Disease and pest management in forages. *Indian farming*, 37 : 26-27.

Alam, M.M., Saxena, S.K. & Khan, A.M. 1979. Suitability of crops to the reniform, stunt and root knot nematodes. In : *Recent researches in plant sciences* (Ed. S.S. Bir) Kalyani Publisher, New Delhi, pp. 630-6.

Alphey, T.J.W. & Robertson, W.M. 1988. Assessement of nematicidal potential of natural plant products (Abstract) *Nematologica* 34 : 253.

Atilano, R.A., Minge, J.A. & Van Gundy, S.D. 1981. Interaction between *Meloidogyne arenaria* and *Glomus fasciculatum* in grape. *Journal of Nematology* 13 : 52- 57.

Ayala, A., Roman, J. & Gonzalex, J. 1967. Pangola grass as a rotation crop for pine apple nematode control. *Journal of Agriculture, University of Puerto Rico*, 51 : 94-96.

Baker, K.F. & Cooke, R.J. 1974. Biological control of plant pathogens, W.H. Freeman, San Fransico, USA.

Bhaskar, R.B., Pandey, K.C. & Hasan, N. 1999. Diseases and pest management in forages. *Proc. Nat. Semin. On Str. For Max. of Forage Prod. By 2000 AD* (Eds. Mukherjee et al.) Dir. of Res. BCKV, Kalyani.

Bingefors, S. 1969. The use of nematode resistant varieties of forage legumes and grasses. *Herbage Abstract*, 39 : 107- 111.

Bingefors, S. 1971. Mutation breeding for disease resistance. In : *Proceedings of the International Atomic Energy Authority*, Vienna pp. 209-235.

Bird, G.W. & Johnson, A.W. 1987. Coastal Bermuda grass rotation for control of root knot nematodes. *Journal of Nematology*, 19 : 138-140.

Caviness, C.S., Walter, H.J., Riggs, R.D. & Beatty, K. 1977. Soybean varietal performance. *Arkansas Farm Research* 26 : 2.

Chwala, A.L. and Prasad, S.K. 1973. Multiple cropping and nematodes, I : Effect of fertility management and intensive rotations on nematode population and crop yield. *Indian J. of nematology*, 3 : 34-39.

Cook, R.J., and Baker, K.F. 1983. The nature and practice of biological control of plant pathogens, *American Phytopathological Society*, Saint paul, Minnesota, pp. 539.

Cooper, K.M. and Grandison. G.S. 1986. Interaction of vesicular arbuscular mycorrhizal fungi and root knot nematode on cultivars of tomato and white clover susceptible to *Meloidogyne hapla. Annals of applied Biology* 108 : 355-365.

Curry, J.P. 1987. The invertebrate fauna of grassland and its influence on productivity. I : composition of the fauna. *Grass and forage Science*, 42 : 103-120.

Daniels, B.A. and Menge, J.A. 1981. Evaluation of the commercial potential of the vesicular arbuscular mycorrhizal fungus *Glomus epigaeus*, *New Phytologists*, 87 : 345-354.

Dasgupta, D.R. and Ganguly, A.K. 1986. Plant resistance to nematodes. In : *Plant Parasitic Nematodes of India, Problems and Progress* (Eds. Swarup, G. and Dasgupta, D.R.). Indian Agricultural Research Institute, New Delhi, India, pp. 458-475.

Dasgupta, D.R. & Ganguly, A.K. 1975. Isolation, Purification and characterization of a trypsin-like protease from the root knot nematode, *Meloidogyne incognita*, *Nematologica*, 21 : 370- 384.

Daulton, R.A.C. 1962. Survey of land use and plant parasitic nematode control practices in the southern flue cured tobacco areas of Rhodesia. *Rhodesian Journal of Agricultural Research*, 59: 216-217.

Davide, R.G. 1985. International *Meloidogyne* project in region VI. In : An Advance Treatise on *Meloidogyne* Vol. I, biology and Control (Ed. Sasser, J.N. and Carter, C.C.). North Carolina State University, USA, PP. 369-372.

Decker, H. 1969. *Phytonematology*. Veb Deutscher Landwirtschafts verlag, Berlin.

Donnelly, E.D. 1981. Registration of All Lotan Sericea lespedeza (Rg. No. 11). *Crop Science*, 21 : 474.

Dunbier, M.V., Pamer, T.P., Ellis, T.J. and Burnett, P.A. 1979. The effect of stem nematode infestation on productivity and persistence of lucerne cultivars. In : *Proceeding 32nd New Zealand Weed and Pest Control Conference*. Palmerston North, New Zealand, pp. 195-198.

Edmunds, J.E. and Mai, W.F. 1967. Population increase of *Pratylenchus penetrans* in alfalfa and celery roots infected with *Trichoderma viride Phytopathology*, 56 : 1320-1321.

Egunjobi, O.A. 1985. The international Meloidogyne project in region IV. In : *An Advance Treatise on Meloidogyne, Vol. I, Biology and Controls (Eds. Sasser, J.N. and Carter, C.C.). North Carolina State University, USA, pp. 353-360.*

Elgin, J.H. Jr. 1982. Registration of W10 multiple pest resistant alfalfa germplasm (Reg. No. GP 116). Crop Science, 22 : 163.

Ericksson, K.B. 1972. Nematode diseases of pasture legumes and turf grasses. In : *Economic Nematology* (Ed. Webster, J.M.) Academic Press, London, New York, pp. 66-69.

Fassuliotis, G. 1985. The role of nematologists in the development of resistant cultivars. In : Advance Treatise on *Meloidogyne* : Biology and Control Vol. I (Sasser, J.N. & Carte, C.C.). North Carolina State University, Releigh, Pp. 233-240.

Fassuliotis, G. 1987. Genetic basis of resistance to plant nematodes. In : *Vistas in nematology* (Eds. Veech. J.A. & Dickson, D.W.) Society of Nematologists, Inc., Hyattsville, Maryland, USA, pp. 364-371.

Ferraz, S. 1985. International Meloidgyne Project in Brazil (Region III). In : *An Advance Treatise on Meloidogyne Vol. I, Biology and Control* (Ed. Sasser, J.N. and Carter, C.C.). North Carolina State University, USA, pp. 351-352.

Gaur, H.S., and Prasad, S.K. 1986. In : Proceedings of National Conference on short term and long term hazards of pesticides and strategies for their safe use. National Academy of Science, India, pp. 66-77.

Goday, G., Rodriguez, R. Kabana, Shelby, R.A. and Jones, G.M. 1983. Chitin amendments for control of *Meloidogyne arenaria* in infested soil, II. Effect on microbial population, *Nematropica* 13 : 63-74.

Gommers, F.J. 1981. Biochemical interactions between nematodes and plants and their relevance to control. *Helminthological Abstract, Series B, Plant Nematology*, 50 : 9-24.

Gooris, J. and Herde, J.D. 1967. Distribution, Host plants and Biology of the grass root knot nematode, *Meloidogyne naasi, Revue Agriculture Brux.* 20 : 1055-1065.

Grandison, G.S. and Cooper, K.M. 1986. Interaction of vesicular arbuscular mycorrhizal and cultivars of alfalfa susceptible and resistant to *Meloidogyne hapla. Journal of Nematology* 18 : 141-149.

Hanna, M.R. & Hawn, E.J. 1975. Treck, alfalfa, *Canadian Journal of Plant Science*, 55 : 847-849.

Hartman, R.W. 1976. Breeding for nematode resistance in vegetables. *Sabrao Journal*, 8 : 1-10.

Hatrwig, E.E. & Epps. J.M. 1977. Registration of centennial soybeans (Reg. No. 114). *Crop Science* 17 : 979.

Hasan, N. 1985a. New host records of the root knot nematode (*Meloidogyne incognita*). *Indian Journal of plant Science*, 3 : 80-82.

Hasan, N. 1985b. Host status of some important cultivated legumes in relation to root knot nematode *Meloidogyne incognita. Proceeding 5th Nematology Symposium*, Udaipur, India, pp. 79.

Hasan, N. 1990.a. Nematode problems of forage grasses and legumes. In : *Progress in Plant Nematology* (Eds. Saxena S.K., Khan, M.W. and Khan, R.M.). CBS Publisher and Distributor Privated Limited, Delhi, India, pp. 433-459.

Hasan, N. 1990b. Major nematode parasites of cultivated forage crops and their management. *Proc. Summ. Inst. on Forage Production Technology*, IGFRI, Jhansi, pp. 141-146.

Hasan, N. 1992. Nematicidal action of essential oils isolated from selected indigenous plants against root knot nematode *Meloidogyne incognita. Acta Botanica Indica*, 20 : 89-92.

Hasan, N. Ahmed, S.T. & Pandey, K.C. 1986. Pest and diseases of rangelands and their importance in forage production. In : Proc. 3rd. Int. Rangl. Cong., Delhi, pp. 393.

Hasan, N. & Jain, R.K. 1984a. Pathogenicity of root knot nematode, *Meloidogyne incognita* to *Clitoria ternatea* L. Agricultural Science Digest, 4 : 189-190.

Hasan, N. & Jain, R.K. 1984b. Effect of soil amendments on fodder production, photosynthetic pigments and nematode associated with berseem followed by bajra. *Agricultural science digest.* 4 : 12- 14.

Hasan, N. & Jain, R.K. 1984c. Biotoxicity of *Parthenium hysterophorus* L. extracts against *Meloidogyne incognita* and *Helicotylenchus dihystera. Nematologia Mediterranea*, 12 : 239-242.

Hasan, N. & Jain, R.K. 1984d. Screening of moth bean germplasm for resistance to root knot nematode, *Meloidogyne incognita. Agricultural Science Digest.* 4 : 198-199.

Hasan, N. & Jain, R.K. 1985a. Preliminary assessment of the response of *Clitoria ternatea* lines to the root knot nematode, *Meloidogyne incognita, Nematologica*, 31 : 236- 238.

Hasan, N. & Jain, R.K. 1985b. Response of some selected *Sesbania* species to root knot nematode, *Meloidogyne incognita, Nematologia Mediterranea*, 13 : 15-19.

Hasan, N. & Jain, R.K. 1987. Parasitic nematodes and vesicular arbuscular mycorrhizal (VAM) fungi associated with berseem in Bundelkhand region. *Indian Journal of Nematology*, 17 : 184-188.

Hasan, N. & Jain, R.K. 1988. Screening cluster bean germplasm for resistance to root knot nematode, *Meloidogyne incognita* and studies on host nematode relationship. *Indian Journal of Botany*, 11 : 33-39.

Hasan, N., Pandey, K.C., Bhaskar, R.B. & Hazra, C.R. 1998. Plant parastic nematodes associated with major forage crops in India. *Indian Journal of Nematology*, 28: 125-131.

Hepher, A. & Atkinson, H.A. 1992. Nematode control with proteinase inhibitors. *European patent application*, number 92301890. 7, Publication number, 0 502 730 Al.

Hayashi, M. Wdia, K. & Munakata, K. 1983. Synthesis and nematicidal activity of phenoxypropionic acid derivatives. *Agricultural Biological Chemistry*, 47 : 2653-2655.

Hawn, E.J. & Kuzubi, G.C. 1978. Influence of *Pratylenchus projectus* on alfalfa sickness in Alberta. *Canadian Plant Disease Survey*, 58 : 1-4.

Hodges, L.R. 1973. *Nematodes and their control.* Union Carbide Corporation, Salinas, California.

Hooper, D.J. & Stone, A.R. 1981. Role of wild plants and weeds in the ecology of plant parasitic nematodes. In : *Pest Pathogen and Vegetation* (Ed. Thresh, M.) Pitman Publisher, London, pp. 199-215.

Hoveland, C.S., Haaland, R.L. & Rodriguez, R. Kabana, 1977. Soil nematode and forage production of some cool season grasses. Proc. 13th. Int. Grassland Cong., Leipzig, GDR, PP. 1257-1259.

Hussey, R.S. & Roncardori, R.W. 1982. Vesicular arbuscular mycorrhizae may limit nematode activity and improve plant growth. *Plant Disease*, 66 : 9-14.

Irwin, J.A.G. & Jones, R.M. 1977. The role of nematodes and fungi as factors associated with death if white clover stolon over summer in Souther Eastern Australian. *Journal of Experimental Agriculture and Animal Husbandry*, 17 : 789-794.

Jain, R.K. & Hasan, N. 1984a. Toxicity of Soo-babool extracts to *Meloidogyne incognita* and *Helicotylenchus dihystera. Indian Journal of Nematology*. 16 : 98-100.

Jain, R.K. & Hasan, N. 1986a. Efficacy of neem cake on fodder production, photosynthetic pigments and nematodes associated with oats and it residual effect on cowpea. *Indian Journal of Nematology*, 16 : 98-100.

Jain,R.K. & Hasan, N. 1986b. Indian Grassland and Fodder Research Institute, Jhansi. In : (Nematology In India) (Institute and Universities), IARI, New Delhi, pp. 28-32.

Jain, R.K. & Hasan, N. 1988. Role of vesicular arbuscular mycorrhizal (VAM) fungi and Nematode activities in forage production. *Acta Botanica Indica*. 16 : 84-88.

Jain, R.K. & Hasan, N. 1995. Nematode Problems in Forage Crops. In : *Nematode Pest Management - An Eco friendly approach.* (Eds. Gopal Swaroop *et al.*). Nematological Society of India. IARI, New Delhi, pp. 217-227.

Jain, R.K. and N. Hasan, 1998a. Nematode pests of Forage crops and their management Current status and future prospects. In : Nematode diseases in Plants (Ed. P.C. Trivedi), CBS Publishers and Distributor Private Limited, New Delhi, India pp. 201-215.

Jain, R.K. & C.L. Sethi, 1987. Pathogenicity of *Heterodera cajani* on cowpea as influenced by the presence of VAM fungi, *Glomus fasciculatum* or *G. epigaeus. Indian Journal of Nematology*. 17 : 165-170.

Jain, R.K. & Sethi, C.L. 1988a. Interaction between vesicular arbuscular mycorrhiza and *Meloidogyne incognita, Heterodera cajani* on cowpea as influenced by time of inoculation. *Indian Journal of Nematology*. 18 : 263-268.

Jain, R.K. & Sethi, C.L. 1988b. Influence of endomycorrhizal fungi *Glomus fasciculatum* and G. epigaeus on penetration and development of *Heterodera cajani* on cowpea. *Indian journal of Nematology*, 18 : 89-93.

Jatala, P. 1986. Biological control of plant parasitic nematodes. *Annual Reveiw of Phytopathology*, 24 : 453-489.

Jonsson, H.A., 1976. Britta- A red clover variety with considerable disease resistance (Abstract). *Plant Breeding Abstract*, 46 : 461.

Jones, F.G.W. and Jones, M.G. 1984. Pest of Field Crops. Arnold Press, London, UK.

Jordaan, E.M., Loots, G.C., Jooste, W.J. and De Waele, D. 1987. Effect of root lesion nematodes (*Pratylenchus brachyurus* Godfrey and *P. zeae* Grahm) and Fusarium monoliforme Sheldon alone or in combination on maize. *Nematologica*, 33 : 213-219.

Kaplan, D.T. 1982. Plant resistance to nematodes : Symposium Introduction. *Journal of Nematology*. 14 : 1-2.

Kali Ram and Gupta, D.C. 1980. A note on the efficacy of fresh Neem leaf extract in the control of *Meloidogyne javanica* infecting Chickpea (*Cicer aretinum*), 10 : 96-98.

Kanwar, R.S., Gupta, D.C. and Walia, K. 1988. Effect of nematode inoculations at different intervals and soil type on interaction between *Meloidogyne javanica* and Rhizoctonia intervals and soil type on interaction between *Meloidogyne javanica* and *Rhizoctonia solani* on cowpea. *Indian Journal of Nematology*, 18 : 112-113.

Kerry, B.R. 1984. Nematophagus fungi and the regulation of nematode population in soil. *Helminthological Abstract* Series B. 53 : 1-14.

Keen, N. T. 1986. Novel approaches to improving disease resistance in plants. In : *Biotechnology for solving agricultural problems* (Eds. Augustin, P.C. ET.AL.) Martinus Nijhoff Pub. Boston, pp. 223-236.

Khan, M.W. 1990. Biocontrol of plant nematodes in closer perspective. In : *Progress in plant Nematology* (Eds. Saxena, S.K., Khan, M.W., Rashid, A. and Khan, M.R.). CBS Publisher and Distributors Private Limited, Delhi, India, pp. 367-382.

Khan, M.W., Alam, M.M. Saxena, S.K. 1974. Effect of water soluble fractions of oil cakes and bitter principles of Neem on some fungi and nematodes. *Acta Botanica Indica*. 2 : 120-128.

Kimura, Y., Mori, M., Hyeons, S., Suzuki, A. and Mitsui, Y. 1981. A rapid and simple method for the assay of nematicidal activity and its application to measuring the activities of dicarboxylic esters. *Agricultural Biological chemistry*, 45 : 249-251.

Kisiei, M., Deubert, K. and Zuckerman, B.M. 1969. The effect of *Tylenchus agricola* and *Tylenchorhynchus claytoni* on root rot of corn caused by Fusarium roseum and Pythium ultimum. *Phytopathology*. 59 : 1387-1390.

Koen, H. and Grobbelaar, N. 1968. The detrimental effect of *Eragrotris curvula* on the *Meloidogyne javanica* population of soil. *Nematologica*, 11 : 573-580.

Kretschmer, A.E. Sonoda, R.M. and Synder, G.H. 1980. Resistance of *Desmodium hetercarpon* and other tropical legumes to root knot nematodes. *Tropical Grassland*, 14 : 115-120.

Kushner, V.D. and Crittenden, H.W. 1967. Influence of Potassium Chloride on the interrelationship of fungi and a nematode in alfalfa roots (Abstract). *Phytopathology* 57 : 464.

Labruyere, R.E. 1979. Resowing problems of old pastures. In : *Soil Borne Plant pathogens* (eds. Schippers, B. and Gams, W.). Academic Press, London, New York, pp. 313-326.

Lal, A., Sanwal, K.C. and Mathur, V.K. 1983. Changes in the nematode population of undisturbed land with the introduction of land development practices and cropping sequences. *Indian Journal of Nematology*, 13 : 133-140.

Larson, R.O. 1987. In : *Proceeding 3rd International Neem Conference* (Eds. Schmutlerer, H. and Ascher, K.R.s.). Eschborn, FRG, Germany, pp. 243.

Lehman, W.F., EDE, L. Marble, V.L. Nelson, M.W. and Radewald, J.D. 1983. Registration of U.C. Cibola alfalfa. *Crop Science*, 23 : 1216.

Litz, R.E. 1986. Germplasm modification and its potential for finding new sources of resistance to diseases. *Journal of Nematology*, 18 : 18-22.

Mahajan, R., Singh, P., Bajaj, K.L. and Kalsi, P.S. 1986. Nematicidal activity of some sesquiterpenoids against root knot nematode *Meloidogyne incognita*. *Nematologica*, 32 : 201- 209.

Malik, M.S., Sangwan, N.K., Dhindsa, K.S. and Bhatti, D.S. 1987. Nematicidal efficacy of some monoterpenes and related derivatives. *Pesticides*, 21 : 30-32.

Mankau, R. 1981. Microbial control of nematodes. In : *Plant parasitic Nematodes* (Eds. Zuckerman, B.M., Mai, W.F. and Rhode, R.A.). Academic Press, London, New York, pp. 475-494.

Manza, B.E. 1974. Effect of *Pratylenchus penetrans* and *Fusarium solani* on growth of alfafa seedlings. *Journal of Nematology*. 6 : 145-146.

Midha, S.K. 1985. Efficacy of *Paecilomyces lilacinus* in controlling root knot infestation on cowpea and mungbean (Abstract). *Indian Journal of Nematology*, 15 : 267.

Minton, N.A., Parker, M.B., Flower, R.A. 1975. Response of soybean cultivars to *Meloidogyne incognita* and combined effects of *Meloidogyne arenaria* and *Sclerotium rolfsii*. *Plant Disease Reporter*, 59 : 920-923.

Mori, M., Hyeons, S., Kimura, Y. and Suzuki, A. 1982. The nematicidal activity of acetylence compounds. *Agricultural Biological chemistry*, 46 : 309-311.

Mukhopadhyay, M.C. and Prasad, S. K. 1968. Population dynamics of *Tylenchorhynchus*. *Nematologica*, 14 : 404-418.

Muller, R. and Gooch, P.S. 1982. Organic amendments in nematode control : An examination of literature. *Nematropica*, 12 : 319- 326.

Nath, R.P., Haider, M.G. and Prasad, S.S. 1974. Combined effect of *Hoplolaimus indicus* and *Fusarium monoliforme* on maize plants. *Indian Journal of Nematology*, 4 :90-93.

Noe, James, P. 1986. Cropping system analysis for limiting losses due to plant parasitic nematodes. In : Guide to Research Methodology, Raleigh, North Carolina, USA, pp. 1-17.

Norton, D.C. 1967. Relationship of *Heterodera trifolii* to some forage legumes. *Phytopathology*, 57 : 1305-1308.

Norton, D.C. 1978. Ecology of Plant Parasitic Nematodes. John Wiley and Sons, New York, pp. 268.

Nusbaum, C.J. and Ferris, H. 1973. The role of cropping system in nematode population management. *Annual Review of Phytopathology*, 11 : 423-448.

Oostenbrink, M. 1961. Nematode in relation to plant growth, III. *Pratylenchus penetrans* (Cobb.) in tree crops, potato and red clover. *Netherland Journal of Agricultural Science*, 9 : 188-209.

Oostenbrink, M. 1964. Hormonious control of nematode infestation. *Nematologica*, 10 : 49-56.

Orton, W.A. and Webber, H.J. 1902. The wilt disease of cowpea and its control. United States Department of Agriculture, *Bureau Plant Industry Bulletin*, 17 : 9-20.

Pandey, K.C., Hasan, N. Bhaskar, R.B. & Hazra, C.R. 1995. Pest and Diseases of Major Forage Crops - Problems and Prospects, Publisher, Porject Coordinator, AICRP (FC), Jhansi, pp. - 1-40.

Peaden, R.N., Evans, D.W., Elgin, J.H. Jr. and Rincker, C.M. 1983. Vernema alfalfa. *Crop Science* 23 : 1009.

Prasad, S.K. and Gaur, H.S. 1974. Relation between nematode population at harvest and yield of wheat. *Indian Journal of Nematology*, 4 : 152-159.

Reddy, K.C. 1983. Nitrogen fixation and nematode resistance of 13 tropical legumes. Dissertation Abstracts International B, 44 : 380.

Reeves, F.B., Wagner, D., Moorman, T. and Kiel J. 1979. The role of endomycorrhiza in revegetation practices in the semi arid West, I. A comparison of incidence of mycorrhiza in severely disturbed natural environment. *American Journal of Botany*, 66 : 6-13.

Reis, R.P., Noel, G.R. and Swanson, E.R. 1983. Economic analysis of a alternative control method for soybean cyst nematode in Southern Illionis. *Plant Diseases*, 67 : 480-483.

Reynolds, H.W., O'Banon, J.H. 1960. Reaction of sixteen varieties of alfalfa to two species of root-knot nematodes. *Plant Disease Reporter*, 44 : 441-443.

Rhoades, H.I. 1976. Effect of *Indigofera hirsuta* on *Belonolaimus longicaudatus*, *Meloidogyne incognita*, on *M. javanica* and subsequent crop yield. *Plant disease Reporter* : 60 : 384-386.

Saleh, H. and Sikora, R.a. 1984. Relationship between *Glomus fasciculatum* root colonization of cotton and its effect on *Meloidogyne incognita*. *Nematologica*, 30 : 230-237.

Sangwan, N.K., Verma, K.K. Verma, B.S., Malik, M.S. and Dhindsa, K.S. 1985. Nematicidal activity of essential oils of *Cymbopogon* grasses. *Nematologica*, 31 : 93-99.

Sasser, J.N. and Freckman, M.W. 1987. *Vistas in Nematology* (Eds. Veech J.A. and Dickson, D.W.). Hayattsville, Mayland, USA, pp. 7-14.

Schenck, N.C., Kinloch, D.W. and Dickson, D.W. 1975. In : *Endomycorrhizas* (Eds. Sanders, E. al.). Academic Press, London, New York, pp. 607-617.

Schonbeck, F. 1979. Endomycorrhiza in relation to plant disease. In : soil Borne plant pathogens (Eds. Schippers, B. and Gams, W.). Academic Press, London, pp. 271-280.

Schonbeck, F. 1984. Endomycorrhizae. In : *Progress in Botany* vol. 46. (Eds. Bochum, et al.). Springer - Verlag, Berhn, Heiderlberg, New York, Tokyo, pp. 382-390.

Seshadri, A.R. and Dasgupta, D.R. 1980. Nematodes. In : Hand book of Agriculture, Indian Council of Agricultural Research, New Delhi, India, pp. 551-577.

Sethi, C.L. and Gaur, H.S. 1986. Nematode management. An over view. In : Plant Parasitic Nematode of India - Problem and Progress. (Eds. Swarup, G. and Dasgupta, D.R). Indian Agricultural Research Institute, NEw Delhi, India pp. 424-445.

Sethi, C.L. and Gaur, H.S. 1990. Population dynamics of plant parasitic nematodes in relation to cropping systems. In : Progress in Plant Nematology (Eds. Saxena *et al.*). CBS Publisher and Distributor Private Limited, Delhi, India, pp. 223- 262.

Sethi, C.L. and Srivastava, A.N. 1986. Management of *Heterodera zeae* on maize through pesticidal seed treatment. *Indian Journal of Nematology* 16 : 82-86.

Shafshak, S.E., El-Debaby, A.S., Sary, G.a. And Salem, M.S 1974. Effect of preceding winter crops on the growth and yield of two suceeding crops, maize and wheat. *Zeitschrift Fuer Ackerund Pflannzenban.* 140 : 44-53.

Shagalina, L.M. 1973. On lucerne resistance to some pathogenic nematode *Lzvestiya Akademii Nank.* 1 : 48-51.

Sharma, J.L. and Trivedi, P.C. 1986. Biological contol of *Meloidogyne incognita* using fungus *Paecilomyces lilacinus* (Abstract). In *Proceeding of National conference* on Plant parasitic Nematodes of India, Problems and Progress. Indian Agricultural Research Institute, New Delhi, India, pp. 86.

Shephard, J.A. 1978. T.R.B. Tests on new love grass strain and recommends it. *Rhodeshian Tobacco Today*, 1 : 15.

Shephard, J.A. & Coombs, R.F. 1979. The effect of four *Meloidogyne* spp. on *Panicum maximum* cv. Umtali, *Rhodesiann Journal of Agricultural Research*, 17 : 155-156.

Sidhu, G.S. & Webster, J.M. 1981a. The genetics of plant nematode parasitic systems. *Botanical Review*, 47 : 387-419.

Sidhu, G.S. & Webster, J.M.1981b. Genetics of plant nematode interaction In : Plant Parasitic Nematodes Vol. III (Eds. Zuckerman, B.M. & Rhode, R.A.) Academic Press, London, pp. 61-83.

Sikora, R.A. & Schonbeck, F. 1975. Mycorrhiza (*Endogone mosseae*) on the population dynamics of the root knot nematodes. 8th. Int. Cong. of Plant Protection, 5 : 158-166.

Singh, B., Bhatt, D.S. & Singh, K. 1974. Resistance to root knot nematodes, *Meloidogyne* spp. in vegetable crops. *Pans*, 20 : 58-67.

Sitaramaiah, K. 1990. Mechanism of reduction of plant parasitic in soil amended with organic materials. In : Progress in Plant Nematology (Saxena *et al.*). CBS Publ. & Distr. Pvt. Ltd., Delhi. pp. 263-295.

Smith, G.S. 1987. Interaction of nematodes with mycorrhizal fungi. In : *Vistas* in Nematology (Eds. Veech, J.A. and Dickson, D.W.). Hayattsville, Maryland, USA, pp. 292-300.

Smith, R.F. and Reynolds, H.T. 1966. Principles, definition and scope of integrated pest control. In : Proceeding FAO symposium on Integrated Pest Control, Rome, Italy, 1 : 11-17.

Smolik, J.D. 1974. Nematode studies at the cotton weed site. US/IBP Grassland Biome Technical Report No. 251, Fort Collins, Colorado, Colorado State University, USA.

Soffes, A.R. Prine, G.M. and Dunn. R.A. 1980. Nitrogen fixation and effect on plant parasitic nematodes by eight legume cover crops (Abstract) 72nd Annual Meeting, American Society of Agronomy, Madison, Wisconsin, USA, pp. 105.

Spaull, A.M. and Clement, R.O. 1982. The effect of root ectoparasitic nematodes upon grass establishments. *Grass and forage science.* 37 : 183.

Spaull, A.M., Clement, R.O. and Newton, P.G. 1983. The effect of root ectoparasitic nematodes on grass establishment and productivity. In : Proceeding of the 9th General meeting of the European Grassland Federation Reading, UK (Ed. Corrall, J.), pp. 310-311.

Srivastava, A.N. and Sethi, C.L. 1986. Seasonal population fluctuation of *Heterodera zeae, Rotylenchulus reniformis* and *Tylenchus vulgaris* in Maize, Cowpea and Wheat rotations. *Indian Journal of Nematology*, 16 : 16-18.

Subramaniyan, S. 1985. Stylo (*Stylosanthes* spp.) a natural host of *Radopholus similis* (Cobb, 1893) Thorne 1949. Proceeding 4th Nematology Symposium, Udaipur, India pp. 78.

Sundaram, R. and Velolayutham, B. 1988. Relative efficacy of some insecticides and Neem cake in the control of *Rotylenchulus reniformis* and *Helicotylenchus dihystera*. *Indian Journal of Nematology*, 18 : 329-331.

Suryanarayana, Y.V. 1990. Botanical pesticides a glance. In : Souvenir National Symposium on Problems and Prospects of Botanical Pesticides in Integrated Pest Management, Indian Society of Tobacco Science, Rajmundry, India, pp. 19-73.

Swarup, G. and Dasgupta, D.R. 1986. Plant Parasitic Nematode of India - Problem and Progress. Indian Agricultural Research Institute, New Delhi, India, pp. 497.

Swarup, G. and Gokte, N. 1986. Biological control. In : Plant Parasitic Nematodes of India Problem and Progress (Eds. Swarup, G. and Dasgupta, D.R.). Indian Agricultural Research Institute, New Delhi, India, pp. 476-479.

Thomason, I.J. Erwin, D.C. and Garber, M.J. 1959. The relationship of root knot nematode, *Meloidogyne javanica* to *Fusarium* wilt of cowpea. *Phytopathology*. 49 : 602.

Upadhyay, K.d. and Swarup, G. 1981. Growth of maize plants in presence of *Tylenchulus vulgaris* Upadhyay *et al.*, 1973, singly and in combination with *Pratylenchus zeae*, Grahm, 1951 and Fusarium monoliforme sheld. *Indian Journal of Nematology*, 11 :29-34.

Vaishnav, M.U. and Sethi, C.L. 1978. Interaction of root knot and stunt nematodes with *Sclerospora graminocola* on bajra. *Indian Phytopathology*, 31 : 497-500.

Vandermeer, J.H. 1990. Inter-cropping. In : Agro-ecology (Eds. Carroll, J.H. *et al.*), Macgraw Hill Pub. Co...NewYork.

Van Gundy, S.D. 1985. Biological control of nematodes : Status and prospects in agricultural IPM system. In : Biological Control in IPM system (Eds. Hoy, M.A., and Herzog, D.C.). Academic Press. New York, London, pp. 467-478.

Vargas, J.M. Jr and Laughlin, C.W. 1972. The role of *Tylenchorhynchus dubius* in the development of fusarium blight of Marion Kentucky Blue grass. *Phytopathology*, 62 : 1311-1314.

Veech, J.A. 1979. Histopathological localization and nematotoxicity of terpenoid aldehydes in cotton. *Journal of Nematology*, 11 : 240-246.

Veech, J.A. 1981. Plant resistance to nematodes. In : Plant parasitic Nematodes vol. III (Eds. Zuckerman, B.M. and Rohde, R.A.). Academic Press, London, New York, pp. 377-400.

Vrain, T.C. 1998. Engineered Resistance In Crops Parasitised By Nematodes. In : Crop Productivity and Sustainability Shaping the Future, Proc. 2nd. Int. Crop Science Cong. (Eds. Chopra, *et al.*), Oxford & IBH Pub. Co. Pvt. Ltd., New Delhi, pp. 274-284.

Warthen, J.D. Jr. 1989. Neem (*Azadirachta indica* Juss), Organism affected and references list uptodate. Proc. Entomological Soc., Washington 91 : 367-388.

Webber, H.J. & Orton, W.A. 1902. A cowpea resistant to root knot *Heterodera radicicola*, US Dept. *Agric. Bureau of Plant Ind. Bull.* 17 : 23-26.

Webster, J.M. 1972. Nematode and biological control. In : Economic Nematology (Ed. Webster, J.M.) Academic Press. London, pp. 469-496.

Welty, R.E. Baker, K.R. & Lindsey, D.L. 1980. Effect of *Meloidogyne hapla* & *M. incognita* on *Phytopathora* root rot of alfalfa. *Plant Disease*, 64 : 1097-1099.

Williams, R.D. 1984. Integrated control of weeds, insects and pathogens. In : Crop Protection Hand Book-Grass and Clover Swards (Ed. Williams, R.D.) BCPC Pub., England.

Winchester, J.A. 1960. Preliminary investigation on the mode of action of pangola grass root in reducing cotton root knot nematode *Meloidogyne incognita acrita* populations. *Proceeding of soil and Crop Science* Florida, 20 : 178-182.

Wyatt, J.E. Fassuliotis, G., Hoffman, J.C. and Deakins, J.r. q1983. Nema Snap, Snap bean. *Horticultural Science*, 18 : 776.

Yeates, G.W. 1979. Soil nematode in terrestrial ecosystem. *Journal of Nematology*, 11 : 213-229.

Yeates, G.W., ross, d.J., Bridges, B.A. and Visser, T.A. 1977. Influence of nematodes *Heterodera trifolii* and *Meloidogyne hapla* and Nitrogen fixation by white clover under green house conditions. New Zealand Journal of Agricultural Research, 20 : 401-413.

CHAPTER - 10

MANAGEMENT OF PLANT PARASITIC NEMATODES WITH LATEX BEARING PLANTS

Mansoor A. Siddiqui and M. Mashkoor Alam

INTRODUCTION

Latex is a milky fluid, often watery or brownish present in plants of some families of angiosperms. It is an emulsion with watery base where various substance like proteins, gums, resins, alkaloids are suspended. Dumb-bell shaped starch grains are frequently noticed in the latex. It is helpful to plants in healing of wounds. This milky fluid is very valuable as the chief source of rubber. The majority of plants yields rubber belong to the families Moraceae, Euphorbiaceae and Apocynaceae.

The laticiferous cells or vessels may contain some fluids which are repulsive to taste or are poisonous. The alkaloids present in the latex sometimes have a disturbing effect on the nervous system of animals.

Earlier studies indicate that the latex bearing plants when used as organic soil amendments have found to be nematode suppressants (Alam, 1986; Siddiqui *et al.*, 1992). It was later claimed that this effect was due to toxic nature of the alkaloids (Siddique *et al.*, 1987, Siddiqui & Alam, 1990). The aim of the present write up is to gather hitherto scattered information.

1. *Artocarpus heterophyllus* Lam.

Syn. : *A. integra* (Thumb.) Merril, *A. integrifolia* Linn. f.

Family : Moraceae

Common names : English Jack tree, Sanskrit & Telugu - Panasa, Hindi - Kathal, Bengali - Kanthal, Marathi - Phanas, Tamil-Pilapalam, Kannada- Halasu, Malayalam-Chakka.

Botany/Uses : The stem is straight, cylindrical and covered with rough bark which exudes milky latex. Fruits and seeds are used as vegetable. Bark is used for tanning. Timber is used for high class furniture, musical instruments, toys, etc.

Nematicidal value : The latex contains high class nematicidal properties. It was found highly deleterious in killing the nematodes, viz. the root-knot nematode, *Meloidogyne incognita* (Kofoid & White) Chitwood, the reniform nematode, *Rotylenchulus reniformis* Linford & Oliveira, the stunt nematode, *Tylenchorynchus brassicae* Siddiqui and the lance nematode, *Hopolaimus indicus* Sher. The latex also significantly inhibited the larval hatching of *Meloidogyne incognita* (Haseeb *et al.*, 1984). The soil amendment with chopped shoot parts of *A. heterophyllus* also gave good control of *M. incognita* and *R. reniformis* on tomato (*Lycopersicon esculentum* Mill.) and eggplant (*Solanum melongena* L.) and *T. brassicae* on cabbage (*Brassica oleracea capitata* L.) and cauliflower (*Brassica oleracea botrytis* L.) (Siddiqui *et al.*, 1987).

2. *Calotropis gigantea* (Linn.) R.Br. ex Ait.

Family : Asclepiadaceae

Common names : English name - Madar, Sanskrit-Arka, Mandara, Hindi Ak, Arbic- Ashur, Persian Khark, Bengali-Akanda, Marathi-Rui, Gujarati Akado, Telugu-Jilledu, Kannada-Akragida, Malayalam-Erikku, Sind- Bijalosha.

Botany/Uses : The latex is present in all parts of the plant. The latex is used in indigenous medicine in combination with *Euphorbia neriifolia* as a strong purgative. The stem, root, bark and leaves are also used in medicine.

Nematicidal value : Seed dressing with the latex of *Calotropis gigantea* significantly inhibited the root knot development caused by *M. incognita* and population of *R. reniformis* and *T. brassicae* on tomato, eggplant, okra (*Abelmoschus esculentus* Moench.), cabbage and cauliflower and resulted in over all improvement of plant growth (Siddiqui & Alam, 1988a, b, c).

3. *Calotropis procera* (Ait). R.Br.

Syn. *Asclepias procera* Willd.

Family : Aslcepiadaceae

Common Names : English Akund, Swallow wark, Sanskrit - Alarka, Hindi - Safed Ak, Madar, Akada, Marathi - Mandara, Tamil-Velleruykku.

Botany/Uses : The plant is an evergreen shrub. Stem fibre is made into cordage and floss from seeds is sued as stuffing material. Stem and leaves are used as green manure. Root bark is used in dysentary, diaphoretic, expectorant and emetic. Powdered flowers are used in cold, cough and asthma.

Nematicidal value : The latex of *C. procera* was found toxic to the nematodes, viz. *Meloidogyne incognita, Helicotylenchus indicus, Siddiqui, Hoplolaimus indicus sher* and *Tylenchus filiformis* Butschli. The toxicity of the latex increased with an increase in the concentration and exposure period. The larval hatching was also significantly arrested by the latex (Siddiqui *et al.*, 1984, Zurreen & Khan, 1984). Soil amendments with pure latex and chopped shoot parts of *C. procera* significantly reduced the root knot development caused by *M. incognita* and the polution of *Hop. indicus, T. brassicae, T. filiformis* improved the plant growth of tomato and eggplant (Alam, 1986, Maqbool *et al.*, 1987). Seed dressing with the latex of *C. procera* significantly inhibited the root knot development caused by *M. incognita* and the population of *R. reniformis* and *T. brassicae* and resulted in overall improvement of plant growth of tomato, eggplant, okra, cabbage and cauliflower (Siddiqui, & Alam, 1988a, b, c).

4. *Caralluma edulis* Benth.

Family : Asclepiadaceae

Common Names : English - Caralluma, Hindi - Pimpa, Rajasthan- Pimpa.

Botany/Uses : A perennial shrub. It is used as vegetable. Plant is used in cooling, alternative and anthelmintic in leprosy and the diseases of blood.

Nematicidal value : The latex was found toxic to the root knot nematode, *M. incognita* (Zurreen & Khan, 1984).

5. *Carica papapya* Linn.

Family : Caricaceae

Common names : English- Papaya, Papaw tree, Hindi Papaya, Papeeta, Bengali- Papaya, Papaya, Marathi- Papaya, Gujarati - Papayi, Tamil - Pappali, Pappayi, Telugu - Boppayi, Kannada - Parangimara.

Botany/Uses : Rapidly growing small tree. Ripe fruits are eaten as desert. Milky latex contains a proteolytic enzyme i.e. Papain, which is extensively used in meat industry as tenderizer. Papain is also used in digestive medicine.

Nematicidal value : The latex of *C. procera* was found toxic to *M. incognita, R. reniformis, Hop. indicus* and *T. brassicae* (Haseeb, *et al.*, 1984). The soil amendment with chopped shoot parts of *C. procera* significantly inhibited the root knot development caused by *M. incognita* and the population of *R. reniformis, T. brassicae* and *Hop. indicus* and resulted in overall improvement of plant growth of tomato, eggplant, okra, cabbage and cauliflower (Siddiqui, *et al.*, 1987).

6. *Euphorbia caducifolia*

Family : Euphorbiaceae

Nematicidal value .: The latex was found toxic to *M. incognita* (Zurreen & Khan, 1984).

7. *Euphorbiax milii* ch. Des Moulins

Syn. *E. splendens* Boj ex Hook,
E. bojeri Hook

Common names : English name : Crown of Thorns

Family : Euphorbiaceae

Botany/Uses : The plant is propagated by cuttings from young growth and it thrives best in a mixture of brick rubbish, leaf mould and charcoal in exposed conditions. Latex is milky. The plant is grown as an ornamental or rockeries in gardens and in hedges.

Nematicidal value : Seed dressing with the latex of *E. milii* significantly inhibited the root knot development caused by *M. incognita* and the population of *R. reniformis* and *T. brassicae* and resulted in overall improvement of plant growth of tomato, eggplant, okra cabbage and cauliflower (Siddiqui & Alam, 1988 a, b, c).

8. *Euphorbia neriifolia* Linn.

Family : Euphorbiaceae

Common names : Sanskrit name - Snuhi, Hindi - Patton ki send, Sehund, thohar Bengali- Mansarij, hiladona, Marathi - Neyadunga, Mingut, Gujarati - Thor tuaria, Telugu - Akiye mudu, Tamil Ilai Kalli Kannada - Yalekalli, Malayalam - Ilakalli, Punjabi - Gangichu, Assam - Hiju.

Botany/uses : A small climbing, spiny shrub. The latex is milky and used as ruberfacient, purgative and expectorant.

Nematicidal value : The latex was found highly toxic to the nematodes viz. *Hop. indicus, Hel. indicus* and *T. filiformis* (Siddiqui *et al.*, 1984). The treatment with the latex significantly inhibited the root knot development caused by *M. incognita* and reduced the population of *R. reniformis* and *T. brassicae* and resulted in he improvement of tomato, eggplant, okra, cabbage and cauliflower (Siddiqui & Alam, 1988 a,b,c).

9. *Euphorbia pulcherrima* Willd. ex Klotz

Syn. *Poinsettia pulcherrima* (Willd. ex Klotz. R. Grali

Family : Euphorbiaceae

Common names : English - Poinsettia

Botany/Uses : Latex is poisnous to livestock, it contains water and water solubes 78.6-84.8% and caoutchouc 4.9-6.9%.

Nematicidal value : The latex of *E. pulcherrima* was found toxic to *M. incognita* (Zurreen & Khan, 1984).

10. *Euphorbia tirucalli* Linn.

Family : Euphorbiaceae

Common names : English - Milk bush, Indian tree spurge, Sanskrit - Trikantaka, Vajradruma, Hindi - Konpal, sehund, Bengali - Lankasy, latadona, Marathi - Shera, Vajradulu, Gujarati - Thor dandalio, Telugu - Chemudu, Kada - Jemudu, Tamil - Tirukalli, Kalli, Kannada - Bonta kalli.

Botany/Uses : Milky latex is vesicant and rubefacient. Wood is white, close grained and fairly hard. It gives a charcoal suitable for use in gun powder. The latex is brittle lustrous resin and used for rheumatism, neuralgia, toothache, cough and asthma. The resin may find use in linoleum, oil skin and leather cloth industries. Roots are used for poisoning fish and birds.

Nematicidal value : The latex of *E. tirucalli* was found highly deleterious in killing the nematodes viz. *Hop. indicus, T. brassicae* and *T. filiformis* (Siddiqui *et al.*, 1984). Seed treatment with the latex significantly inhibited the root knot development caused by *M. incognita* and suppressed the population of *R. reniformis* and *T. brassicae* and resulted in the overall improvement of plant growth of tomato, eggplant, okra, cabbage and cauliflower (Siddiqui & Alam, 1988 a, b, c).

11. *Ficus benghalensis* Linn.

Family : Moraceae

Common names : English The banayan tree, Sanskrit - Bahupada, Vata, Hindi - Bar, Bargad, Bor, Bengali - Bar, Bot, Gujarati - Vad, Vadlo, Vor, Marathi - Vada, Wad, War, Telugu - Marri, Peddamarri, Vati, Tamil - Al, Alam, Kannada - Ala, Alada, Mara, Vata. Malayalam - Ala, Vatam.

Botany/Uses : A large, spreading, evergreen tree extending laterally by sending down aerial roots from branches. The milky latex is externally applied for paines and bruises and as an anodyne in rheumatism and lumbago. It is also used as remedy for toothache. The bark is astringent and is used in dysentary. The young tips of hanging roots are given for obstinate vomiting.

Nematicidal value : The latex was toxic to the root-knot nematode, *M. incognita* (Zurreen and Khan, 1984).

12. *Ficus carrica* Linn

Family : Moraceae

Common names : English -Common fig., Hindi, Bengali, Gujarati and Marathi - Anjir, Telugu - Anjuru, Manjimedi, Simayatti, Tamil - Simaiyatti, Tenatti, Kannada - Anjura, Malayalam - Simayatti.

Botany/Uses : A small tree or bush. The leaves are used as fodder. The latex contains caoutchouc (2.4%), resin, albumin, cerin, sugar and malic acid. Fig latex is used as anthelmintic. The antehlmintic action has been traced to ficin. The fincin is effective against Helminths.

Nematicidal value : The latex was highly deleterious in killing he nematodes viz., *M. incognita, R. reniformis, Hop. indicus* and *T. brassicae.* The latex also inhibited the larval hatching of *M. incognita* (Haseeb *et al.*, 1984). The soil amendment with chopped shoot parts significantly reduced the population of *M. incognita* and *R. reniformis* on tomato, eggplant and *T. brassicae* on cabbage and cauliflower (Siddiqui *et al.*, 1987).

13. *Ficus elastica* Roxb.

Family : Moraceae

Common names : English - Assam rubber tree, India rubber tree, Bengal and Assam - Bor, Attah, Bor.

Botany/Uses : A small or medium sized evergreen tree. The fresh latex has a faint ethereal, odour and is chalky white in colour turning pink on exposure. Coagulation takes place when the latex is exposed to sunlight. The tree is considered to be a good source of fodder in Bengal. It is a commercial source of natural rubber. But the tree is valued mostly for shade and ornament. The wood is white or light brown in colour and in used as fuel.

Nematicidal value : The latex was toxic to *M. incognita* (Zurreen and Khan, 1984). Soil amendment with chopped shoot parts of *F. elastica* significantly suppressed the population of *M. incognita* and *R. reniformis* on tomato, eggplant and *T. brassicae* on cabbage and cauliflower (Siddiqui *et al.*, 1987).

14. *Ficus glomerata* Roxb.

Syn. *F. gooleraeea* Roxb., *Covellia glomerata* Miq., F. racemosa Linn.

Family : Moraceae

Common names : English - Cluster fig. Hindi - Gular, Umar, Bengali - Dumur, Jagyadumbar, Marathi - Umbar, Gujarati - Umar, Gular, Telugu - Ati, Udumbaramu, Tamil, Kannada and Malayalam - Athi, Oriya- Dimiri.

Botany/Uses : A deciduous tree; leaves are frequently used as fodder and fruits are edible. Bark is used as tan. The latex contain, 4.0-7.4% caoutchouc. The coagulum of latex is used in manufacture of ground sheets water proof bonded paper. The fruits are astringent, stomachic and carminative. The latex is given in piles and diarrhoea.

Nematicidal value : The latex was found toxic to *M. incognita, R. reniformis, T. brassicae* and *Hop. indicus*. It also inhibited the larval hatching of *M. incognita* (Haseeb *et al.*, 1984). Soil amendement with chopped shoot parts significantly reduced the population of *M. incognita* and *R. reniformis* on tomato, eggplant and *T. brassicae* on cabbage and (Siddiqui *et al.* 1987).

15. *Ficus religiosa* Linn.

Syn. *Urostigma religiosum* Gasp.

Family : Moraceae

Common names : English - The Peepal tree, Hindi - Pipal, Sanskrit - Ashwathaha, Bengali - Ashwatha, Gujarati - Jari, Pipro, Pipul, Marathi - Ashwatha, Pimpala, Telugu - Ashvathamu, Bodhi, Tamil - Arasu, Ashwattham, Kannada- Arali, Ashwattha, Malayalam - Arachu, Arayal, Ashwattham.

Botany/Uses : A large or medium sized tree grown as an avenue tree. A sacred tree of Hindus and Buddhists. Latex contain 0.7 - 5.1% caoutchouc. An infusion of it is used for ulcers and skin diseases. Leaves and tender shoots are used as purgative and in skin diseases. The fruit is laxative and the seeds are considered to be cooling, alternative and laxative.

Nematicidal value : The latex was found highly toxic to *M. incognita* (Zurreen and Khan, 1984).

16. *Ipomoea fistulosa* Mart. exch.

Syn. I. *crassicaulis* (Benth.) B.L. Robinson. I., *Carnea auct* (non Jacq.), Batatas crassicaulis Benth.).

Family : Convolvulaceae

Common names : Hindi - Sada Suhagan, Besharm, Rajasthani - Nagarpan

Botany/Uses : A stout straggling shrub, native to S. America, and is grown in gardens for dense and foliage flowers. It is used as a green manure crop in Madras. The plant is toxic to live stock. Leaves contain a polysaccharide, ipomase, an anthracene glucoside, a gum, jalapin and saponins.

Nematicidal value : The latex was found highly deleterious in killing the nematodes viz., *M. incognita, R. reniformis, T. brassicae* and *Hop. indicus*. It also inhibited the larval hatching of *M. incognita* (Haseeb, *et al.*, 1984). Soil amendment with chopped shoot parts significantly suppressed the population of *M. incognita* and *R. reniformis* on tomato, eggplant, and *T. brassicae* on cabbage and cauliflower (Siddiqui *et al.*, 1987).

17. *Nerium indicum* Mill.

Syn. *N. odorum* Soland, *N. Oleander* Blanco.

Family : Apocynaceae

Common names : English- Indian oleander, Sweet Scented oleander, Hindi - Kaner, Karber, Kuruvira, Bengali - Karabi, Marathi - Kanher, Kaneri, Gujarati-Kagaer, Telgu Ganneru, Kastoori, Pathelu, Tamil -Arali, Kannada Kangalu, Oriya - Kanero, Korobiro, Mundari Kanailiba, Santal - Rajbaka.

Botany/Uses : A shrub. All parts are poisonous. Roots, bark and seeds contain cardio-active glycosides i.e. olenadrin. The bark contain scopoletin and scopolin. Root is externally applied as a paste against ring worm. Dried leaves are an effective cardiac stimulant.

Nematicidal value : The latices of *N. indicum* and *N. odorum* were found highly toxic to *M. incognita, R. reniformis, T. brassicae, Hop. indicus, Hel. indicus* and *T. filiformis* (Haseeb *et al.*, 1984, Siddiqui *et al.*, 1984). It also significantly inhibited the larval hatching of *M. incognita* (Haseeb *et al.*, 1984; Zurreen and Khan, 1984). Soil amendment with chopped shoot parts of *N. odorum* significantly suppressed the population of *M. incognita* and *R. reniformis* on tomato, eggplant and *T. brassicae* on babbage and cauliflower (Siddiqui *et al.*, 1987).

18. *Pedilanthus tithymaloides* Poit.

Family : Euphorbiaceae

Common names : English name - Slipper plant, Bengali - Belatisii, Maharashtra - Vilaytisher, M.P. - Nagphani, Nagdaman, Mundari - Airi.

Botany/Uses : An ornamental laticiferous shrub grown in hedges. The root is powerful emetic. The latex is emetic, irritant and caustic. The latex is said to be used in venereal diseases. It is also applied to warts and leucoderma patches.

Nematicidal value : The latex was found toxic to *Hop. indicus Hel. indicus* and *T. filiformis* (Siddiqui *et al.*, 1984).

19. *Plumeria acutifolia* Poir

Syn. *P. acuminata* Ait, *P. rubra* Linn. var. *acutifolia* Bailey.

Family : Apocynaceae

Common names : English name - Temple or Pagoda tree, Sanskrit - Kshira champa, Hindi - Golainchi, Bengali - Sonchampa, Gujarati - Rhadachampo, Telugu - Nuru varahaalu, Vaada ganneru, Tamil - Arali, Kalimandharai, Kannada - Devaganigile, Kadu sampage, Malayalam -

Ezha - champakam, Arali, Oriya katchampa, Golochi, Torato, santal Champa Pungar, Golanj baha, Assam - Golanchi, Gunach.

Botany/Uses : An evergreen or partly deciduous; tree Bark is stimulant. Its decoction is used as a purgative, febrifuge, and emmenagogue. Latex rubefacient and purgative, enters into applications for itch. rheumatism and gum troubles. The root is cathartic. The wood is for making drums and other musical instruments.

Nematicidal value *:* The latex was found toxic to *M. incognita* (Zurreen & Khan, 1984).

20. *Plumeria oblongifolia*

Family : Apocynaceae

Nematicidal value : The latex was found toxic to *M. incognita* (Zurreen & Khan, 1984).

21. *Tabernaemontana coronaria* R.Br.

Syn. *Ervatamia divaricata* (Linn. (Alston, *Ervatamia coronaria* Stapf.

Family : Apocynaceae

Common names : English name - East Indian Rosebay, Sanskrit - Nandivriksha, tagara, Hindi - Chandni, Tagara, Bengali - Tagar, GujaratiSagar, Tagar, Marathi - Ananta, Gondetagara, Sagar, Tagar, Telugu - Gandhitagarapu, Nandivardhanamu, Tamil - Adukkunandiyavattai, Kurudup-palai, Nandiyavatta, Kannada - Kottuhale, Nandibatlu.

Botany/Uses : A small evergreen shrub, the root is acrid and bitter. It is employed as a local andodyne and chewed for the relief of tootache. It is vermicide. The milky juice of leaves is used in opthalmia. Juice of flowers is employed for relieving the burning sensation in sore eyes. It is also applied in skin diseases.

Nematicidal value : The latex was found highly deleterious in Killing *M. incognita, R. reniformis, T. brassicae* and *Hop. indicus*. It also inhibited the larval hatching of *M. incognita* (Haseeb, *et al.*, 1984). Soil amendment with chopped shoot parts significantly suppressed the population of *M. incognita* and *R. reniformis* on tomato, eggplant and T. brassicae on cabbage and cauliflower (Siddiqui *et al.*, 1987).

22. *Thevetia peruviana* (Pers.) Merrill

Syn. *T. neriifolia* Juss. ex Steud.

Family : Apocynaceae

Common names : English names - Luck nut tree, yellow oleander, Sanskrit - Ashvaghna Divyapushapa, Haripriya, Hindi - Pili Kaner, Zard Kaner, Kulkephul, Bengali - Koklaphul, Haldi Korubi, China Karab, Marathi - Pilvalakanhera, Gujarati - Kanera, Telugu - Pachaganneru, Kannada - Kadukasi, Kanogalu, Malayalam - Maya areli, Pachcha areli, Orissa - Konyar phul.

Botany/Uses : All parts of the plant produce latex which is highly poisonous. Flower is yellow in colour. Latex is sometimes applied to sores and cotton soaked in the juice is used for the relief of toothache. The plant also yields a dark gum. Seeds are used as an abortifacient and as a purgative in reheumatism and dropsy. They are also used as an alexiteric. Kernels are very bitter. They are used for insecticidal purposes.

Nematicidal value : The latex was found toxic to *M. incognita, T. brassicae, Hop, indicus* and *Hel. indicus* (Siddiqui *et al.*., 1984; Zurreen & Khan, 1984).

Appendix 1. Components of latex and its coagulum

Plants	Coagulum			Original latex	
	Caoutchouc conc. %	Resin %	Insoluble %	Water & Water solubles %	Caoutchouc %
Artocarpus heterophyllus	6.0-10.0	82.6-86.4	3.9-8.1	65.9-76.0	2.3-2.9
Calotropis gigantea	5.1-18.6	73.6-87.8	4.5-13.8	86.0-93.0	0.6-1.9
C. procera	–	52.8-85.0	11.4-22.9	88.4-93.0	0.8-2.5
Euphorbia neriifolia	–	–	–	69.4-93.3	0.2-2.6
E. pulcherrima	–	–	–	78.6-84.8	4.9-6.9
E. tirucalli	–	0.9	–	53.8-79.9	2.8-3.8
Ficus benghalensis	2.7-15.9	77.3-94.9	0.7-20.0	46.0-90.6	0.3-7.7
F. carica	7.2	55.4	37.4	66.1	2.4
F. elastica	92.2	7.1	0.7	56.7	39.9
F. glomerata	16.3-25.7	72.3 - 77.6	0.8-8.3	70.2-76.4	4.0-7.4
F. religiosa	5.2-7.1	74.9, 77.6	4.8-18.0	23.5-86.4	0.7-51
Plumeria acutifolia	19.1, 16.1	62.7, 75.6	18.2-8.5	–	–
Thevetia peruviana	13.3-9.7	69.7, 67.4	17.0-22.9	–	–

Appendix 2. Analysis of latex and other products occurring in different plants

Plant	Plant parts	Chemical/Chemical constitution
Atrocarpus heterophyllus	Latex	Crystalline steroketone, artostenone, $C_{30}H_{50}O$, m.p. 109.
	Fruit	Moisture 77.2, protein 1.9, fat 0.1, carbohydrates 18.9, fibre 1.1, mineral matter 0.8, Ca, 0.02 P 0.03%, Fe 0.5 mg; vitamin A, 540 I.U., vitamin C, 10 mg/100g
	Seeds	Starch moisture 51.6, protein, 6.6, fat, 0.4 carbohydrates, 38.4, fibre 1.5, mineral matter, 1.5, Ca 0.05, P, 0.13%, Fe, 1.2 mg/100 g
Calotropis gigantea	Latex	Isomeric resinol ($C_{30}H_{50}O$, α-Calotropeol (m.p. 204°-5°) β-calotropeol (m.p. 216°-17°)
	Seeds	Moisture 7.4, protein 27, ether extract, 32.4, ash, 6.55%
	Oil extract from seeds	Palmitic 15, oleic 52, linole, 32 and linolenic acid, 0.9%
	Flowers	Esters of α & β-calotropeols, β-amyrin
	Bark	α & β-calotropeols, β-amyrin, giganteol
	Root bark	β-amyrin, 2-isomeric crystalline alcohols, giganteol, m.p. 223°-24°, isogiganteol, m.p. 117°-78°
	Ash (12.7%)	Potash (K_2O, 20.8%).
Calotropis procera	Latex	Trypsin, α-lactuceryl isovalerate (m.p. 181) α- lactuceryl acetate, m.p. 252°, α-lactuceryl, $C_{30}H_{50}O$, m.p. 224.50, calotoxin, $C_{29}H_{40}O_{10}$, m.p. 224° (decomp.) Calotropagenin, $C_{23}H_{32}O_6$, m.p. 240°
	Leaf & stalk	Calotropin, $C_{29}H_{40}O10$ m.p. 221° (decomp.)
Carica papaya	Fruit	Moisture 89,6, proteins, 0.5, carbohydrate, 9.5, ether extract 0.1, mineral matter, 0.4 calcium, 0.01, phosphorus, 0.01%, iron, 0.4 mg/100g
	Seed	Protein 24.3, carbohydrate, 15.5; fatty oil 25.3, crude fibre 17.0; ash, 8.8; volatile oil, 0.09% a glucoside i.e. caricin
	Leaf	Vitamin C, 286 mg per 100 g, Vitamin E, 36 mg/per 100 g.
	Root	A singirin-like glycoside
Euphorbia milli	Latex	Tarpenic alcohol

Appendix 2 contd...

Plant	Plant parts	Chemical/Chemical constitution
E. nerriifolia	Latex	Tarpenic alcohol
E. pulcherria	Latex	Tarpenic alcohol
E. tirucalli	Latex	Tarpenic alcohol, isoeuphorol $C_{30}H_{50}O$; m.p. 117-18°
Ficus benghalensis	Latex	Ficosterol, $C_{20}H_{50}$; m.p. 135° [α]D, + 63.59° in chloroform, glutathione (34 mg/100g)
F. elastica	Latex	Caoutchouc 77.5 resin 19.3, protein 1.5% insoluble matter, 1.7% and Ash, 0.5%
	Leaf	Caoutchouc 0.3, sugar, 1.1, tannins 1, and Ash 0.27%
F. glomerata	Latex	Caoutchouc 4.0-7.4%
	Fruit	13.6, albuminoids, 7.4 fat, 5.6, carbohydrates, 49.0 colouring matter, 8.5 fibre, 17.9 ash, 6.5,
	Leaf	Siliqua (SiO2) 0.25 phosphorous (P_2O_5) 0.91%
F. religiosa	Fruit	Moisture 9.9% albuminoids, 7.9, fatty matter 5.3 carbohydrates 34.9, coloring matter 7.5, ash 8.3 silica, 1.85 phosphorous (P_2O_5) 0.69%
	Leaf	Crude protein 12.99 ether extract, 2.71, crude fibre, 22.36 total ash, 15.06 lime (CaO), 4.64 phosphorous (P_2O_5) 0.52%
		Polysaccharide ipomose, anthracene, glycoside, gum, saponin
Nerium indicum	Seed	Neriodorin ($C_{22}H_{32}O_7$, m.p. p. 86-87°), Neriodorein ($C_{23}H_{34}O_{11}$ m.p. 106-7°), Karabin ($C_{21}H_{49}O_6$)
and *N. odorum*	Leaf	Oleandrin ($C_{32}H_{48}O_9$, m.p. 250° decomp.), Neriodin (m.p. 238-39°), Nerium D (m.p. 235-38°), Nerium E (16-deacetylanhydro- oleandrin, $C_{30}H_{44}O_7$ m.p. 22-24°)
	Bark	Odoroside A ($C_{30}H_{46}O_7$ m.p. 183°/198°), Digitoxienin β-D diginoside
		Odoroside B ($C_{30}H_{46}O_7$ m.p. 150°/200°), uzarigenin β-D diginoside
		Odoroside D ($C_{30}H_5O_{12}$ m.p. 219°/254°), Digitoxigenin β - D - glucoside β-D-diginoside
		Odoroside F ($C_{36}H_{56}O_{13}$ m.p. 298°), Digitoxigenin β - D - glucoside β-D-diginoside
		Odoroside G ($C_{44}H_{68}O_{19}$ m.p. 282), Digitoxigenin β- D-glucoside β-D-diginoside-monoacetyl β-D-Digitatoside
		Odoroside H ($C_{30}H_{46}O_8$, m.p. 236°), Digitoxienin β-D-diginoside β-D - Digitatoside
		Odoroside K ($C_{42}H_{66}O_{17}$, m.p. 196°/246-65°), uzarigenin - β - D-Digitatoside. Glucosido β-D-Digitatoside. Glucosido-β-D-Digitato-side-glucosido - β-D-diginoxido
		Odorobioside K ($C_{36}H_{56}O_{12}$, m.p. 178°/222-55°), uzarigenin - β - D-glucosido - β-D-glucosido - β-D-diginosido (monoacetate; $C_{34}H_{48}O_{10}$ m.p. 178°) Decomposes to D - digitalose and a derivative of 16-anhydrodigitoxigenin
		Odoroside M (Monoacetate : $C_{34}H_{48}O_{10}$ m.p. 219°/230°) Anisomer of L.
	Root	Phenolic compound (m.p. 140-141°) essential oil
		(d^{30} $_{40}$,O. 8666 nD, 1.40315; [α]-408), resinous matter 7.5%, Alcohol isolated from resinous matter ($C_{10}H_{50}O$, m.p. 184-85°)
Pedilanthus tithymaloides	Latex	Total solids 0.4%, euphorbin, cerin, myricine, resin, fatty oil.

Appendix 2 contd...

Plant	Plant parts	Chemical/Chemical constitution
Plumeria acutifolia	Flower	Essential oil (0.04-0.07) d_{280}^{280} $0.9^{6.8}{}_{5}$, $n^{280}{}_{D}$, $1.4^{88}{}_{5}$, acid val, 20.2 sap. val. 123
	Leaf	Pectin 5.6%
	Bark	Glucoside plumieride ($C_{12}H_{26}O_{12}$, m.p. 224-25°, yield 4%), Fulvoplumierin (5' - carbomethoxy-3', 4'-1,2-cumalino-w- propenyl fulvene, $C_{14}H_{12}O_4$ m.p. 151-52° (decomp.) α-amyrin acetate
Tabernaemontana coronaria	Latex	Moisture, 84.27, total solids, 15.73, alcohol extract, 6.94, chlorofrorm extract, 2.54 and residue 5.75%
	Bark & stem	Tabernaemontanine ($C_{20}H_{26}O_3N_2$; m.p. 208-10°), Coronarine ($C_{44}H_{56}O_6N_4$ 2.5 H_2O m.p. 196-98°), A crystalline resin alcohol (m.p. 180-81°)
		Caotchouc, resins, sugars and fatty matter yielding on saponification, palmitic acid, ceratic and oleic acids.
Thevetia peruuviana	Different parts	Glycoside aglycone sugar (s) mol. formula m.p. Cerberoside Digitoxigenin L-thevetose + $C_{42}H_{66}O_{18}$ 196 (thevetin B) D - glucose (2 mols.)
		2'-O-Acetyl Digitoxigenin L-thevetose+ $C_{44}H_{68}O_{19}$ 162-85° derberoside D-glucose (2 mols.)
		Neriifolin Digitoxigenin L-thevetose $C_{30}H_{46}O_8$ 175/230°, 239°
		2'-O-Neriifolin digitoxigenin L-thevetose $C_{32}H_{48}O_9$ 148/230° 212° Cerberin Thevetin A
		19 (OxO-cerberoside) Cannogenin L-thevetose+$C_{42}H_{64}O_{19}$ 219° D-glucose (2 mols.)
		Peruvoside
		(19-Oxoneriifolin) Cannogenin L-thevetose $C_{30}H_{44}O_9$ 164/216°
		Theveneriin Cannogenol L - thevetose $C_{30}H_{46}O_9$ 234° (ruvoside) (19-Oxyneriifolin)
		Peruvosidic acid Cannogenic L-thevetose $C_{30}H_{44}O_{10}$ 226° (perusitin) acid

REFERENCES

Alam, M.M. 1986. Possible use of weeds as soil amendment for the management of root knot and stunt nematodes attacking eggplant. *Agric. Wastes.* 16 : 97-102.

Haseeb, A., Singh, B., Khan, A.M. and Saxena, S.K. 1978. Nematode population as influenced by soil amendments. *Geobios.* 6 : 152-155.

Haseeb, A. *et al.* 1982. Toxicity of leaf extracts of plants to root knot and reniform nematodes. *Indian J. Parasitol* 6 ; 119-120.

Haseeb, A., Siddiqui, M.A. and Alam, 1984. Toxicity of latex bearing plants to phytonematodes. In : "Environment & Biotic Interaction" (eds. - A.K. Datta gupta & R.P. Maleyvar) : 67 - 71.

Maqbool, M.A. Hashmi, S. & Ghaffar, A. 1982. Effect of latex extract from *Euphorbia caducifolia* and *Calotropis procera* on root knot nematode *Meloidogyne incognita* infesting tomato and eggplant. *Pak. J. Nematol.* 5 : 43-47.

Siddiqui, M.A. & Alam, M.M. 1988a. Effect of latex seed dressing on *Rotylenchulus reniformis* and Plant growth of some vegetables. *Nematol medit.* 16 : 129-131.

Siddiqui, M.A. & Alam, M.M. 1988b. Effect of seed dressing with plant latex on *Tylenchorhynchus brassicae* and plant growth of cabbage and cauliflower *Pak. J. Nematol.* 6 : 65-71.

Siddiqui, M.A. & Alam, M.M. 1988c. Effect of seed treatment with plant lactices on the root knot development and plant growth of some vegetables. *Proc. Nat. Acad. Sci. India* 58 : 321-328.

Siddiqui, M.A., Haseeb, A. & Alam, M.M. 1984. Toxicity of plant latex to some plant parasitic nematodes. *National Academy Science letters* 7 : 1-2.

Siddiqui, M.A., Haseeb, A.& Alam. M.M. 1987. Evaluation of nematicidal properties in some latex bearing plants. *Indian J. Nematol.* 17 : 99-102.

Siddiqui, M.A. & Alam, M.M. 1990. Potential of plant latex as a botanical nematicide. *Integrated Pest Management Practitioner* 12 : 6-8 (USA).

Zurreen, S. & Khan, M.I. 1984. Nematicidal activity in some plant latices. *Pak. J. Nematol,* 2 : 69-71.

CHAPTER - 11

MANAGEMENT OF PHYTONEMATODES BY BIOAGENTS

M.S. Sheela, Hebsybai and P. Sivaprasad

INTRODUCTION

Nematodes constitute a group of primitive microscopic invertebrate animals which occur in all conceivable habitats in amazingly large numbers. Besides bacteria, viruses, fungi, insects etc. the nematodes are regarded as important pests of our agricultural and horticultural crops. Their success in all types of biosphere is related to their varied feeding mechanism e.g. microphagous, saprophagous, carnivorus and predaceous. The nematodes are now well known to cause severe diseases not only to plants but also to animals including man and the losses may run into billions of rupees annually.

The control of nematodes is far more difficult than any other kind of pests because, they inhabit soil and usually attack the underground parts of plants. Though, many methods for controlling nematodes have been devised only some have proved effective, others being particularly effective or ineffective. Chemical control of nematodes proved generally effective, but the majority of these chemicals are general biocides and are highly expensive, often causing health hazards and pollution problems.

Generally, the nematode control measures are either corrective (e.g., chemical) or preventive (e.g., cultural). However, the biological control has both these possibilities. It may be preventive in the sense that it can help in evading the disease. If the disease has already set in, the melody can be corrected by reducing the population density of pest. Biological control with reference to plant parasitic nematodes is thus a means of promoting organisms that are harmless to the plants but attack and destroy those organisms which are harmful to the plants. Many biocontrol agents can be employed for this purpose.

On account of the ecofriendly plant protection drive, great emphasis has been given to the exploitation of potential bioagents for controlling nematodes. Soil being a complex envrionment housing various flora and fauna, nematodes are naturally exposed to many enemies of nematodes. But the widely exploited once are fungi and bacteria. These bioagents have several advantages. Not only are they ecofriendly, but also result in delayed development of resistance. Besides, biotechnological interventions for evolving efficient strains are possible in these organisms as they possess a simple genome. Eventhough fungi is considered as one of the most important agent, in the past decade bacterial pathogen and their metabolites gained prime importance.

A. FUNGI

Fungi, one of the most important microbial agents attacking nematodes have been extensively investigated for their potential in the bio control of nematodes. The fungal bioagents can broadly be grouped into 3 categories viz., predaceous fungi, endozoic fungi and egg parasitic fungi (opportunistic fungi).

1. Predaceous fungi

The predaceous fungi also known as nematode trapping fungi capture nematodes by adhesive mycelial networks, sticky knobs, non constricting and constricting rings. They are good saprophytes, thriving well in soil with high organic matter, utilising nematodes as alternative food source. Some species routinely have trapping devices on the mycelium, while many other species produce the devices only when the nematodes are present in the vicinity. Works of Drechster (1937, 1941) and Duddington (1957, 1960) generated interest in this group of fungi as potential agents for the biocontrol of nematodes. The scope of utilising nematode trapping fungi as biocontrol agents was first examnined in Hawaii for the control of nematodes on pineapple (Linford *et al.*, 1938). Literature survey showed that 57 per cent of research efforts on utilization of fungal pathogen for nematode management (Goswami and Uma Rao, 1995) are concentrated on predaceous fungi. Among the predaceous fungi, *Arthrobotrys* spp. and *Dactylella* spp. have proved promising for nematode management.

1.1. *Arthrobotrys* spp.

Successful reduction in the invasion of plants by *Heterodera* could be obtained in pot trials. However, little success was achieved in the field when trapping fungi were introduced with organic matter of various kinds (Hans and Wilkin, 1961). The number of *H. avenae* in roots of wheat was significantly reduced when *A. thanmaria* was incorporated in microplots (Duddington and Duthoit, 1960; Duddigton *et al.*, 1961). *A. oligospora* and *A. superba* have been recorded to reduce nematode population and increase yields of mushroom (Kermerec, 1973). A formulation of *A. robusta* (Royal 300) protect the mushroom, *Agaricus bisporus* from *Ditylenchus myceliophagus* (Cayrol *et al.*, 1978). Upadhayay and Dwivedi (1988) showed the use of predaceous fungi as biocontrol agents against plant parasitic nematodes in India. *A. irregularis* protected plants against light and moderate infections (Cayrol, 1988). *A. dactyloides* was found effective for controlling root knot nematodes (Nakasono and Gaspard, 1991).

Tunlid and Janson (1992) reported that lectin carbohydrate interaction was responsible for adhesion of nematodes to *A. oligospora* (first step of interactions). Dackman and Nordbring-Hertz (1992) found that the fungus developed traps in response to diffusing substances from the faeces of cows rather than in response to living nematodes.

1.2 *Dactylella* spp.

Linford and Yap (1939) added 5 species of nematode trapping fungi to soil infested with *Meloidogyne* sp. and found that *Dactylella ellipsospora* had significantly reduced nematode injury in plants after 15 months.

Deschiens *et al.* (1943) tried *A. oligospora* and *D. thaumasia* for the control of *H. schachtii* which gave encouraging but inconclusive results. In microplot trials, *H. avenae* was controlled to a great extent when green manure or fungal mycelium were added to the soil (1960). Nakasoma and Gaspard (1991) determined the effectiveness of *D. haptotyla* for root knot and free living nematodes.

2. Endozoic fungi

Endoparasitic fungi produce either simple spores or flagellate spores as infectious agents. Germination of the spores are within the nematode body. The spores present in soil remain viable but dormant until they make contact with the host. The simple spores are ingested by the nematodes, reach the oesophagous or buccal cavity where they germinate, penetrate the oesophagus and colonise the body cavity eg. *Harposporium anguillulae, H. bysmatosporum.* The sticky spores stick to the nematodes at any point but mostly they cause infections in the head regions. They germinate and penetrate directly through the nematode cuticle and produce infective hyphae within the body cavity e.g. *Meria coniospora, Nematoctomus bisporus.* The flagellated spores stick to the nematodes and encyst before germination, penetration and colonisation of the host eg. *Catenaria anguillulae, C. vermicola.*

The non trap forming endozoic (endoparasitic) fungi usually enter the nematode by a germ tube that penetrates the cuticle from a sticky pore. Most of these fungi form vegetative hyphae within the nematode body. Once inside the nematode, the hyphae ramify throughout the body, absorb the contents and then spores bearing hyphae emerge from the nematode cadaver.

Catenaria spp. and *Nematophthora* spp. are two important flagellate fungi involved in biocontrol of plant parasitic nematodes.

2.1 *Catenaria* spp.

C. auxilaris parasitized females of *H. avenae* and *Globodera rostochiensis* (Kerry, 1975; Kerry *et al.*, 1976). *C. auxilaris* was recorded from Australia (Stirling, 1980) as parasitizing females of *H. avenae* in late winter and early spring and also on *H. glycines* females from a sandy loam soil at Woburn, U.K. (Kerry and Muller, 1981). Crump *et al.*, (1983) surveyed several soybean growing areas of Tennessee, USA and collected females of *H. glycines* infected with *C. auxilaris*, Choudhary and Dhawan (1984) reported infection of larvae and eggs of *H. avenae* by *C vermicola.*

Voss and Wyss (1990) determined the potential of *C. anguillulae* as a biocontrol agent against several plant parasitic nematodes. Variations between strains of the fungus and its attact on *H. schachtii* was also studied (Voss *et al.*, 1992). *In vitro* observations on the infection of *M. incognita* eggs by *C. anguillulae* was reported by Wyss *et al.*, (1992), where embryo was observed to be killed within a few minutes following mass aggregation and encystment of flagellate spore of *C. anguillulae.*

2.2 *Nematophthora* spp.

N. gynophila parasitises *H. avenae, H. trifolii, H. cruciferae, H. carotae, H. goetingiana* and *H. schachtii* but not *G. rostochiensis* (Kerry and Crump, 1977). Kerry *et al.*, (1980) found that the ability of *N. gynophila* to infect nematode is dependent upon adequate soil moisture.

Kerry (1981a), reviewed the biological control of soil nematodes and discussed the role of *N. gynophila* in decline of *H. avenae* populations. The fungus was found to control *H. avenae* population in fields, enough to allow continuous growing of susceptible crops on nematode infested land (Sayre, 1980; Kerry, 1981b). In a field trial on sandy loam soil at Woburn (U.K.), high levels of fungal parasitism were associated with low rates of *H. avenae* multiplications (Kerry, *et al.*, 1982b). In pot tests, 97 per cent females were killed by the fungus but the rate of parasitism was much reduced in soil which were kept dry (Kerry *et al.*, 1980), since the motile zoospores of the fungus were less active in dry soils.

In a survey at Woburn, Tidworth and Sydomonton (U.K.). Kerry *et al.*, (1982c) observed a decline in *H. avenae* cyst nematodes on susceptible barley. Less than 40 per cent of females formed cysts containing eggs. The fungus was active at all the three sites, but was most active at Woburn where there was irrigation. The fungus has been recorded from the females of *H. glycines* (Kerry and Muller, 1981; Crump, *et al.*, 1983) and *M. arenaria* (Coetzee, 1956) from USA and also from Germany (Knuth and Dietermann, 1983).

2.3 Others

The adhesion and infection of *Drechmeria coniospora* to the plant parasitic nematodes were studied and protein was found to be involved in the adhesion process (Jansson *et al.*, 1987). Colonisation and digestion was investigated by Dijksterhuis *et al.*, (1991).

3. Egg parasitic fungi

Several fungal species have been found to be parasitic on eggs and cysts of plant parasitic nematodes. These are soil fungi and they colonise the reproductive strucutures of the nematodes particularly, *Meloidogyne, Heterodera* and *Globodera* spp. where as sedentary stages are most vulnerable to fungal attack. These fungi include *Verticillium, Cylindrocarpon, Fusarium, Exophiala, Torula, Gliocladium, Cephalosporium, Phoma* and *Paecilomyces*. Destructive activity of the fungi is mainly due to the enzymatic disruption of nematodes structural elements, such as egg shells and larval cuticle of nematodes and physiological disturbances brought about by the biosynthesis of diffusible toxic metabolites (Morgan Jones *et al.*, 1983 and 1984).]

3.1 Verticillum

The fungus *V. chlamydosporium* was designated as the principal egg pathogen of *H. schachtii*, The contents of an invaded egg get converted into a mass of hyphae which are apparently non persistent as they undergo lysis in due course. The young white females on roots are not attacked but those in which eggs have been formed become susceptible to invasion (Bursnall and Tribe, 1974).

In southernEngland, it was recorded that 76 per cent of soils contained fungal parasites of *H. avenae* and were considered to be the most important egg parasite, capable of infesting all developmental stages (Kerry, 1975; Morgan Jones *et al.*, 1981). It penetrated the eggs, colonised them and prevented hatching. Further, it disrupted the larval cuticle and egg shell and the hyphae readily proliferated endogenously. The invaded larvae become totally necrotic and disintegrated (Morgan Jones *et al.*, 1983).

Similarly *Verticillium* was observed to be the most frequent parasite of eggs, young cyst and larvae of *H. schachtii* (Crump *et al.*, 1983; Thomas, 1982). It was also a frequent parasite on *H. avenae* cysts in Australia (Stirling and Kerry, 1983). Integrated management of the citrus nematode *T. semipenetrans* infecting acid lime using *V. chlamydosporium*, *V. lecanii*, oil cakes, endomycorrhizae and nematicides has been developed (Parvatha Reddy *et al.*, 1991a, 1995, 1996a, b, 1997).

Integrated management of root knot nematode infecting crossandra was achieved by rational combination of *V. chlamydosporium*, a nematicide or an oil cake (Khan and Parvatha, Reddy, 1994).

3.2 Paecilomyces

P. lilacinus is a cosmoplitan fungus having wide distribution in the tropics, including India. Extensive data have been generated on its usefulness against the potato cyst nematode, *Globodera* spp., the root knot nematode, *Meloidogyne* spp., the reniform nematode, *Rotylenchus reniformis* and the burrowing nematode, *Radopholus similis*. *P. nostocoides* was reported to parasitise eggs of *H. zeae* (Koshy, Swamp and Sethi, *et al.*, 1971) in USA. The use of *P. nostocoides* as a biocontrol agent of *H. zeae* has not been tested so far, but *In vitro*, colonization of eggs does occur (Dunn, 1983).

P. lilacinus was found to parasitise eggs of *M. incognita* acrita (Jalata *et al.*, 1979) in Peru. Seventy to ninety per cent of eggs of *Globodera pallida* were also infected by this fungus.

The efficiency of *P. lilacinus* in controlling *M. incognita* and *G. pallida* was demonstrated by France *et al.*, (1981) and Jatala *et al.*, (1981). The eggs were deformed by the fungus with the help of diffusible toxic metabolites. Mature females of root knot nematodes were usually penetrated by the fungus through the anus or vulva while cysts of *G. pallida* through the vulva and exposed or broken neck region (Jalata *et al.*, 1985; Jalata, 1986). They reported that hyphal penetration of egg shell is brought about by mechanical pressure and or enzymatic activity. Rate of fungal penetration into egg shell of *Meloidogyne* eggs is faster than in those of *Globodera* and *Naccobus* where the egg shell shows more complexing (Jalata, 1986).

The fungus was also found to be parasitic on the cyst and eggs of *H. glycines* in USA (Morgan Jones *et al.*, 1984; Morgan Jones and Rodriguez Kabana, 1984). The parasitic fungus *P. marguandi* grown on paddy seeds at 2 g per kg soil gave effective control of root knot nematodes infesting crossandra (Khan and Parvatha Reddy, 1994).

P. lilacinus and *Trichoderma* spp. alone or in combination suppressed *Meloidogyne* spp. when incorporated in solarized and non-solarized cardamom nursery beds (Eapen and Venugopal, 1995).

Combined effects of extracts of leaves of neem and calotropis and *P. lilacinus* reduced infection of *M. incognita* in tube rose significantly compared to their individual effect (Nagesh and Reddy, 1995).

Application of *P. lilacinus* at 4 and8 kg/ soil singnificantly reduced the root knot nematode population in the rhizosphere of betel vine (Natal *et al.*, 1995). *M. incognita* and *R. reniformis* were found to be less damaging to *M. incognita* in the presence of *P. lilacinus* (Anver and Alam, 1997).

Integration of neem cake, *Pasteuria penetrans* and *P. lilacinus* gave maximum increase in plant growth and number of seedlings per bed (Reddy *et al.*, 1997). Root dip treatment of egg plant seedlings in formulation of *P. lilacinus* resulted in drastic reduction of population of *M. incognita* (Rao *et al.*, 1998).

The culture filtrate of medium isolates of *P. lilacinus* possessed nematicidal action on juveniles of *M. incognita*. Seventy seven per cent mortality was obtained within 24 hours (Khan and Goswami, 1999).

3.3 Others

The fungus *Cylindrocarpon destructans* (Syn. C. radicicola) has been recorded in cysts of *H. schachtii* (Cursnacs and Tribs, 1974). Eggs of *G. rostochiensis* was also infected *in vitro* by the fungus (Kondakova, 1973). The fungus was found to be most effective in destroying all three stages namely eggs, larvae and cysts of *H. avenae* in a minimum time period (Jain, 1979), Vinduska (1979) recorded *C. destructans* to be an important parasite of *H. schachtii* cysts. In field applications against *H. avenae* a single dose led to 76.7 to 81.6 per cent reduction in the number of infected plants and was further reduced to 93.2 per cent with 2 applications (Kondakoka and Tikhonova, 1981). Choudhary and Kaushal (1984) reported the infections of eggs of *H. avenae* by *C. uniseptatum* in wheat.

Khan and Hussain (1986) reported that females, eggs and juveniles of *M. incognita* were infected by *Fusarium solani*. Goswami and Uma Rao (1997) also reported *Fusarium oxysporum, Aspergillus terreus, Paecilomyces lilacinus, Phoma sorghina* etc infecting eggs of *M. incognita*. They were deformed by thickly covered fungal mycelium.

Aspergillus versiuler, A. fumigation, Cylindrocarpon olidium, F. solani, F. equisetti, F. oxysporum, Acremonium sclerotigmum, A. persicum were reported to infect eggs cysts of *M. javanica* and *H. schachtti* (Salesh, 1990; Al-Hazmi and Razik, 1991). Goswami (1994) also found inhibitory effect of culture filtrates of *A. miner, Trichoderma* spp. and many other fungi.

4. Arbuscular Mycorrhizal Fungi (AMF)

Mycorrhiza, a symbiotic association of fungi and plant roots, is a universal phenomenon throughout the plant kingdom. It is beneficial and even indispensible for the healthy growth of host plants (Sharma and Trivedi, 1994).

The mycorrhizal endophytes as well as phytonematodes in the soil environment are part of a complex interacting system. The dynamics of such microbial systems in the rhizosphere exert measurable effects on the plants by altering the morphology and physiology of the root system. Various aspects of mycorrhizae and nematode interaction have been reviewed by many workers (Hasan and Jain, 1991; Fitter and Sandus, 1992; Jain and Hasan, 1994). The research on biocontrol potentialities of this microbial model system is highlighted in this articles.

1. Importance in relation to nematological work

Mycorrhizal fungi and plant parasitic nematodes are commonly found inhabiting the rhizosphere and colonising the roots of their host plants. These two groups of microorganisms exert a characteristic but opposite effect on plant heatlh. The potenital role of mycorrhizal fungi as biocontrol agents for nematode infection has received considerable attention (Jalali & Chand, 1988; Jalali and Jalali, 1991; Fitter & Sandus, 1992; Franc, 1993, Jain and Hasan, 1994, Sharma and Trivedi, 1994). It has been observed that the plants colonised with mycorrhizal fungi generally grow well inspite of the presence of damaging levels of plant parasitic nematodes and this may be due to improved plant nutrition especially phosphorus (Hussey & Roncadori, 1982) or altered biochemical constituents in the host plant (Sikora & Schonbeck, 1975).

In a nematode susceptible cotton presence of VA mycorrhizae offsets the damage caused by *M. incognita* (Roncadori & Hussey, 1977). Similarly, Soybeans inoculated with *Glomus macrocarpus* and *M. incognita* exhibited fewer galls per gram root in comparison to plants inoculated with nematode alone (Kellam & Schenck, 1980). In another study, tomato roots colonised by *G. fasciculatum* and infested by *M. incognita* or *M. hapla* exhibited fewer and smaller galls than did nematode infected non mycorrhizal plants (Bagyaraj *et al.*, 1979). Small giant cells with few nuclear and retarded nematode development are also observed in mycorrhizal tomato roots infected with *M. incognita* (Sikora, 1979). In addition to nematode penetration and development, there are indications that nematode reproduction may also be suppressed on mycorrhizal plants. Such studies suggest that AMF fungi have potential in reducing plant diseases caused by certain plant parasitic nematodes. However, contrary to the observed beneficial effects of VAM fungi, there are also reports that suggest that presence of VAM fungi may actually help the nematode in reproduction or they may not have any effect on the nematode development. Thus, the nature of interaction between nematodes and mycorrhizae may be antagonistic, neutral or synergistic. Saleh & Sikora (1984), however, attributed these differences to nematode-host combination.

The research on biocontrol potentialities of AMF in different crops is reviewed in this article.

1.1. Vegetables

Bagyaraj *et al.*, (1979) and Manjunath and Bagyaraj (1981) reported that increased spore production by *G. fasciculatum* reduced the number and size of galls produced by *M. incognita* and *M. javanica* in tomato plants.

Sitaramaiah and Sikora (1982) indicated that endotrophic mycorrhizal fungi decreased the penetration of *R. reniformis* on mycorrhizal tomato plants. They also observed delayed development of the gelatinous matrix with few eggs per egg sac on the plants with mycorrhizal inoculation. Sitaramaiah & Sikora (1982) conducted green house testing to evaluate the influence of *G. mosseae* on *R. reniformis* penetration and development on bushbean, cucumber and muskmelon. The reduction in number of nematodes that penetrated mycorrhizal bushbean was 35-41 per cent lower than the control, 4 ans 8 days after nematode attack. A similar reduction in larval penetration was observed on mycorrhizal cucumber and muskmelon 8 days after nematode introduction. These studies have shown that the endomycorrhizal fungus (*G. mosseae*), pre-disposes the host plants causing an increase in plant resistance to nematode infection.

In a study, Suresh & Bagyaraj (1984) reported that mycorrhizal inoculation caused reduced root knot infestation. Such plant had increased quantities of phosphorus, potassium, calcium, total and reducing sugars and amino acids (Phenyl alanine and serine) which alone or collectively played a role in suppressing the development of *M. incognita* in mycorrhizal tomato plants.

Culture of *G. fasciculatum* when inoculated in the nursery protected brinjal plants from root knot nematode infestation. When transplanted in the main field, yield of the plant too was increased significantly (Asha, 1996).

Integration of *G. fasciculatum* with castor cake resulted in significant reduction in root galling and fecundity of *M. incognita* and increased root colonisation in tomato (Rao *et al.*, 1997).

The cultures *G. fasciculatum, G. etunicatum, G. constrictum, G. mosseae, G. monosporum* and *Acaulospora morroweae* resulted in significantly lower number of galls in amaranthus. The population of the nematode in the soil and in the root system was significantly reduced. Among the six cultures, *G. fasciculatum* was superior in protecting the plants from the attack of root knot nematode (Asha, 1998).

1.2. Pulses

Schenck *et al.*, (1975) compared the interaction of three species of VA mycorrhizal fungi *G. heterogama, G. macrocorpus* and *G. margarita* with *M. incognita* on resistant and susceptible soybean cultivars. Generally an increase in root weight resulted in increased nematode populations. However, on susceptible cultivar (Picket) root weight and seed weight were the greatest when plants were inoculated with *G. macrocorpus*, but nematode populations remained low.

Effect of AMF on cowpea has been studied by La Taorraca (1979) who found that VA mycorrhizal inoculation increased yield in cowpea and decreased nematode infestation.

Carling and Brown (1980) observed that colonization of soybean roots by most of the isolates of *Glomus* species significantly increased yield and plant dry weight in low fertility soils and reduced nematode infestation.

Islam *et al.*, (1980) reported that among transplanted seedlings with and without *G. fasciculatum* the growth parameters were highest in the cowpea plants inoculated with VAm fungi in combinations with rock phosphate. Kellam and Schenck (1980) found that *G. macrocorpus* inoculated soybean plants had five galls per g of root weight and increased yield.

Zaki and Bhatti (1987) reported that pigeon pea plants infected with *H. cajani* showed a reduction in the level of all the aminoacids except for aspartic acid which increased in both shoots and roots whereas alanine, glycine and serine increased in the roots only.

Jain and Sethi (1988) reported that inoculation of *V. unguiculata* seedlings with *G. fasciculatum* or *G. epigaeus* one to two weeks prior to inoculation with *M. incognita* or *H. cajani* alleviated the deleterious effect of both nematodes to a considerable extent.

Jain and Sethi (1989) studied the influence of endomycorrhizal fungi *G. fasciculatum* and *G. epigaeus* on penetration and development of *H. cajani* on cowpea. It was found that the introduction of mycorrhizal fungi 15 days before the nematodes, adversely affected root penetration to greater extent than simultaneous inoculation. Over 60 per cent colonization of the root system by VAM fungi considerably hampered root invasions.

Devi and Sitaramaiah (1990) reported that inoculation with *G. fasciculatum* in combination with 'P' increased per cent mycorrhizal root colonization, dry weight of root and shoot, total chlorophyll conent and chemical content of blackgram over uninoculated with 'P' alone.

Glomus fasciculatum significantly reduced juvenile penetration and development of *R. reniformis* on mycorrhizal cowpea plants. Development of gelatinous matrix was delayed and fewer egg mass were produced on the plants (Deepthi, 1993).

Sankaranarayanan and Sundarababu (1994) reported that inoculation of blackgram with *G. fasciculatum* 15 to 20 days earlier than inoculation with *M. incognita* controlled the nematode population and also increased biomass production.

Santhi and Sundarababu (1995) reported that pot mixture soil increased the growth as well as VAM spore production and colonization thereby reducing the nematode population. They also observed the effect of different levels of moisture on the interaction of *G. fasciculatum* with *M. incognita* on cowpea. Maximum plant growth occurred at 70 per cent moisture, growth decreasing at higher levels. A reverse trend was noticed with root weight. Nematode populations were positively correlated with moisture levels while *G. fasciculatum* spore population and colonization showed a negative correlation.

1.3. Fruits

Umesh *et al.* (1988) reported low number of *Radopholus similis* in mycorrhizal treated banana plant.

Multiplication of *M. javanica* was suppressed in peach by *G. mosseae* (Baghel and Bhatti, 1990).

Integration of neem cake @ 400 g/plant, carbofuran @ 20 g/plant, *G. fasciculatum* @ 50 g cultural plant and *P. penetrans* @ 100 g soil was most effective in reducing the population of *R. similis* in banana (Channabasappa *et al.*, 1995).

Vesicular Arbuscular mycorrhizal fungus, *Glomus mosseae* was found effective in reducing the nematode populations especially that of *M. incognita* and *R. similis* infesting banana besides enhancing growth and bunch weight under field conditions. There was 18 per cent increase in height, 2.5 per cent increase in the number of leaves, 30.7 per cent increase in bunch weight, 30 per cent increase in the number of hands, 39.6 per cent increase in number of fingers and 30.1 per cent increase in shoot weight in the mycorrhizae colonized plants when compared to that of control banana plants in the field (Sosamma *et al.*, 1999).

1.4. Spices

A general reduction in galling and nematode population due to AMF colonization was observed in cardamom. However number of root knot nematode population was observation in *G. fasciculatum* treatment. Maximum dry weight of plants and tillering were observed in *A. morrowae* treatment (Sivaprasad, 1995). Root knot nematode infestation and multiplication was generally low in AMF inoculated ginger and turmeric plants. *G. fasciculatum* was most effective followed by *G. mosseae* (Sivaprasad, 1995).

Extensive survey conducted in three major pepper growing districts of Kerala (Kannur, Waynad and Idukki) revealed high AMF colonization and spore count coupled with low population of root knot nematode, *M. incognita* and burrowing nematode, *R. similis* in different pepper cultivars. Pot culture studies indicated improvement in the biometric characters like plant height and dry weight of shoot and root of pepper plants due to inoculation of *G. etunicatum*, *G. mosseae* being more effective in suppressing the nematode population in root (Sivaprasad and Sheela, 1998).

1.5. Others

The reproduction of *H. solanacearum* on tobacco plants was suppressed by inoculation of *Andogene gigantes* (Sox and Spasoff, 1972). The mycorrhizal fungi *G. intradices* and *G. margarita* reduced the severity of yield losses in the susceptible cotton cultivar, Stonevill 213 due to *M. incognita* infestation (Smith *et al.*, 1986).

Chitinase activity was higher in mycorrhizal roots of *Allium porum* L. than in the uninfected controls resulting in effective nematode control. However, 60 to 90 days after inoculation, when the symbiosis was fully established the mycorrhizal roots contained much less chitinase than control (Panu *et al.*, 1989). Varaprasad and Kumar (1991) observed that two hundred juveniles of *H. cajani* and 2 g of *Fasciculatum solani* mycelial pieces in single inoculation significantly reduced nitrate reductase and the glutamine synthetase in cowpea (*V. unguiculata* cv. Pusa).

Combined application of *G. fasciculatum* and Achook, a neem product was very effective in reducing root knot nematode population in fodder cowpea (jain and Hasan, 1995). Sitaramaiah (1995) also reported significantly decrease in the population of *R. reniforms* on cotton when inoculated with *G. fasciculatum*.

2. Commercialisation of nematofagous fungi

Only a few attempts have been done for commercialisation of nematopathogenic fungi. However, in France, two commercial agents, Royal 300, Royal 350, are sold for controlling. *Ditylenchus myceliophagous* on mushrooms and *Meloidogyne* spp. on vegetables respectively. More recently, the facultative parasite, *P. lilacinus* has been produced in Philippines as 'Biocon' which can be applied for the control of several nematode species including root knot nematode.

B. BACTERIA

Among the various antagonists of nematodes known to occur in the rhizosphere of soil, bacteria fall next to fungi. Their potentiality as a biocontrol agent in reducing the population of phytoparasitic nematode is well explored.

On the basis of their mode of action, bacteria have been categorized mainly into two groups. The first groups includes those bacteria which liberate metabolites that have toxic effect on phytoparasitic nematodes. The second group includes those bacteria which parasitise directly on nematodes thereby affecting the penetration, entry and reproduction of nematodes.

1. Bacteria releasing metabolites toxic to nematodes

Several bacteria like *Bacillus* spp., *Agrobacterium radiobacter, Pseudomonas* sp. etc. belong to this group. The first report on bacterial metabolites was published by Johnson (1954). He reported that *Clostridium butryic* produced a mixture of formic, acetic, propionic and butyric acid in culture filtrates that were toxic to *T. martini*. Other compounds like hydrogen sulphate and ammonia produced by bacteria have also been found to have deleterious effect on *Hirschmanniella oryzae* in rice fields and root knot nematodes respectively (Jacq and Fortuner, 1979; Mankau and Minter, 1962; Zavaleta, 1985).

1.1 *Bacillus* spp.

Various formulation of *Bacillus thuringiensis* normally used for insectpest control, have also been evaluated and found toxic to eggs and larvae of *Meloidogyne* species (Prasad *et al.*, 1972).

Ignoffo and Dropkin (1977) reported that a thermostable toxic of *B. thuringiensis* was found toxic to population of *Meloidogyne, Ponagrellus* and *Aphelenchus* and Prevented *M. incognita* juveniles from forming galls on tomaton roots.

Similarly culture filtrates of *Bacillus* species like *B. cereus, B. subtilis* and *B. pumilus* have been found to be toxic to larvae of *Anguina tritici* (Gokte and Swarup, 1986). The effect of *Bacillus* spp. on different host plants of root knot nematode representing the families Piperaceae (betel vine), Malvaceae (bhindi), Solanaceae (tomato), Zingiberaceae (Ginger), Cucurbitaceae (bittergourd), Leguminosae (erythrin-a), Moringaceae (drum-stick) and Labiatae (coleus) were studied. None of these plants showed any symptoms of phytotoxicity, withering or other bacteria diseases (Sheela, 1991). The pathogenicity of the bacteria on higher animals, if any was assessed using chick as the test organism. The tests (subcutaneous injection and oral dose) revealed that

Bacillus spp. were not injurious to birds and hence could be used in INM programmes (Sheela, 1991).

Survey for bacterial antagonists of root knot nematode in nematode infested traditional pepper growing tracts in Kannur, Ernakulam, Idukki, Kollam, Kottayam, Kozhikode, Pathanamthitta, Thiruvananthapuram, Thrissur, and Wynad districts revealed the presence of six species of *Bacillus* viz., *B. circulans, B. macerans, B. coagulans, B. licheniformis, B. subtilis* and *B. pumilus*, The first four were observed in India for the first time and *B. subtilis* and *B. pumilus* were reported for first time in Kerala. Their percentage of occurrence at different stages of life (egg larva and adult female) was estimated. The bacterial spores were seen externally in the eggs and females of the nematode. Nematode count was minimum in bacteria infected sample, the percentage reduction being 35 under natural conditions (Sheela and Venkitesan, 1992).

Detailed *in vitro* studies showed the larvicidal and ovicidal effects of different isolates of *Bacillus* spp. obtained from the egg masses and females of root knot nematode. The larvicidal effect ranged from 70 to 90 per cent and ovicided effect ranged from 60 to 75 per cent. Studies on infected tomato plants revealed a reduction in population of *M. incognita* in soil (60 per cent) and root (70 per cent) (Sheela and Venkitesan, 1992).

The pathogenicity of *Bacillus* spp. was established through pot culture studies. The nematode population characteristics viz., reproductive rate, nemostatic value compared at different periods and the overall performance of the pathogens revealed *B. macerans* and *B. circulans* as potent pathogens. The number of egg per egg mass (egg production) was reduced by 50 per cent (Sheela *et al.*, 1993).

An inoculum level of 1.2 x 10^8 cells of *B. macerans* per pot significantly reduced root knot nematode population in soil and root of pepper significantly improving the length of vine number of leaves, weight of shoot (fresh and dry) and root (Sheela and Venkitesan, 1994). Toxic effect of *B. cereus* on juveniles and eggs of *Meloidogyne javanica* was also reported by Oka *et al.*, (1995). He isolated several bacterial form *M. incognita* egg masses and *Heterodera zeae* cyst and reported that liquid cultures of *Bacillus* sp., *Corynebacterium* sp., *Serratia* sp. and *Arthrobacterium* spp. at a concentration as low as 0.1 per cent inhibited hatching of *M. incognita* eggs. The cultural filtrates at a concentration of 0.6 per cent was highly toxic to juveniles of *M. incognita, R. reniformis* and *T. semipenetrans*. The same bacterial species were found to have suppressive effect and it ranged between 46 to 96 per cent for *M. javanica* and 36 to 100 per cent for *R. reniformis, infesting* sunflower. The carryover effect of the bacterial isolates continued upto 15 months at which the reduction in nematode population was upto 91 per cent for *R. reniformis* and 5 to 88 per cent for *M. javanica*. Bacteria (*Bacillus* sp.) could be reisolated from soil 15 months after inoculation (Ali, 1996).

Two bacterial insecticide strains of *B. thuringiensis* (Dipel and SAN 415) were evaluated for their bioefficacy against *M. javanica* and *T. semipenetrans* in tomato. Both the strains suppressed the total nematode population and egg masses of the two nematode genera. Besides, the two strains of *Bacillus*. remarkably reduced the percentage of hatchability of eggs of the two nematode genera. SAN 415 was more effective than Dipel and the effect was approximately similar to the standard nematicidal nemacur (Osman *et al.*, 1988). Another isolate of *B. thuringiensis* (CR-371) was found effective in reducing galling by root knot nematode on tomato under green house and field environment (Zuckerman *et al.* 1993). They also reported smaller populations of *Pratylenchus penetrans* in roots of strawberry plants treated with CR-371. Sharma (1994) compared the efficacy of two strains of *B. thuringiensis* viz., Btt and Bti against *M. incognita*. He found 66 per cent nematode control by Btt and 53 to 65 per cent with Bti respectively.

1.2 *Pseudomonas*

The plant growth promoting rhizobacteria *Pseudomonas fluorescens* also controls nematodes effectively. A reduction in penetration by *H. schachtii* and significant increase in yield of sugar beet as seed treatment with selected strains of *P. fluorescens* was observed by Ostendrop and Sikora (1989). It also reduced the root-knot nematode population in soil as well as roots of brinjal (Sheela, *et al.*, 1999).

Application of *P. flurorescens* as seed treatment at a dosage of 10 g/kg of seed was effective in reducing the infestation of *H. gracillis* in rice. Maximum bacterial colonisation and nematode suppression was observed in plants treated with *P. fluorescens* as seed treatment and it increased yield of rice by 13 per cent (Ramakrsihnan *et al.* 1998).

Severity of root knot infection in grape vine root was significantly reduced by application of *P. fluorescens*. The extent of root colonisation was significantly better at all dosage levels on the yield also. The effect of treatments when studied in the second crop, 8 months after treatment also showed effective suppressions of nematode multiplication (Shanti *et al.*, 1998).

1.3 Others

Plant health promoting rhizobacteria (PHPR) namely *Agrobacterium radiobacter* and *Bacillus sphaericus* have also been found to roduce toxic metabolites affecting penetration of *Globodera pallida* by 41 per cent in potato roots and improving yield (Racke and Sikoru, 1992).

2. Bacteria parasitising nematodes

2.1 *Pasteuria* spp.

Among the various bacteria parasitising nematode, *Pasteuria penetrans* is the most specific obligate parasite. Two species namely *P. penetrans* (Thorne, 1940, Sayre, 1988) and *P. nshizawae* (Sayre *et al.*, 1991) have been identified to parasitise different groups of nematodes. Initially described as microsporadian, *P. penetrans* has showed great promise as a potential significant biocontrol agent. Its life cycle remarkably synchronises with the development of its host (Mankau and Imbriani, 1975) except in a few isolates where the bacterium completes its life cycle in second stage of juveniles itself.

The effectiveness of *P. penetrans* in controlling *Meloidogyne* spp. is quite well established and its ability to prevent nematode reproduction is an ideal characteristic of any biocontrol agent. Various degrees of control of root knot and cyst forming nematodes exerted by the bacterium have been achieved by several investigators (Mankau 1980; Nishizawa, 1984; Sharma, 1992; Singh and Dhawan, 1994; Ciancio and Baurijate, 1995).

Pot cultures of *Meloidogyne* populations came near to extinction within 4-5 generations when treated with *P. penetrans* (Mankau, 1980). Infection of *Meloidogyne* appears to be directly proportional to the rate of application of the bacterium and in a dose of 100 mg/kg of soil all juveniles were infected in two hours (Stirling and Wachtel, 1980).

Cucumbers grown in *M. incognita* infested microplots gave consistently higher yields in spore treated than in non spore treated plots. Parasitism was so effective that yield differences were insignificant between treated and untreated controls (Sayre, 1984). Stirling (1984) also recorded no difference between bacteria treated and nematicide treated soil. Field trials with the bacterium also gave significant increase in the yield of crops (Brown *et al.*, 1985). Page and Bridge (1985) reported complete destruction of glass house *M. acronea* cultures by the bacterium.

P. penetrans is effective in natural suppressions of nematodes especially the root knot nematodes where monoculture or continuous cropping with nematode susceptible crops. Thus, effective natural suppression of *M. incognita* in old plantations of grape vine has been reported from Australia and India. Similarly, appreciable natural suppression of *Meloidogyne* spp. took place in pockets growing the nematode susceptible vegetables continuously in Africa, preventing yield loss.

Repeated cultivated of susceptible crop tend to increase the inoculum of bacterium over time from the relatively low levels of spore added initially in the soil (Ostendrop *et al.*, 1991; Singh and Dhawan, 1992; Kausumimoto *et al.*, 1993). The potential of *P. penetrans* for managing root knot nematode in different crops have been reported (Daudi and Gowen, 1992; Rao *et al.*, 1998; Rangaswamy *et al.*, 1999).

2.2 Others

Azotobacter significantly reduced root knot nematode infestation in Okra (Amalendu Chatterjee, 1995).

C. PREDATORY NEMATODES

Nature has several inbuilt mechanisms to regulate and maintain the population of different forms of life. Predation is one such mechanism operating to maintain the stability of the ecosystem. In nematode control too predation is in no way a unique phenomenon. The predatory nematodes can play a significant role in nematode management if given suffcient attention.

Most predatory nematodes depending on their mode of feeding and type of feeding apparatus belong to three groups viz., *Mononchida* (e.g., *Clarkus papillatus*, *Lotonchus blachylaimus*, *Mononchus aquaticus* etc.), Dorylaimida (e.g., *Actinolaimus* sp., *Monochoides potohikus* etc.). The obligatory form *Seinera* sp. belongs to *Aphelenchoida. Mononchus* feed on nematodes, rotifers, protozoans, algae and fungal spores while members of Dorylaimida are largely omnivorus and feed on fungi, algae, nematodes, enchytraeids and other microfauna.

Predatory nematodes have either a large open stomata armed with teeth or denticles used in seizing and tipping open the prey or swallowing them whole e.g., *Mononchus* sp., *Butlerius* sp., *Anatonchus* sp., *Diplogaster* sp., *Tripyla* sp.) or a pharynx equipped with on onchiostylet or stomato stylet for piercing the prey (e.g., *Seinura* sp.). *Seinura* spp. can kill even larger nematodes than their own body size due to its ability to paralyse the prey with the help of secretions from the dorsal oesophageal gland. The chemical is so strong that the prey is immediately immobilized after insertion of the stylet into its body.

1. Mononchus

Cobb (1917) in his classical work on mononchus had suggested the possibility of using predatory nematodes in controlling populations of plant parasitic nematodes. Steiner and Heinley 91922) had found that a single *C. papillatus*. killed 1322 juveniles of *H. radicicola* in one week and accrodingly thought of using them in controlling the cyst. Similarly, Thorne (1927) suggested the use of *C. papillatus*, against the sugar beet nematode. Cassidy (1931) observed that the mononch species, *L. blachylaimus, C. papillatus* and *Hnatonchus tridentatus* predated upon the eggs and adults of nematodes and hence may be of use in their biological control. Esser (1963) had also speculated on the possibilities of using Mononchs for controlling nematodes. *Mylonchulus sigmaturus* was used by Cohn and Mordechai (1974) to keep in check the population levels of citrus nematodes *T. semipenetrans*. Severals other workers have made contributions to the study of predation by Mononchus (Mulvey and Jonson, 1961; Esser, 1963; Nelmo, 1974; Grootaert and Maurtens, 1976, Small and Grootaer, 1983).

Small (1979) found a significant reduction in the population of *G. rostochiensis* in the presence of *Prionchulus panctatus*. Jairajpuri (1979) demonstrated that predation by *M. detritatus* depended on the type of prey and the physiological state of predator and that at low prey density or in the absence of prey and predator often restorted to cannibalism, an observation that has been corroborated by studies in *M. aquaticus* by Bilgrami and Jairajpuri (1984). Further studies on *M. aquaticus* showed the predation by this nematode is determined by the activity of prey, temeprature and other experimental factors (Bilgrami *et al.*, 1983). Azmi (1983) has clearly indicated that with increase in the population of *F. monhystera*, a gradual reduction took place in the population of *H. dihystera* and after some time an equilibrium was maintained between the prey and predator population.

Very significant and interesting ecological study was made in the relationship between the population levels of *Parabadranchus shakili* and *Trichodorus* sp. and *Hemicriconemoids* sp. (Ahmed and Jairajpuri, 1984). They found that higher predator populations were associated with low populations of the species of *Trichodorus*, *Thornenema* and *Hemicriconemoides*. The predatory nematodes *T. tenuicaudatus* was observed to be a potential biocontrol agent in the management of *T. semipenetrans* and *H. dihystera*.

2. Dorylaimid

The Dorylaimid nematodes dominate all other soil inhibitory nematodes, both in number of species and in population levels. Based on the studies conducted on the behaviour of Dorylaimids it has been found that comparatively only a few species of this group belonging to the Longidoridae are phytoparasitic, some are suspected parasites such as *Enchodelus, paraenchodelus, Basirotyleptus, Colifornidorus*, while majority of Dorylamida especially those belonging to the superfamilies Actinolaimidea, Nygolaimoidea, Dorylaimoidea are predaceous (Jairajpuri and Ahmed, 1992). Recent studies particularly those carried out by Jairajpuri and his co-workers at AMU, have proved that there is good prospect for employing these animals in the bio control of plant nematodes.

Thorne (1930) observed the predaceous nature of feeding by species of Dorylaims. The studies by Linford and Oliviera (1937) showed predation by *Aporcelaimus, Nygolaimus, Sectonema, Labronema, Dorylaimus* and *Actinolaimus*. Esser (1963) also found species of Dorylaims to be predaceous including those of *Carcharolaimus*. More studies on Dorylaimid predators were made by Linford and Oliveira (1937), Hechler (1963), Esser (1963 and 87), etc. Feeding behaviour of *Labronema vulvapapillatum* was observed by Wyss and Grootaert (1977). In India, the predatory behaviour of *Aquatides thornei* was studied for the first time by Bilgrami *et al.*, (1985). Aggregation of predators was observed in *Dorylaimus stagnalis* and *Aporcelaimellus nivalis* (Shafqat *et al.*,

1987; Khan *et al.*, 1991) around prey. The predation was affected by various factors such as temperature, prey density type of prey, starvation period etc. Jairajpuri and Bilgrami (1990) and Jairajpuri (1990) have expressed the opinion that these groups have many more species which would be efficient predators particularly those belonging to the genera *Eudorylaimus, Mesodorylaimus, Aporcelaimus, N. ygolaimus, Sectonema* etc. Recently a comprehensive study in the predation abilities of two species of *Dorylaims* was made, viz., *Allodorylaimus americanus* and *Discolaimus silvicolus* (Khan *et al.*, 1995). The predators aggregated around injured prey and showed chemoattraction towards excised prey.

3. Diplogasterid

The rapid rate of multiplication ability of perceive prey attactants, as also the indication that the high rate of predation, high strike rate against plant parasitic nematodes, short life cycle, has made them an ideal candidate for the biocontrol of nematodes. Yeats (1969) carried out excellent work on the predatory abilities of *Monochoides potolhikus*. Grootaert (1976), Goortaertfeted (1977) and Small and Grootaert (1983) did same work on the predation by mesodiplogaster and *Butlerius* sp. At AMU valuable research was conducted on two species of *Mononchoides* viz., *O. longicaudatus* and *D. fortidens* (Bilgram and Jairajpuri 1988, 89). These studies conclude that diplogasterid nematodes are very effective predators.

4. Other agents

Framentary evidence indicate that several other agents attack nematode. Several reports are there on tradigrades devouring nematodes (Esser, 1963; Esser and Sobers, 1964; Hetchinsonan Streu, 1960). Tardigrades are usually polyphagous and in certain microhabitats, nematodes serve as a partial food source for tardigrades. Observations indicate that tardigrades feed on nematodes nonselectively. Hutchinson and Streu (1960) found tardigrades feeding on *Trichodorus acqulis* and *Tylenchus* sp.. *Hypsibius myrops* significantly reduced populations of *M. incognita, D. dipsaci* and *Panagrellus redivius* (Sayre, 1969).

Mites too have been seen feeding on *Meloidogyne* (Linford and Oliveira, 1938) and *Heterodera cysts* (Murphy and Doncaster, 1957). Rodriquez *et al.*, 1962) found that three species of nematodes were prey to the mite *Macrochetes muscaedomesticae*. Schaerffenberg and Tendl (1951) reported control of *H. schachtii* by enchytraeids. An undescribed species of turbellaria was found to ingest nematodes (Sayre and Sowers, 1966).

Scattered informations are also available on insects feeding on nematodes. Esser (1963) observed a dipterous larva devouring *Belonolaimus*. Brown (1954) reported that a spring tail, *Isotoma* sp. ingested nematodes. Murphy and Doncaster (1957) reported that the Collembola *Onychiurus aramtus* was a voracious feeder of *H. cruciferae*. The insect was also seen to feed on *Dorylaimus* and *Mononchus*. Other collembola seen to damage cysts of *H. cruciferae* to varying degrees were *Hopogastrura* sp. *Isotoma veridis* and *Orchesella villosa*.

The only published report of nematodes being attacked by viruses is that of Loewenberg *et al.*, (1959). These authors found that juveniles of *M. incognita* became sluggish in their movements and galls were not subsequently produced.

Several types of protozoans are known to parasitize nematodes. Some species of protozoan parasitize nematodes by forming resistant cysts (Coxidicene) while others by aggregation around the excretory pore and vulval opening and culturing finally into the body of the nematode (Flagellates). A flagellated protozoan, *Bodo caudatum* which was seen on dead *D. dipsaci* (Canning, 1973) holds some promise as an agent of biocontrol in that it can enter the nematodes through the excretory pore. *Theratomyxa weberi* was observed feeding on *G. rostochiensis* larvae by Weber *et al.*, (1952).

Probably more of such other natural control mechanism is in operation in the soil. Relatively, little is known about such agents. Though it is doubtful if these agents can be exploited for practical use, work along this like could prove interesting.

D. FUTURE PROSPECTS

Over reliance on chemicals leads to problems of residues in crops. Hence priority should be given to developing sustainable methods like biological methods to manage nematodes. Intensive and systemic surveys should be conducted to identify potential biocontrol agents for nematode pests. Research efforts need to be directed in standardizing techniques for the mass multiplication of the identified biocontrol agents.

Methods and strategies for introducing and monitoring a population in the soil and exploiting its antagonistic potential needs to be investigated. Formulation of these into suitable products that allow easy application to soil should also be given importance.

Possibility of altering host specificity and improving virulence of suitable biocontrol agents through biotechnological interventions can be explored. More of the studies on biocontrol agents are confined to green house conditions. The possibility of using these against nematodes should be tested under field conditions. If found effective, use of this should be popularized among farmers through demonstration trials and training programmes.

Organic amendments/inorganic wastes are ideal components in INM. Not only will the organic amendments reduce the nematode population through its inherent properties, but also will act as an ideal medium for the multiplication of the bioagents, thus imparting long term protection. So specific organic amendments that may stimulate the microflora antagonistic to nematodes should be identified. Thus an ideal bio approach for managing the phyto nematodes would be the application of a potential bioagents along with its specific organic amendment.

REFERENCES

Anver, S. and Alam, M. 1997. Control of *Meloidogyne incognita* and *Rotylenculus reniformis* singly and concomitantly on Pigeon pea with *Paecilomyces lilacinus*. Indian J. Nematol. 27 : 209-213.

Asha John, 1996. Phytochemicals and VAM for management of nematodes in Brinjal (*Solanum melongena* L.) M.Sc. (Ag.) thesis, Kerala Agricultural University.

Deepthi, N.P. 1993. Controls of collar rot and root knot of cowpea with VA mycorrhizae. *M.Sc. (Ag.) thesis* Kerala Agricultural University, p. 109.

Khan, M.R. and Goswami, B.R. 1999. Nematicidal effect of culture filtrates of *Paecilomyces lilacinus isolates* on *Meloidogyne incognita. Indian J. Nematol.* 29 : 145-148.

Lue, M., Sikora, R.A. and Bridge, J. (eds.) 1990. *Plant Parasitic Nematodes in Subtropical and Tropical Agriculture.* C.A.B. International. pp. 1-615.

Nematological Society of India 1995. Nematode Pest management. An Appraisal of Eco friendly Approaches. Edts. Swarup, G., Dasgupta, D.R. and Gill, J.S.

Ramakrishnan, S., Sivakumar, C.V. and Poornima, K. 1998. Management of Rice Root nematode, *Hirschmaniella gracilis* (deman/LISC and Goodey with *Pseudomonas fluorescens* Migula. *Proc. Third Intl. Symp. Afro Asian Soc. Nematologists (TISAASN)*, p. 194-198.

Rangaswamy, S.D., Reddy, P.P. Nagesh, M. and Nanjegowda, D. 1999. Integrated management of *M. incognita* using *Pasteuria penetrans, Trichoderma viride* and oil cakes in tomato. *Pest Magmnt. Hortl. Eco-system* 5 : 122-126.

Rao, M.S. Reddy, P.P. and Nagesh, M. 1998. Evaluation of *Paecilomyces lilacinus* cultured on neem cake extract for the management of root knot nematode on egg plant. *Pest Magmnt. Hortl. Eco system* 4 : 116-119.

Reddy, P.P., Nagesh M. and Devaffa, V. 1997. Effect of interaction of *Pasteuria penetrans, Paecilomyces lilacinus* and neem cake for the management of root knot nematode infecting tomato. *Pest Magmnt. Hortl. Eco system* 3 : 100-104.

Saxena, S.K., Khan, M.W., Rashid, A. and Khan, R.M. (eds.) 1990. *Progress in Plant Nematology*. CBS publishers and Distributes Pvt. Ltd. pp. 1-583.

Shanthi, A., Rajeswari, S. and Sivakumar, C.VB. 1998. Soil Application of *Pseudomonas fluorescens* (Migula) for the control of root knot nematode (*M. incognita*) on grapevine (*Vitis vindera* Linn). Proc. Third Intl. Symp. Afro Asian Soc. Nematologists (TISAASN) p. 203-206.

Sharma, R. and Swarup, G. 1998. Pathology of Cyst Nematodes. Malhotra Publishing House, New Delhi pp. 1-84.

Sheela, M.S., Jiji, T. and Premila, K.S. 1999. Management of root knot nematode through the plant rhizosphere bacterium *Pseudomonas fluorescence* Migula. *Nat. Seminar Nematol Res. in India. Abstr.* p. 33-34.

Sivaprasad, P. and Sheela, M.S. 1998. Distribution of phytonematodes and VAM fungi in the rhizosphere of pepper and its impact on nematode management. *Proc. Third Intl. Symp. Afro Asian Soc. Nematologists (TISAASN)* p. 102-105.

Sosamma, V.K., Koshy, P.K. and Samud, R. 1998. Effect of VAM, *Glomus mosseae* on nematode infesting banana under coconut based high density multi species cropping system. *Proc. Third Intl. Symp. Afro-Asian Soc. Nematologists (TISAASN)*, p. 94-97.

CHAPTER - 12

CURRENT TRENDS IN BIOLOGICAL CONTROL OF NEMATODE PESTS IN INDIA

M. Nagesh and P. Parvatha Reddy

INTRODUCTION

According to recent WHO estimates approximately 7.5 lakh people are taken ill every year with pesticide poisoning and up to 14,000 of these die in agony. Although third world uses one sixth of the total pesticides produced globally, at least 37,500 people are poisoned yearly and 10,000 of them fatally. In the USA, some three lakh farm workers are affected by pesticides related illness such as dizziness, vomitting, pinpoint pupils and severe skin rashes.

There is a considerable pressure from public and environmental scientists to reduce emphasis on chemical control and the need to switch over to biological control of plant pests and diseases. Pesticides are necessary at present, but are not a long term solution to crop health. Besides, their effects on non target organisms and hazardous nature, petroleum based pesticides will become more expensive. Moreover, high residues of pesticides are increasingly being found in products of agriculture. Biological control alone cannot replace the use of pesticides entirely since the effectiveness of biocontrol is dependent on several factors. However, research has shown that biological control can be very efficiently combined with pesticides in an integrated manner so as to bring down the use of pesticides by half or even lower. In many instances pesticides are lossing their effectiveness because of development of resistance and resurgence in target organisms. Breeding for resistance, which continues to be the most practical and feasible method for controlling plant diseases, pests and plant parasitic nematodes, is not able to keep pace with the development of more virulent strains of these crop pests. Another disturbing factor is the presence of spurious chemicals/ pesticides or formulations of bioagents in the market. Since no proper quality control regulations or guidelines are existing at present, it is imperative that the government enact rules for quality control parameters to be followed strictly by producers of biocontrol agents and synthetic chemicals.

Biological control of plant parasitic nematodes has become most significant and intensive area of research of late because of environmental concerns and non availability of effective nematicides. Among the various antagonists of plant parasitic nematodes, the parasitic fungus, *Paecilomyces lilacinus*, is the most important and promising bioagent that has wide nematode host range, wide occurrence, and amenability to survive and parasitize in different agro-climatic situations. Further formulations of *P. lilacinus* have a long shelf life which makes the production of the fungus commercially viable.

In this chapter, the current trends and future needs of biological control using potential biocontrol agents such as bacteria (including actinomycetes) and fungi (including VAM fungi) in the sustainable management of plant parasitic nematodes has been discussed along with the possibilities of their integration with eco-friendly components into overall nematode management strategies.

A. BACTERIA

A large number of rhizobacteria are known (Table 1) to reduce nematode populations and important genera include *Agrobacterium, Alcaligenes, Bacillus, Clostridium Desulfovibrio, Pseudomonas, Serratia* and *Streptomyces*. Application of some of these bacteria has given very promising results (Siddiqui & Mahmood, 1999).

The bacterial antagonists studied were mainly *Pseudomonas fluorescens, P. stutzeri, B. subtilis, B. thuringiensis, B. cereus, B. licheniformis* and *P. mindocina* against root knot cyst, citrus, reniform, rice white tip nematodes in vegetable, citrus, grape vine, spices, black gram, chick pea, pigeon pea, wheat and paddy crops (Table 2). These antagonists effectively reduced nematode populations in different crop rhizospheres. These bacteria were either given as soil amendments or as seed treatment before planting. Further different doses in terms of colony forming units were employed in the form of talc or aquaceous formulations.

The bacterial antagonists depending on mode of action may be classified into two groups. First group includes the bacteria which release metabolites, inhibitory to nematodes (*Bacillus, Clostridium, Pseudomonas, Azotabacter* etc.). The metabolites released by soil microorganisms may be toxic, antibiotic, or inhibitory to nematodes. The second groups includes the bacteria which directly parasitise the nematodes and their life cycle stages such as *Pasteuria*.

The exact mode of action against nematodes by these bacteria were not studied extensively although, it is speculated that these bacteria generally induce pathogenicity related proteins in (bacteria) colonized roots, root exudates antagonistic to nematodes and other plant pathogens, lectin related compounds or siderophore productions.

Bansal *et al.*, (1999) examined eight strains of *Azatobacter chroococcum* to explain the mechanism of its antagonistic effect. Six strains showing different properties like nitrogen fixation, ammonia excretion indole acetic acid (IAA) production and siderophore production, reduced cyst formation from 6 to 60 per cent. Strain 1 (1) and Mac 27 having high siderophore production did not control nematode infection but high temperature

Table 1. Rhizobacteria evaluated for the control of phyto parasitic nematodes.

Biotic agent	Nematode spp.	Crop	Location	References
Bacillus licheniformis, Pseudomonas mindocina	*M. incognita*	—	AMU, Aligarh	Siddiqui & Husain, 1991
Bacillus subtilis, B. pumilis, B. cereus, Pseudomonas spp.	*M. incognita, H. cajani, H. zeae, H. avenae*	—	IARI, New Delhi	Gokte & Swarup, 1988
P. fluorescens	*H. avenae*	Wheat	IARI, New Delhi	Anju Kamra et al., 1999
P. fluorescens	*M. incognita*	Tomato	—	Verma *et al.*, 1999
P. stutzeri, B. thuringiensis	*M. incognita*	Tomato	—	Khan & Tarrum, 1999
B. subtilis, P. fluorescens	*M. incognita*	Tomato	TNAU, Coimbatore	Santhi & Siva Kumar, 1995
P. fluorescens	*M. incognita*	Black pepper	IISR, Calicut	Eapen *et al.*, 1997
P. fluorescens	*H. cajani*	Pigeonpea	AMU, Aligarh	Siddiqui *et al.*, 1998
Bacillus subtilis, Bradyrhizobium japonicum	*H. cajani*	Pigeonpea	AMU, Aligarh	Siddiqui & Mahmood, 1995a
P. fluorescens	*Hirchmaniella gracilis*	Paddy	TNAU, Coimbatore	Ramakrishnan & Siva Kumar, 1999
B. subtilis	*M. incognita*	Chickpea	AMU, Aligarh	Siddiqui & Mahmood, 1995b.
P. fluorescens	*H. cajani*	Blackgram	TNAU, Coimbatore	Latha & Siva Kumar, 1998
P. fluorescens	*Tylenchulus semipenetrans*	Sweet orange & lime	TNAU, Coimbatore	Santhi et al., 1999
P. fluorescens	*H. avenae*	Wheat	—	Basal et al., 1999
P. fluorescens	*M. incognita*	Potato	TNAU, Coimbatore	Mani et al., 1998
P. fluorescens	*Tylenchulus semipenetrans*	Acid lime	IIHR, Bangalore	Parvatha Reddy and Nagesh, 2000

Table 2. Rhizobacteria evaluated against plant parasitic nematodes and employed in field.

Nema spp.	Crop	Biotic agent	Dose	Application method	Reference
M. incognita	Tomato	*B. subtilis, B. thuringiensis, Pseudomonas stutzeri*	—	Soil amendment	Khan & Tarannum, 1999
Hirschmanniella gracilis	Paddy	*Pseudomonas strain* Pf-1	10 g/kg seed	Seed treatment	Ramakrishnan & Sivakumar, 1999
Tylenchulus semipenetrans	Citrus sinensis, Citrus limon	*P. fluorescens*	3 x 10^9 cfu/tree, 6 x 10^9 cfu/tree talc formulation	Soil amendment	Santhi *et al.* 1999
Hirschmanniella gracils	Paddy	*P. fluorescens* strain Pf-1	10 g/kg seed	Seed treatment	Ramakrishnan *et al.*, 1998
Tylenchulus semipenetrans	Acid lime	*P. fluorescens*	3 x 10^9 cfu/ plant	Soil amendment	Parvatha Reddy *et al.*, 2000
Golden Cyst nematode	Potato	*P. fluorescens*	10-20 kg/ha (15 x 10^8 cfu/g)	Soil amendment	Mani *et al.*, 1998

resistant mutant HT54, possessing the highest ammonia excretion ability and negligible IAA production, reduced nematode infection by 48 per cent. They concluded that ammonia but not siderophores or IAA had a role in the control of *H. avenae* infection in wheat.

The cell free culture filtrate of *P. fluorescens* was found to be toxic to *H. avenae*. Emergence was completely inhibited in the standard filtrate (S). It was significantly reduced in the S/10 and S/100 dilutions of the filtrate where the cumulative number of juveniles after 96 h were 33 and 120, respectively compared to 340 in distilled water. The S/1000 dilution has a stimulatory effect on emergence; 570 juveniles emerged after 96 h. Emergence resumed in cysts that were placed in S/100 and S/1000 dilutions but was significantly reduced in the cysts kept in the S and S/10 dilution, when replaced with distilled water. The standard filtrate (S) showed 100% mortality within 24 h and the subsequent dilutions of S/4, S/8, S/12, S/16 and S/20 caused an average mortality of 94.7, 76.6, 72.3, 57.0 and 56.6%, respectively. (Anju Kamra *et al.*, 1999).

1. Combinations

Endomycorrhiza, *Glomus mosseae* and antagonists - *P. lilacinus* and *Pseudomonas fluorescens* were used alone or in combination to control the wilt disease combination pigeonpea plant growth, nodulation, phosphorus contents and reduced nematode multiplication and wilting in infected plants. Treatment with *G. mosseae* or *P. lilacinus* caused similar increases in plant growth and nodulation. The use of *P. fluorescens* against *F. udum* caused a greater increase in plant growth and nodulation compared with its application against *H. cajani*. Greater growth and nodulation were observed when all the three antagonists were used together. The use of antagonists on plants inoculated with both pathogens caused a greater increase in plant growth and nodulation compared with plants inoculated with a single pathogen. Treatment with *G. mosseae* increased phosphorus content both in pathogen - inoculated and uninoculated plants whereas other antagonists increased the phosphorus content only in pathogen-inoculated plants (Siddiqui *et al.*, 1998).

2. *Pasteuria penetrans*

Pasteuria penetrans is known to parasitize 205 nematode species belonging to 96 genera from 10 orders of the phylum nematoda. It occurs widely (recorded from 51 countries in 5 continents) and is a potential biocontrol agent. It is a mycelial endospore forming bacterium which is an obligate parasite of large number of nematodes with a life cycle remarkably synchronizing with the development of its host. Infection occurs through direct penetration of the cuticle by germinating spores that encumber to the surface of the nematodes. The life cycle begins with the vegetative mycelial colonies which eventually fill the nematode body with a large mass of bacterial endospores. Sporangia are liberated in the soil from the body of the nematode upon decomposition.

So far three species belonging to the genus *Pasteuria* viz., *P. penetrans* (ex-thorne, 1940) Sayre & Starr, 1985, *P. thornei* star & Sayre, 1988 and *P. nishizawae* Sayre *et al.*, 1991 have been known to parasitize nematodes, *Pasteuria penetrans* has been subject of intensive studies. It was first described as a sporozoan. *Dubosquia penetrans* (Thorne 1940). Mankau (1975) transferred it to *Bacillus penetrans* and finally redesignated as *P. penetrans* by Sayre & Starr (1985). Infection occurs through direct penetration of the cuticle by germinating spores. The vegetable mycelial colonies fill the nematode body with large mass of endospores. Sporangia are liberated from the body of the nematode into soil. Duration of life cycle of parasite from endospore to endospore varies with species and temperature. A number of strains of *P. penetrans* are reported. The populations dynamics in soil is not clearly understood. Bacteria are ubiquitous and destroy nematodes in virtually all soils because of their constant association in the rhizosphere. Bacteria like *Pasteuria penetrans* destroy nematodes by their parasitic behaviour while the non-parasite rhizobacteria reduce nematode populations by colonizing the rhizosphere of the host plant.

In India several studies were carried out on *P. penetrans* against several economically important nematodes in different crops (Table 3). The occurrence of a new strain of *Pasteuria penetrans* on *H. cajani* was reported by Bansa Singh & Dhawan (1990). Spore attachment was only observed on the cuticle of second stage juveniles of *H. avenae*, while *H. mothi*, *R. reniformis*, *T. semipenetrans*, *P. thornei* were free of spores. Significantly greater attachment was at 25°C than at 15 or 35°C. The life cycle of the bacterium was completed in 49 days at 10-17°C (Dolly Bhattacharya & Gopal Swarup, 1988). Bio efficacy studies showed that it is effective against root knot, citrus, cyst and burrowing nematodes (Table 4).

Pasteuria penetrans reduces root penetration by the nematode juveniles, inhibits egg formation, reduces nematode population in soil and results in increase in yield. The bacterium being resistant to heat, desiccation its favorable compatibility with nematicides and its ability to survive for more than 2 years in soil, qualifies it as

Table 3. Reports of *Pasteuria penetrans* studies in India

Antagonist	Host nematode Species	Crop	Reference
Pasteuria penetrans	*M. incognita*	Brinjal	Nagesh et al. 1996
P. penetrans	*M. javanica*	Tomato	Maheswhari et al. 1988
P. penetrans	*M. incognita*	Tomato	Parvatha Reddy et al., 1997
P. penetrans	*M. incognita*	Blackpepper	Sosamma and Koshy 1995
P. penetrans	*M. incognita*	Brinjal	Singh et al., 1998
P. penetrans	*Heterodera cajani*	Cowpea	Singh and Dhawan, 1996
P. penetrans	*Heterodera cajani*	Pigeonpea	Mohinder Singh et al., 1996
P. penetrans	*M. incognita*	Sunflower	Devappa et al., 1997
P. penetrans	*H. aveane*	Wheat	Singh and Dhawan, 1990
P. penetrans	*Rotylenchulus reniformis*	—	Sharma and Sharma, 1989
P. penetrans	*Heterodera spp.*	—	Bhattacharya and Swarup, 1989
P. penetrans	*Meloidogyne spp.*	Grapevine	Mani *et al.*, 1998
P. penetrans	*Radopholus similis*	Banana	Vidya and Reddy, 1998
P. penetrans	*T. semipenetrans*	Acid lime	Parvatha Reddy and Nagesh, 2000

Table 4 : Bioefficacy of *Pasteuria penetrans* in different crops

Nematode species	Crop	Dose	Glass house/ Field/Pot	Reference
M. incognita, *M. javanica*	Tomato	2.5 g/kg soil	Glass house	Soma Shekhar and Gill, 1991
M. incognita	Brinjal	—	Glass house	Singh et al., 1998
M. incognita	Pigeonpea	—	Glass house	Mohinder Singh et al., 1996
M. incognita	Sunflower	924 g/m^2	Field	Devappa *et al.*, 1997
M. incognita	Tomato	1 x 10^5 spores/g soil	Pot	Walia and Dalal, 1994

the most potential bacterial biocontrol candidate which could be used on a large scale under field conditions. However, the obligate nature of the organism, its lack of mobility and dependence on dispersal agents and host specificity are its major drawbacks. Nagesh and Parvatha Reddy (1998) modified the method of mass production of *P. penetrans* under glass house conditions and achieved higher recovery of bacterial spores.

B. FUNGI

Interaction between fungal antagonists and nematodes have been known to occur in agricultural soils for many years (Mankau, 1980). Nematode destroying fungi play an important role in regulating nematode population dynamics. Recent studies have indicated that the existence of opportunistic soil fungi which are capable of colonizing females and eggs of nematodes and deleteriously affecting them. Nematodes belonging to such genera as *Meloidogyne*, *Tylenchulus* and *Rotylenchulus*, at sedentary stages of their life cycle, are vulnerable to attack by fungi either within host plant roots or when females and egg masses are exposed on the root surface or whithin soil. Fungal growth is known to be enhanced in the rhizosphere, increasing the likelihood of fungal invasion of nematode structures present in the ecological niche. Once in contact with eggs, larvae or females, the fungi grow rapidly and eventually parasitizes all the eggs that are in the early embryonic development stages. Several fungi such as *Paecilomyces lilacinus*, *Verticillum lecanii*, *V. chlamydosporium*, *Trichoderma harzianum*, *T. viride* and *Gliocladium virens* have been reported to parasitize eggs of plant parasitic nematodes.

The nematophagous fungi cause pathogenicity to phyto-parasitic nematodes and the stages of their life cycle either by directly parasitising the eggs, egg masses, females etc., or by releasing toxins/ substances/secondary metabolites inhibitory/ antagonistic to egg hatching, juveniles, fecundity, root penetration etc. The perusal of literature indicated that culture filtrates from different fungi exhibited antagonistic effects on the nematodes and the stages of their life cycles. The studies carried out on the culture filtrates in relation to plant parasitic nematodes in the country are discussed as follows.

Paecilomyces lilacinus is by far the most promising and practicable biocontrol agent for the management of root knot and cyst nematodes. The fungus grows abundantly in sterilized paddy seeds, oil cakes, mashed potato, chopped water lily plants, ipil ipil (*Leucaena glauca*) leaves and corn grits. Eggs of *Meloidogyne incognita* on potato roots were found to be infected by *P. lilacinus*. This fungus was also capable of invading mature females and cysts of *Globodera pallida* (Jatala *et al.*, 1979). *Paecilomyces lilacinus* increased growth of brinjal plants and reduced root galling, egg masses and number of eggs per egg mass of *M. incognita* (Sharma and Trivedi, 1989). Morgan-Jones *et al.*, (1984) showed that hyphae of *P. lilacinus* readily penetrate egg shells of *M. arenaria* through small pores dissolved in the vitelline layer. The effect seems to involve disruption of embryonic development and death of larvae, resulting in reduction of nematode populations. *Paecilomyces lilacinus* was most effective in the management of root knot nematodes in black pepper vines (Sosamma and Koshy, 1995). *Paecilomyces lilacinus* suppressed *Meloidogyne* spp. population by 58.3-86.9% in cardamon nursery. The number of quality seedlings for transplanting was greater in beds treated with this biocontrol agent (Eapen and Venugopal, 1995). *Paecilomyces lilacinus* at 4 and 8 g per kg soil significantly reduced *M. incognita* population in soil and root knot index and

Table 5. Reports of oppurtunistic fungi effective against plant parasitic nematodes in India

Antagonists	Nematodes	Crop	Reference
Paecilomyces lilacinus	*Meloidogyne incognita*	Tomato	Sharma et al., 1988
P. lilacinus	*M. javanica*	Tomato	Khan and Esfahmi 1990
P. lilacinus	*M. javanica*	Okra	Walia et al., 1991
P. lilacinus	*M. javanica*	Brinjal	Zaki and Maqbool, 1991
P. lilacinus	*M. incognita*	Chickpea	Siddiqui and Muhmood 1992
P. lilacinus	*M. incognita*	Chilli	Pandey and Trivedi 1992
P. lilacinus	*M. incognita*	Brinjal	Trivedi 1992
P. lilacinus	*Rotylenchulus reniformis*	Pigeonpdea	Mahmood and Siddiqui 1993
P. lilacinus	*M. incognita*	Tomato	Goutam et al. 1995
P. lilacinus	*M. javanica*	Chickpea	Siddiqui et al. 1996
P. lilacinus	*M. incognita*	Brinjal	Rao et al., 1998
Veriticillium chlamydosporium	*Heterodera cajani*	Pigeonpea	Siddiqui and Mahmood 1995
V. chlamydosporium	*M. javanica*	Chickpea	Siddiqui et al. 1996
V. chlamydosporium	*M. incognita*	Tomato	Rao et al., 1998
V. chlamydosporium	*M. incognita*	Tomato	Sankaranarayanan et al., (Unpublished)
V. chlamydosporium	*H. cajani*	Cowpea	Bhardwaj, 1997
P. lilacinus, T. harzianum, T. viride	*M. incognita*	Gladiolus	Nagesh et al., 1999
P. lilacinus	*M. incognita*	Tuberose	Nagesh et al., 1997
P. lilacinus, T. harzianum, T. viride	*M. incognita*	Carnation, Gerbera	Nagesh & Reddy, 1996
P. lilacinus	*M. incognita*	Corssandra	Nagesh & Reddy, 1997
P. lilacinus	*M. incognita*	Chickpea	Bhat et al., 1999.

increased number of leaves per plant, weight of 100 leaves, shoot and root weight of betelvine (Nakat *et al.*, 1995). *Paecilomyces lilacinus* was also effective for the management of reniform infecting tomato and brinjal (Parvatha Reddy and Khan, 1988, 1989).

Testing culture filtrates of fungi for their nematicidal action have been attempted by several workers (Table 6). Khan *et al.*, (1984) tested filtrates of 8 species of *Aspergillus* at different concentrations. The S and S/2 concentrations of *A. niger, A. candidus, A. flavus* and *A. fumigatus* caused 100% mortality of *M. incognita*. Hatching was totally inhibited at S concentration of all four species of *Aspergillus*. Culture filtrates (c.f.) of *P. lilacinus* effectively inhibited hatching of *M. incognita* juveniles in 100 and 50% c.f. Tomato seedlings dipped in c.f. for 2-3 h reduced root galling significantly (Khan *et al.*, 1988).

1. Mass production of antagonistic fungi

The antagonistic fungi reported from India mainly belong to the category of saprophytic oppurtunistic soil inhabiting fungi. Several substrates have been tried (Table 7).

Use of oil cakes will have an additive effect, as they are already known to be toxic to nematode (Sharma & Trivedi, 1987).

In a preliminary experiment 5% raw molasses amended with N+P significantly enhanced fungal (624.0 mg dry mycelial wt) and was at par with 10% molasses + N + P (627. 5 mg) and the potato dextrose broth control (684.25 mg). Higher concentrations of molasses supported very poor growth. Of the agricultural wastes tested, rice husk

Table 6. Effect of Culture filtrates from antagonistic fungi.

Nematode sp.	Biotic agent	Concentration	Effect on	Referneces
M. incognita, *M. javanica*	*T. harzianum*, *T. koningii*	No dilution	Juvenile mortality	Sankaranarayanan et al., 1998
M. graminicola	*T. harzianum*	50% and no dilution	Juvenile mortality	Pathak & Kumar, 1995
M. incognita	*R. solani*, *T. viride*	Full strength	Hatching	Sharma & Saxena, 1992
M. javanica	*A. niger*, *T. viride*, *Chaetomium cochliodes*, *R. solani*	Full strength	Root penetration by juveniles Root galling Dev. of nematodes	Khan & Saxena, 1976
H. cajani	*T. viride*, *T. harzianum*, *P. fluorescens*, *B. subtilis*	Soil drenching	Fecundity	Pathak & Kumar, 1998

Table 7. Mass production of various fungal biotic agents on different substrates

	Nematode species	Biotic agent	Substrate	Reference
1	—	*P. lilacinus*	Leaves of *Leucaena leucocephala*, *Azadirachta indica*, *Cassia aurculata*, *Pongamia glabra*, *Gliricidia maculata*	Mani & Anandam, 1989
2	*Tylenchulus semipenetrans*	*P. lilacinus*	Wheat, Bajra, jowar, paddy, grain	Mani et al., 1989
3	*M. incognita*	*P. lilacinus*	Paddy grain	Trivedi 1992
4	—	*P. lilacinus*	*Azadirachta indica*, *Cisium arvense*	Bhatti et al., 1989
5	*M. javanica*	*P. lilacinus*	Leaf extracts of *Peristrophe bicalyculata*, *Dalbergia sissoo*	Siddiqui & Mahmood, 1994
6	—	*P. lilacinus*	Wheat, maize, bajra, paddy, sorghum, grains	Patel et al., 1992
7	—	*P. lilacinus*	Oil cakes - *Sesamum indicum*, Cotton, Linseed mustard, groundnut, husk of mungbean, cotton seed, guar (*Cyamopsis tetragonoloba*)	
8	—	*P. lilacinus*, *T. viride*	Sugarcane by products (bagasse, trash, press mud)	Hari & Somasekhar, 1998
9	—	*P. lilacinus*	5% raw molasses + inorganic N and P fertilizers	Bansal et al., 1988

produced the least number of spores while wheat bran produced the maximum and was comparable to the rice grain control. The addition of 5% molasses amended with N and P significantly enhanced sporulation only in the case of rice husk and rice bran. (Bansal *et al.*, 1988).

From the literature it is clear that most of the media/substrates tried for mass production included solid media such as plant leaves, oil cakes, whole grains of cereals, brans, flours, etc. To a small extent liquid media including aqueous leaf extracts, oil cake extracts, tapioca broth and molasses, were used for mass production of these fungi.

Several workers have tested bioefficacy under pot culture and green house (Table 8).

The main types of destructive activity of oppurtunistic fungi are thought to be enzymatic disruption of nematode egg shells and larval cuticles and physiological disturbances brought about by biosynthesis of diffusible toxic metabolites (Morgan Jones and Rodriguez kabana, 1985). The feasibility of using selected

Table 8. Bioefficacy of *Paecilomyces lilacinus* tested against plant parasitic nematodes under pot/glass house conditions

	Nematode sp.	Crop	Substrate used	GH/ Field/Pot	Reference
1	*M. incognita*	*Trifolium foenumgraceum*	Goat dung, Sesame oilcake	Pot	Sharma & Trivedi, 1989a
2	*M. incognita*	Eggplant	Paddy grain	Pot	Sharma & Trivedi, 1989b
3	*M. incognita, R. reniformis*	*Vigna unguiculata*	Mycelium	Pot	Khan & Hussain, 1990
4	*M. javanica*	Tomato	Castor leaves Gram seed	Pot	Zaki & Bhatti, 1990
5	*M. incognita*	Chickpea	Organic/FYM		Sivakumar & Vidhyasekharan, 1990
6	*T. semipenetrans*	Pigeonpea	Oilcakes-castor karanj neem		Reddy et al., 1991
7	*M. incognita* race 3	Chickpea	—		Siddiqui & Mahmood, 1993
8	*M. javanica*	Betelvine	Aq.solution as Drench		Walia & Bansal, 1992
9	*M. incognita.* race 3 + *M. phaseolina*	Eggplant	—		Siddiqui & Mahmood, 1992
10	*R. reniformis*	*Capsicum annum*	—	Field	Mahmood & Siddiqui 1993
11	*M. incognita* + *M. phaseolina*	Pepper	Culture filtrate As seed treatment	Pot	Siddiqui & Mahmood, 1993
12	*H. cajani*	Groundnut	Paddy grain amendment	Pot	Rana & Dalal, 1995
13	*R. similis*	Okra	Seed treatment At planting	Pot	Sosamma *et al.*, 1994
14	*M. incognita*	*Geranium*	Neem cake extract	Pot	Rao & Reddy, 1994
15	*M. incognita*	Pigeonpea	—	Pot	Pandey & Trivedi, 1992
16	*M. incognita*	Tomato	Paddy grain	Pot	Jonathan et al., 1995 Patel *et al.*, 1995
17	*M. javanica*	*Cowpea*	Neem Cake	Pot	Walia et al., 1991
18	*M. javanica*	Betelvine	Wheat bran as seed	Pot	Anita & Vadivelu, 1997
19	*M. incognita*	Chickpea	Neem and mustard oil cakes at 1 g/kg soil with *P. lilacinus* at 3.5 x 10^8 spores	Pot	Yaqub Bhat *et al.*, 1999
20	*M. incognita*	Carrot	Zino, neem, castor oil cakes 1 ton/ha with *P. lilacinus* at 10^7 spores/m^2	Pot	Sivakumar, 1999

opportunistic soil fungi such as *P. lilacinus, V. chlamydosporium, V. lecanii, Trichoderma harzianum, T. viridae* and *Gliocladium virens* have shown some effect in controlling *M. incognita, T. semipenetrans* and *R. reniformis*. Species of *Verticillium, Gliocladium* and *P. lilacinus* were recovered from females obtained in galls of plants.

2. Mycorrhizae

Mycorrhizae fungi have shown antagonistic influence on the population of plant parasitic nematodes which may be due to increased amino acid contents (especially arginine) in the mycorrhizal roots (Nemec and Meridith, 1981) or the

Table 9. Bioefficiency of *Trichoderma, Gliocladium* sp. tested against plant parasitic nematodes under pot/glass house conditions

Nematode sp.	Crop	Combination/ integration with	Dose	Appln. method	References
M. incognita	Okra	*T. harzianum / A. niger / Fusarium oxysporum* in combination with oil cakes of Koranj / mustard /cotton seed / mahua	—	Soil amendment	Goswami *et al.*, 1998
M. incognita	Tomato	Neem cake with *T. harzianum*	—	Neem cake amendment	Rao et al., 1997
M. incognita. F. solani	Papaya	*T. harzianum* and *P. lilacinus*	1g/kg soil	Soil amendment	Khan et al., 1997
M. javanica	Tomato	*A. niger. P. lilacinus. P. viride*	Culture filtrates	Root dip treatment	Khan & Saxena, 1997a
H. cajani. F. udum	Pigeonpea	*Colonus mosseae. T. harzianum. V. chlamydosporium*	—	Simultaneous application of all three to soil	Siddiqui & Mahmood, 1996
T. semipenetrans	Citrus aurantifolia	*T. harzianum* in combination with neem/castor/karanj oil cakes	Oil cakes @ 20 g with *T. hazarianum at 2 g/plant*	Soil amendment before planting	Reddy *et al.*, 1996
M. incognita	Sunflower	*T. harzianum/T. viride/ G.virens/G. deliquescens*	—	Soil amendment	Sankaranarayanan et al. 1999
M. incognita	Tomato	*T. harzianum. Trichoderma roseum*	T. harzianum 5g/kg soil *T. roseum* 7 g/kg soil	Soil amendment	Rekha *et al.*, 1998
M. incognita	Tomato	*V. chlamydosporium* and neem or *Calotropis procera* leaves	—	Soil amendment	Reddy *et al.* 1999
M. incognita	Tomato	*V. chlamydosporium* and neem or pongamia oil cakes	V. c at 4 x 106 spores/plant	Soil amendment	Reddy *et al.*, 1997
H. avenae	Wheat	*V. chlamydosporium*	Oilcake 10g plant V.c 8 g substrate/kg soil	Soil amendment	Bhardwaj et al. 1996
T. semipenetrans	Citrus aurantifolia	*V. chlamydosporium / V. lecani*, oil cakes castor / karanj /neem	V.c/V.1 2g/ plant Oil cakes 10 g/plant	Soil amendment	Reddy *et al.*, 1996
H. cajani. F. udum	Pigeonpea	*Gigaspora margarita. V. chlamydosporium. P. lilacinus*	—	All three simultaneously as soil amendment	Siddiqui & Mahmood, 1998

presence of additional amount of phosphorus in roots. VAM can markedly alter the plant response to nematodes. For example, tomato roots colonised by *Glomus fasciculatum* and infected by *M. incognita* or *M. hapla* exhibited fewer and smaller galls than did nematode infected non mycorrhizal plants (Bagyaraj, *et al.*, 1979). Small giant cells with few nuclei and ratarded nematode development are observed in mycorrhizal tomato roots infected with *M. incognita* (Sikora, 1979). In addition to adverse effect on nematode penetration and development, there are indications that nematode reproduction is also suppressed on mycorrhizal plants.

Table 10. Field trials with *Trichoderma* spp.

Nematode spp.	Crop	Combination or integration with	Dose	Application method	References
M. incognita	Tomato	Glomus fasciculatum, neem cake, *T. harzianum*	—		Reddy et al., 1998
Nematodes, *Fusarium solani*	—	—	—	Glomus fasciculatum & neem cake in nursery and *T. harzianum* in microplot at planting	Sathpathi, 1998
M. incognita	Tomato	—	V.c at 2 x 10^8 spores/m^2 *P. penetrans* at 10^7 spores/m^2	Soil amendment	Rao *et al.*, 1998a
M. incognita	Egg plant	—	10% w/v	Drench	Rao et al., 1998b

Umesh *et al.* (1988) recorded low number of *Radopholus similis* in mycorrhizal treated banana plants. Establishment of *G. fasciculatum* two weeks prior to *M. incognita* resulted in increased fresh and dry shoot weight of okra and reduced larval penetration, development, number of galls and egg masses per plant (Sharma and Trivedi 1994).

Sitaramaiah and Sikora (1982) indicated that the penetration of *R. reniformis* decreased on mycorrhizal tomato plants. They also observed delayed development of gelatinous matrix with few eggs per egg sac. The reduction in larval penetration of *R. reniformis* on mycorrhizal (*G. mosseae* bush bean, cucumber and muskmelon was recorded by Sitaramaiah and Sikora (1982). Suresh and Bagyaraj (1984) reported that mycorrhizal inoculation caused reduced *M. incognita* infestation in tomato plants which had increased quantities of P, K, Ca, total and reducing sugars and amino acdis (phenyl alanine and serine).

The establishment of *G. fasciculatum* and *G. epigaeus* in the root system of cowpea plants caused an increase in plant resistance to *Heterodera cajani* infection (Jain and Sethi, 1988a). They suggested that the reduction in penetration may be due to the fact that mycorrhizal roots were undesirable sites for larval penetration. Jain and Sethi (1988b) reported reduction in number of galls of *M. incognita* and number of cysts of *H. cajani* on cowpea roots with the prior colonisation of *G fasciculatum.*

G. mosseae was the most effective VAM fungi in reducing *M. incognita* population in roots and promoting growth of crossandra (Nagesh *et al.*, 1997). Four VAM fungi viz., *Acaulospora laevis, Glomus fasciculatum, G. intraradices* and *G. mosseae* in combination with neem cake were evaluated for their comparative efficacy to colonise crosandra roots and suppress *M. incognita* infection. *Glomus fasciculatum* and *G. mosseae* colonised crossandra roots better compared to *G. intradices* and *A. laevis* and comparatively lower nematode infection and multiplication, thus promoting better root health and plant growth in former species Although all the four VAM fungi in combination with neem cake gave better control of nematodes over carbofuran treatment, two species namely *G. fasciculatum* and *G. mosseae* were found to be more efficient (Nagesh *et al.* 1999). The VAm colonisation of roots mechanically may prevent nematode penetration and establishment, increased nutritional status, increased sugar content of roots (Smith, 1987), increased levels of P in roots or increased host tolerance. Although most of the research conducted under greenhouse conditions indicated that mycorrhizal fungi have the potential to nullify the damage due to phytonematodes, but intensive investigations need to be carried out under field conditions to assess their potential utilization as biocontrol agents. Since VAM fungi are obligate parasites, they can not be cultured on large scale which hinders their effective utilization in integrated nematode management programmes under filed conditions.

3. Actinomycetes

The avermectins are a new class of macro cyclic lactones isolated from the soil actinomycetes. *Streptomyces avermitilis*. These compounds were reported to be possessing insecticidal, acaricidal and nematicidal properties (Putter *et al.*, 1981). They are commonly distributed in most of the cultivated soils. The available literature on avermectins indicates that *M. incognita*, *M. javanica*, *M. arenaria* and *T. semipenetrans* were the commonly used test organisms to assess the nematicidal activity of these metabolites on tomato and citrus (Garabedian and Van Gundy, 1983). In citrus, a monthly rate of 1.1 kg a.i. per ha of avermectins for 7 months gave maximum increase in yield and reduction of *T semipenetrans* population. While avermectins B1 and B2a have shown high toxicity against *M. incognita* in greenhouse conditions when incorporated into soil, avermectin B2a was found to be more potent than B1 (Putter *et al.*, 1981). In a microplot study, avermectin B1 at 0.17 kg a.i. per ha was equal to ethoprop, aldicarb, fenamiphos, oxamyl and carbofuran at 6.7 kg a.i. per ha (Normeyer and Dickson, 1981). The avermectins have a direct action on neurotransmission. They are antagonists of gama amino butryic acid.

C. ROLE OF BIOLOGICAL CONTROL IN INTEGRATED NEMATODE MANAGEMENT (INM)

The extensive research on nematode management in India and abroad arrived at the time tested conclusion that no single method of control, be it the chemical, host resistance, cultural or biological method, could achieve long term and sustained nematode control with a safety to environment. To alleviate nematode problems, integrated nematode management (INM) procedures are implemented based on the principles of prevention, population reduction and tolerance. INM seeks to stabilize populations of target nematodes at acceptable levels resulting in favourable long term socio-economic and environmental consequences. In practice, in integration of different methods of nematode control is the rule than the exception. In most cases, practical control of a nematode disease involves integration of several diverse control measures. In this context, biological control forms an ideal and essential component in the INM.

1. Integration of bio-agents and botanicals

A strategy was standardised for effective use of neem cake extract for the management of *Meloidogyne incognita* on okra (*Abelmoschus esculentus*) under field conditions. Combination of the use of aqueous extract of neem cake for seed treatment and soil drenching under field conditions was found as effective as application of carbofuran at 2 kg a.i. per ha or neem cake at 2 tonnes per ha for the management of *M. incognita* on okra (Rao *et al.*, 1997).

Management of the citrus nematode, *Tylenchulus semipenetrans* based on the application of neem cake and castor cake both at 10 g per plant along with an antagonistic fungus *Paecilomyces lilacinus* at 4 g (grown on paddy seeds) per plant was found to be extremely effective in reducing the citrus nematode population with a consequent increase in the growth of acid lime seedlings (Parvatha Reddy and Khan, 1991). *Trichoderma harzianum* in combination with neem, Karanj and castor oil cakes was effective in increasing the growth of acid lime seedlings and reducing the citrus nematode population both in the soil and roots. The parasitization of citrus nematode females with *T. harzianum* increased in the presence of the oil cakes (Parvatha Reddy *et al.*, 1996c). *Veriticillium lecanii* in combination with the oil cakes was effective in increasing the growth of acid lime and in reducing the citrus nematode population both in soil and roots. Between the two parasitic fungi tested in combination with oil cakes, *V. lecanii* performed better than *V. chlamydosporium*. The parasitisation of nematode eggs was also maximum when *V. lecanii* was used in combination with the oil cakes (Parvatha Reddy *et al.*, 1996b). This strategy forms the basis for the effective managment of citrus nematode under field conditions with minimum amount of botanicals and bio-agents.

Integration of *P. lilacinus/Trichoderma harzianum* at 500 ml/sq m (aqueous spore suspension containing 2 x 104 spores per ml) with neem cake at 0.5 kg/sq m or fenamiphos at 2 g ai/sq m increased plant growth parameters and flower yield of carnation and gerbera. The above treatments also increased root knot egg parasitisation by the parasitic fungi (Nagesh and Parvatha Reddy, 1996).

Integration of neem cake at 400 g per plant with *p. penetrans* at 100 g soil (300 spores) was found effective in reducing *R. similis* population both in soil and roots of banana by more than 50% and increased plant growth parameters (Channabasappa *et al.*, 1995). Application of 5 per cent inoculum of *Arthrobotrys conoides* to the pot soil amended with FYM, effectively reduced the larval penetration of *M. incognita* and root galling was reduced by 30-40% in okra (Srivastava and Swarup, 1986). Integration of neem cake with *Trichoderma harzianum* gave effective control of *M. incognita* infecting tomato and increased root colonisation with biocontrol agent (Rao *et al.*, 1977d).

Integration of *Paecilomyces lilacinus* with 5% neem leaf extract increased plant growth parameters and flower yield of tuberose. The above treatment also reduced root galling and increased per cent egg parasitzation by *P. lilacinus* (Nagesh *et al.*, 1996). Further, it was observed that split application of *P. lilacinus* in combination with oil cakes viz., castor, pongamia or neem (at planting and 45 days later) under field conditions significantly reduced root gall index, root and soil populations of *M. incognita* and its multiplication in tuberose, compared to single application of *P. lilacinus*, oil cakes and their combinations at planting (Nagesh, *et al.*, 1998).

After standardising a formulation of neem cake aqueous extract which supported the growth of *Paecilomyces lilacinus*, experiments were conducted to evaluate this formulation as bare root dip treatment for the management of *Meloidogyne incognita* on egg plant (*Solanum melongena*). Root dip treatment of egg plant seedlings in this formulation of *P. lilacinus*, resulted in significant increase in plant growth parameters and drastic reduction in root knot index and final popualtion of *M. incognita*. Further, significant increase in root colonisation, propagule density in soil and per cent infection by *P. lilacinus* revealed the complementary effect of these two components (*P. lilacinus* and neem cake extract) required for their effective use in sustainable management of *M. incognita* on egg plant (Nagesh *et al.*, 1998).

Combined application of *P. penetrans*, *T. viride* and neem or castor cake at 1/3rd dosage was significantly superior compared to their individual applications in terms of increase plant growth and reduced root galling, egg mass production and final population of *M. incognita*. Similarly total parasitisation by bioagents was higher in plants treated with these treatments, in terms of number of junveniles encumbered and adult females infected with the bacterium, and egg masses parasitised by *T. viride*, compared to their individual application (Rangaswamy *et al.*, 1999).

Incorporation of neem cake/castor cake/neem leaf/castor leaf/pongamia leaf in nursery beds at 400 g/m^2 and application of spore suspension of *P. lilacinus* or *V. chlamydosporium* in tomato and brinjal nursery beds amended with the above plant products facilitated the effective management of root knot nematode, *M. incognita* and reniform nematode, *R. reniformis*. Further, root dip treatment of seedlings in 5% aqueous suspension of neem leaf/cake mixed with the spores of *P. lilacinus* of *V. chlamydosporium* for 20-30 minutes before transplanting was effective in the management of root knot and reniform nematodes in the main field. Neem cake suspension (5%) mixed with the spores of *P. lilacinus* is also effective for the treatment of okra seeds for the managment of *M. incognita* under field conditions (Rao and Parvatha Reddy, 1992a,b, 1993; 1994; Rao *et al.*, 1994, 1997b,c). Transplanting of tomato seedlings raised in nursery beds treated with neem cake and *P. penetrans* in pits incorporated with *P. lilacinus* gave effective management of *M. incognita* and increased tomato fruit yield (Parvatha Reddy *et al.* 1997). Combination of *V. lecanii* and *P. lilacinus* with neem leaf extract resulted in increased flower yield of crossandra. Root galling was least when neem leaf extract was integrated with *P. lilacinus* (Nagesh and Parvatha Reddy, 1995a, b). Treatment of tuberose bulbs with neem and calotropis leaf extracts mixed with *P. lilacinus* spores significantly reduced *M. incognita* infection and multiplication besides stimulating the plant growth (Nagesh *et al.*, 1997). Three antagonistic fungi viz., *T. harzianum*, *V. lecanii* and *P. lilacinus* at 2 x 10^4 spores/ml in combination with neem cake reduced *M. incognita* population in both soil and roots of carnation (Nagesh and parvatha Reddy, 1996a). Identification of unique properties of the botanical extracts to support the growth of the promising bio-agents for the first time, has not only helped in evolving integrated methods by rational combination of bio agents and botanicals, but also in standardising the protocol for mass production of these bio- agents.

2. Integration of Mycorrhizae and botanicals

Inoculation of endomycorrhiza, *Glomus fasciculatum* in the soils amended with neem cake was found to be effective in the management of *T. semipenetrans* on acid lime. This strategy can help in combating the menace of citrus nematode at nursery stage and also provide highly mycorrhizal seedlings of acid lime for transplanting in the main field for the management of *T. semipenetrans* under field conditions (Parvatha Reddy *et al.*, 1995).

Integration of karanj cake with *G. mosseae* gave maximum banana fruit yield, while neem cake + *G. mosseae* was most effective in reducing *Radopholus similis* population both in soil and roots (Parvatha Reddy *et al.*, 1997). Integration of neem, Karanj and castor cakes with VAM, *Glomus mosseae* significantly enhanced plant growth parameters and flower yield of crossandra, root colonization and sporulation of VAM. The above treatments also reduced root knot nematode multiplication and root galling (Nagesh and Parvatha Reddy, 1997). Inoculation of *G. mosseae* or *G. fasciculatum* in the nursery beds amended with neem cake/castor cake/neem leaf/calotropis leaf reduced the infestation of root knot and reniform nematodes to the maximum extent on tomato. Amendment of botanical in the nursery beds increased the multiplication of endomycorrhizae and colonization of tomato root which in turn could protect the crop from these nematode in the main field resulting in increased yield (Rao *et al.* 1995b; 1996, 1997e). Similarly, inoculation of *G. mosseae/G. fasciculatum* in the brinjal nursery beds and subsequent application of 5% aqueous extracts of neem cake/ castor cake/neem leaf/calotropis leaf in the nursery beds resulted in the effective management of root knot and reniform nematodes and yielded healthy brinjal seedlings which would withstand the attack of these nematodes after transplanting in the main field (Rao *et al.*, 1993a,c). The root colonisation of VAM fungi increased significantly in the presence of neem cake which in turn improved their efficacy in reducing *M. incognita* population in crossandra roots. (Nagesh and Parvatha Reddy, 1996b).

Three oil cakes viz. Castor, Karanj and neem each at 400 g per plant and *G. mosseae* at 200 g inoculum per plant (containing 25 chlamydodspores per g of inoculum) were evaluated either individually or in combination (at half the dosages) for the management of *R. similis* under field conditions. Karanj cake + *G. mosseae* was the most effective treatment in increasing plant growth parameters (plant height, pseudostem girth, leaf area and number of leaves) and fruit yield (bunch weight, number of hands per bunch, number of fingers per hand and finger weight), while neem cake + *G. mosseae* was most effective in reducing the nematode population both in soil and roots (Parvatha Reddy *et al.*, 1997).

Interestingly, integration of bio-agent (*P. lilacinus*) and endomycorrhizae (*G. mosseae/G. fasciculatum*) have culminated into the successful management of *M. incognita* infecting tomato. This phenomenon facilitated standardisation of a strategy wherein inoculation of mycorrhizae and bio agent in the nursery beds protected the seedlings of tomato from the attack of *M. incognita*. Further, these mycorrhizal seedlings (colonised either with *G. mosseae* or *G. fasciculatum*) can be given a root dip treatment with spore suspension of *P. lilacinus* for 5-10 minutes for the effective management of these nematodes in the main field after transplanting (Rao *et al*, 1993b, d).

3. Integration of bio-agents, mycorrhizae and botanicals

After investigating the interactive effects of useful components of management (botanicals, bio-agents and endomycorrhizae), strategies to integrate these components for the management of root knot and reniform nematodes infecting tomato and brinjal have been developed. These strategies include inoculation of *G. fasciculatum* in the castor cake amended nursery beds followed by root dip treatment of mycorrhizal seedlings of brinjal in spore suspension of *P. lilacinus* for the managment of *R. reniformis* (Rao *et al.*, 1993b). Similarly, inoculation of *G. mosseae* in neem leaf/neem cake amended nursery beds followed by the root dip treatment of mycorrhizal seedlings of tomato in spore suspension of *P. lilacinus* was effective for the management of *M. incognita* under field conditions (Rao *et al.*, 1995a). Integration of a bioagent (*V. lecanii*), endomycorrhiza (*G. mosseae*) with botanicals improved the growth of crossandra and reduced the population of *M. incognita*, *G. mosseae* reduced the requirement of phosphatic fertilizer and favoured the antagonistic potential of *V.*

lecanii against *M. incognita* (Nagesh and Parvatha Reddy, 1995b).

4. Integration of bio-agents and chemicals

An integrated management of the citrus nematodes, *T. semipenetrans* infecting acid lime based on application of the parasitic fungus, *P. lilacinus* at 4 g per plant (grown on paddy seeds) along with carbofuran at 1 kg a.i./ha has been developed. It is found to substantially reduced the nematode population both in soil and roots and to increase plant growth parameters (Parvatha Reddy *et al.*, 1996a). Integration of a bio agent, *P. lilacinus* with carbofuran at 1 kg a.i./ha was found effective in the management of reniform nematode, *R. reniformis* infecting tomato and brinjal (Parvatha Reddy and Khan, 1988, 1989). Integrated management of root knot nematode infecting crossandra was obtained by rational combination of bio control agents *V. chlamydosporium*/ *Trichoderma harzianum* with aldicarb at 0.5 kg a.i./ha (Parvatha Reddy and Nagesh, 1995). Maheshwari *et al.* (1987) reported that application of *P. penetrans* in combination with carbofuran, aldicarb, miral, sebifos and phorate significantly improved plant growth of tomato by greatly reducing galling due to *M. javanica*. *Paecilomyces lilacinus* at 4 g per plant + carbofuran at 0.75 kg a.i. per ha significantly reduced *M. incognita* population in soil and root knot index and increased plant growth parameters of betelvine (Nakat *et al.*, 1995).

5. Integration of mycorrhizae and chemicals

Integration of *G. mosseae*/*G. fasciculatum* with aldicarb at 0.5 kg a.i./ha gave effective management of root knot nematodes infecting tuberose (Parvatha Reddy and Nagesh, 1995).

6. Integration of bio-agents

Integration of two bio-agents, *P. lilacinus* and *V. chlamydosporium* resulted in combined and complimentary effects for the successful management of *M. incognita* infecting tomato (Rao and Parvatha Reddy, 1992a). Combinations of *Verticillium lecanii* and *Paecilomyces lilacinus* (2×10^4 spores/ ml each) with 5% neem leaf extract resulted in significantly higher plant growth parameters and crossandra flower yield (Nagesh and Parvatha Reddy, 1995). Root galling, nematode multiplication factor were least and the parasitisation of eggs and egg masses was highest in *V. lecanii* + 5% neem leaf extract. Combined application of *P. lilacinus* and *Pasteuria penetrans* enhanced plant growth and flower yield of crossandra besides reducing root galling due to *M. incognita* (Nagesh *et al.* 1995). Integration of *Glomus deserticola* with *V. chlamydosporium* gave effective management of *M. incognita* on tomato seedlings and increased the egg parasitisation by the bioagent (Rao *et al.*, 1997a). Three antagonistic fungi *Paecilomyces lilacinus*, *Trichoderma harzianum*, *T. viride*, neem cake and their combinations were evaluated to manage the root knot nematode and *Fusarium* wilt in *Gladiolus* under glasshouse conditions. The plants treated with *P. lilacinus* + *T. harzianum* + neem cake and *P. lilacinus* + *T. viride* + neem cake combinations not only controlled nematode infection but also the wilt till the harvest of flower spikes. The corms and cormels obtained from these treated plants were free from *Fusarium* fungus compared to that in other treated plants (Nagesh *et al.*, 1999). Maheshwari and Mani (1988) observed that population of *M. incognita* and *M. javanica* were suppressed effectively and yields of tomato were greater when *P. penetrans* and *P. lilacinus* were applied together. Black pepper vines combinedly inoculated with *P.*

lilacinus and *P. penetrans* had put out 23-112 per cent increase in plant growth over control and were very effective in the management of root knot nematodes (Sosamma and Koshy, 1995).

7. Integration of bioagents and physical methods

Soil solarisation of nursery beds and application of bioagents such as *P. lilacinus* or *Trichoderma* spp. improved growth of cardamom seedlings by suppressing root knot nematode population (Eapen and Venugopal, 1995).

Most of the bioagents have exhibited excellent versatility in integrated use with other methods of control without losing their efficacy.

D. IMPORTANT RESEARCH ISSUES IN BIOLOGICAL CONTROL OF PLANT PARASITIC NEMATODES

1. Exploration of Nematode suppressive soils

Nematode suppressive soils not only are likely sources of potentially useful biological control agents, but also they could yield invaluable general information on the ecology of antagonists in soil. Whether biological control is to be achieved by introducing selected antagonists into the nematodes natural environment or by manipulating the antagonists that are already present, a detailed understanding of the interrelationships between nematodes and their antagonists will be required. Ideally, this information should be available for all the cultivated soils, but since it is time consuming to obtain, perhaps the best place to start is in soils known to be suppressive to nematodes. For example, thesc soils could provide data on level of mortality caused by antagonist in the natural environment and give some indication of how this varies with changes in the soil physical environemt and nematode density.

2. Cataloguing of Pathogenic diversity

The fungal and bacterial antagonists of nematodes exhibit considerable diversity. The diversity of fungi (as isolates of the same taxon) is generally expressed and observed in culture, often in terms of colour and general appearance, degree of sporulation and growth rate. The selection of the most virulent isolate from the material available must be on the first steps in any biological control programme. Basic studies are needed on the underlying reasons why isolates differ in virulence, because such studies may identify some specific morphological, biochemical or genetic trait which can be identified readily and which is correlated with virulence. In the absence of such studies, a pathogenicity test must be used to identify isolates with good biological control potential.

3. Evolving screening methods

It is often difficult to identify the right type of screening test to be used. The same isolates do not necessarily show consistent levels of parasitic activity in different screening tests, viz., water agar test, filter paper test, agar plate test and millet substrate test. The results obtained with each of the tests probably differed becasue different attributes of the fungi (e.g. the virulence of the isolate, its ability to establish from macerated mycelium in the absence of a nutrient source and its capacity to survive saprophytically on a millet substrate) may have been the major factors determining levels of parasitism in each situation and we need a clear idea of the attribute we are trying to identify in the screen.

4. Search for new antagonists and novel mechanism of action

Although several species of nematodes, fungi and bacteria are reported to parasitise or prey on nematodes, there are many undescribed species which could be useful biological control agents. In the last 20 years a number of new fungal parasites of females and eggs have been recognised (Kerry and Crump, 1980; Morgan - Jones and Rodriguez Kabana, 1988) and several new strains of *Pasteuria* capable of attacking root knot and cyst nematodes have been found (Nishizawa, 1987; Davies *et al.*, 1990; Sturhan and Winkelheide, 1990). The tropical soils and regions which are considered the centres of origin of root knot and cyst nematodes have not been surveyed in detail. There is an urgent need to develop germplasm collections and national repositories for potentially useful antagonists.

It is also important that we continue to seek antagonists with novel mechanisms of action. Some fungi and bacteria produce highly nematicidal chemicals (Burg *et al.*, 1979; Mishra *et al.*, 1987; Devidas *et al.*, 1990), while a number of rhizosphere -inhabiting bacteria have recently been found to reduce the number of *Heterodera and Meloidogyne* larvae invading roots (Becker *et al.*, 1988; Oostendrop and Sikora, 1989; Spiegel *et al.*, 1990; Stirling *et al.*, 1990). Although the mechanisms of action of the latter bacteria are not yet known, it is thought that they may alter exudation patterns from roots and therefore affect the capacity of nematodes to locate roots. Their main role might be to complement the activity of some of the parasites and predators which have traditionally been used as biological control agents. Many of these organisms act through density dependent mechanisms and therefore have little impact on the low nematode populations that usually occur at the time crops are planted.

5. Selection of rhizosphere-competent antagonists

In most experiments with biological control agents, the agent being tested has been applied by mixing propagules thoroughly into the soil surrounding the plant. Although such application methods may be necessary for some organsims, they usually necessitate the use of large amounts of inoculum. Since plant parasitic nematodes tend to aggregate in and around roots, we need to concentrate much more on application methods which target the root system rather than the bulk soil mass. Seed treatments, and procedures which establish biological control agents in transplants, must therefore be tested. In conjunction with these studies, it will be necessary to determine whether a biological control agent has the capacity to survive in the rhizosphere, as this is the main region where intimate contact between nematodes and their antagonists can occur. De Leij and Kerry (1990) demonstrated the importance of rhizosphere competence for *V. chlamydosporium*. All the isolates which were introduced into soil became established at high populations densities, but the isolate which colonised roots and galls was the only one to give any nematode control.

6. Nutritional relationships of each antagonist

Many potentially useful biological control agents are not obligate parasites and their parasitic activity is therefore likely to be influenced by the presence of food sources other than nematodes.

7. Genetic improvement of biocontrol agents

Although wild type strains of biocontrol agents especially antagonistic fungi have shown considerable promise in the biological control of plant parasitic nematodes, research carried out to improve their desirable characters are meager. The desirable characters include tolerance to commonly used agro- chemicals, adverse environmental conditions such as extreme moisture, temperature, pH and competition from other microbes in plant rhizosphere, vigorous growth, sporulation and long shelf life. The new technologies of protoplast fusion, transformation and gene cloning can develop new strains of bacteria, whereas mutagenesis and protoplast fusion can develop variants in fungi.

E. CONCLUSION

Although, most of the research conducted under greenhouse conditions indicated that bio-control agents have the potential to nullify the damage due to phytonematodes, intensive investigations need to be carried out under field conditions to assess their practical utilization. The biocontrol agents with the highest potential for the management of plant nematodes are likely to be those that are endemic in agricultural soils and can be induced to proliferate by methods compatible with conventional farming practices.

Among bacteria, *Pasteuria penetrans* provides ample scope in exploiting its potential as biocontrol agent against root knot, citrus and lesion nematodes. Since *Pasteuria penetrans* can not be cultured on a large scale which hinders its effective utilization in integrated nematode management programmes.

The opportunistic fungi which infect eggs and females belonging to the genera, *Paecilomyces*, *Trichoderma* and *Verticillium* seem to be most promising. These fungi spread rapidly, are ubiquitous and are present abundantly in the rhizosphere. They are well adapted to the environment, being successful competitors and are able to survive host free periods. Besides, they also have the added advantage of being easily grown in axenic culture and introduced into soil. In France, two commercial agents, 'Royal 300' and 'Royal 350', are available for control of *Ditylenchus myceliophagous* on mushrooms and *Meloidogyne* spp. on vegetables. More recently, the parasitic fungus, *Paecilomyces lilacinus* has been produced in Philippines as 'Biocon' which can be used for the control of several nematode species. In view of the above advantages, opportunistic fungi form ideal biological control components which can be effectively utilized in integrated nematode management programmes under field conditions.

F. FUTURE THRUSTS

Research on biological control of nematodes needs focus on certain important areas as given below for its better exploitation academically as well as at farmer's level.

- Quantification of microbial bio-diversity in a given agro-climatic region.
- Selection/development of superior or virulent strains.
- Development of simple, efficient and economical methods of field application and advice for the users.
- Development of economically viable mass production systems and proper storage techniques.
- Development of suitable formulations, safety testing and quality control protocols.
- Registration protocols for most effective and safe microbial products.
- Use of genetic engineering techniques for creating transgenics and new combinations of multipurpose pathogens.

- Training and human resource development.

  Creation of village level crop protection units involving unemployed youth, farm women and artisans in production, packing and transport of biocontrol agents.

REFERENCES

Ahmed, A. and Alam, M.M. 1997. Utilization of fly ash and *Paecilomyces lilacinus* for the control of *Meloidogyne incognita. Internationa Journal of Nematology*, 7 : 162-164.

Alam, M.M. 1990. *Paecilomyces lilacinus*-a nematode bio control agent. Aspects ; *Prospects.* In *Nematode Biocontrol.* (Jairajpuri, M.S., Alam, M.M. and Ahmad, I. eds.) pp 71-82.

Alam, M.M. and Jairajpuri, M.S. 1990. Natural enemies of nematodes. *Aspects : Prospects.* In *Nematodes Biocontrol.* (Jairajpuri, M.S., Alam, M.M. and Ahmad, I. eds.) 17-40.

Anandaraj, M. 1996. Biological control in species. (Peter, K.V., ed.), vii + 52 pp.

Anita, B. and Vadivelu, S. 1997. Management of root knot nematode, *Meloidogyne hapla* on scented geranium, *Pelargonium graveolens L. Indian Journal of Nematology* : 27 : 1, 123-125.

Anju Kamra and Dhawan, S.C. 1999. Effect of *Pseudomonas fluorescens* culture filtrate on the emergence and mortality of juveniles of *Heterodera avenae.* In : *Proceedings of national symposium on National approaches in nematode management for sustainable agriculture,* (Dhawan, S.C. and Kausahl, K.K. eds.) Anand, India 23-25 November, 1998, 116-118.

Anju Kamra, Dhawan, S.C., Kamra, A. 1998. Observations on host range, vertical movement and soil pH on spores of *Pasteuria penetrans* infecting *Heterodera cajani.* Indian Journal of Nematology 28 : 2, 192-202.

Anwar, L. Bilgrami, Irfan Ahmed, and M. Shamim Jairajpuri, 1987. Interaction between mononchs in non sterile agar plates. *Indian Journal of Nematology* 17 : 339-340.

Bagyaraj, D.J., Manjunath, A. and Reddy, D.D.R. 1979. Interaction of vesicular arbuscular mycorrhiza with root knot nematodes in tomato. *Plant and soil* 51 : 397-403.

Bansa Singh, and Dhawan, S.C. 1992. Effect of soil texture on attachment of bacterial spores of *Pasteuria penetrans* to second stage juveniles of *Heterodera cajani.* Indian Journal of Nematology 22 : 72-74.

Bansa Singh, Dhawan, S.C. and Singh, B. 1998. Compatibility of *Pasteuria penetrans* with carbofuran in controlling *Heterodera cajani* on cowpea. *Indian Journal of Nematology* 28 : 2, 110-113.

Bansal, R.K., Walia, R.K. and Bhatti, D.S. 1988. Evaluation of some agro industrial wastes for mass propagation of the nematode parasitic fungus *Paecilomyces lilacinus. Nematologia Mediterranea* 16 : 135-136.

Bansal, R.K., Walia, R.K. and Bhatti, D.S. 1992. Wood charcoal powder, a carrier of *Paecilomyces lilacinus*, spores. *Nematologia Mediterranea* 20 : 5-7.

Bhardwaj, P. and Trivedi, P.C. 1996. Biological control of *Heterodera avenae* on wheat using different inoculum levels of *Verticillium chlamydosporium. Annals of Plant Protection Sciences* 4 : 111-114.

Bhattacharya, D. and Swarup, G. 1988. *Pasteuria penetrans,* a pathogens of the genus *Heterodera,* its effect on nematode biology and control. *Indian Journal of Nematology*

18 : 61-70.

Bhatti, D.S. Verma, K.K. and Goel, S.R. 1989. Growth responses of some soil inhabiting fungi to leaf extracts of nematicidal plants. *Current Nematology, 1992, 3: 53-60.*

Bilgrami, A.L. 1993. Analysis of the predation by *Aporeclaimellus nivalis* on prey nematodes from different prey trophic categories. *Nematologica* 39 : 356-365.

Bilgrami, A.L., Kulshreshtha, R. and Pervez, R. 1997. Community analysis of predaceous and free living nematodes of Aligarh district. *Indian Journal of Nematology* 27 : 1, 104-110.

Bilgrami, A.L. 1992. Resistance and susceptibility of prey nematodes to predation and strike rate of the predators. *Mononchus aquaticus, Dorylaimus stagnalis* and *Aquatides thornei. Fundamental and Applied Nematology* 15 : 265-270. Biological suppression of Plant diseases, phytoparasitic nematodes and weeds. Project Directorate of Biological Control, Bangalore p. 154-163.

Bourne, J.M. and Kerry, B.R. 1999. Effect of the host plant on the efficacy of *Verticillium chlamydosporium* as a biological control agent of root knot nematode densities and fungal application rate. *Soil Biology and Biochemistry* 39 : 75-84.

Chahal, P.P.K. and Chahal, V.P.S. 1993. Effect of thuricide on the hatching of eggs of root knot nematode (*Meloidogyne incognita*). *Current-Nematology* 4 : 247.

Chahal, P.P.K. and Chahal, V.P.S. 1988. Biological control of root knot nematode on brinjal (*Solanum melongena* L.) with *Azatobacter chroococcum. In advances in plant Nematology*". (eds. M.A. Maqbool, A.M. Golden, A. Gaffar and L. Q., Krusberg) National Nematological Research Centre, Univ. of Karachi, Karachi, Pakistan, pp. 257-263.

Channabasappa, B.S., Krishnappa, K. and Reddy, B.M.R. 1995. Utilization of eco friendly biological agents and biocomponents in the integrated management of *Radopholus similis* on banana. In : *National symposium on nematode problems of India - An appraisal of the nematode managment with eco friendly approaches and biocomponents*, Indian Agril. Res. Inst., New Delhi.

Choudhury, B.N. and Kumar, C.V.S. 1999. Seasonal fluctuations of a commonly occurring predatory mononch. *Mylonchulus minor* (Nematoda : Mononchida). *Indian Journal of Nematology* 29 : 1, 69-73.

Das, T.K. and Mukhopadhyaya, M.C. 1990. Preliminary observations on the population fluctuations of some plant parasitic, predatory and freeliving nematodes in cultivated soils and grassland. *Indian Journal of Nematology* 20 : 108-111.

Deschiens, R. 1943. La prophylaxie des infestations vermineuses a 'Nematodes par les agents biologiques. *France Medical.* 2 : 573-574.

Devappa, V., Krishnappa, K. and Reddy, B.M.R. 1997. Management of root knot nematode *Meloidogyne incognita* in sunflower. *Mysore Journal of Agricultural Sciences* 31 : 155-158.

Dhanam, M. 1998. Predatory potential and host preference of *Clarkus elongatus* under laboratory conditions. In : *Proceedings of the First national Symposium on Pest Management in Horticultural Crops*, Bangalore, 1998. p. 315- 317.

Dhawan, S.C. and Anju Kamra, 1995. Bacterial antagonists of nematodes. IN *Nematode Pest Management : An Appraisal of Eco Friendly Approaches.* (Gopal Swarup, D.R. Dasgupta and J.S. Gill eds.). Nematological Society of India, New Delhi pp. 88-97.

Dohroo, N.P. and Meera Sharma, M. 1992. New host records of fungi from India. *Indian Phytopathology* 45 : 280.

Dohroo, N.P., Sharma, O., Sharma, M., Sarlach, R.S. 1994. Effect of organic amendments of soil on rhizome rot. nematodes and rhizosphere mycoflora of ginger (*Zingiber officinale* Roscoe). *Annals of Biology Ludhiana* 10 : 208-210.

Duddington, C.L. 1956. Biological control-Predacious Fungi. In : *Nematology* (Sasser, J.N. and Jenkins, W.R. eds.). The University of North Carolina Press, Raleigh, U.S.A. pp. 461-465.

Duddington, C.L. 1957. *The friendly Fungi* : Faber and Faber, London, U.K.

Duddington, C.L. and Duthoit, C.M.G. 1960. Green manuring and cereal root eelworm. *Plant pathology* 9 : 7-9.

Eapen, S.J. and Venugopal, M.N. 1995. Field evaluation of *Paecilomyces lilacinus* and *Trichoderma* spp. in cardamom nurseries for the control of root knot nematodes and rhizome rot disease. In : *National symposium on nematode problems of India - An appraisal of the nematode management with eco friendly approaches and biocomponents*, Indian Agril. Res. Inst., New Delhi.

Eapen, S.J., Ramana, K.V. and Sarma, Y.R. 1997. Evaluation of *Pseudomonas fluorescens* isolated for control of *Meloidogyne incognita* in black pepper (*Piper nigrum* L.) In *Biotechnology of spices, medicinal & aromatic plants.* Proceedings of the National Seminar on Biotechnology of Spices and Aromatic Plants. Calicut India. (Edison, S., Ramana, K.V., Sasikumar, B. Babu, K.N. and Eapen, S.J. eds.). pp. 129-133.

Fauzia Jairajpuri, M.S. and Khan Z. 1998. Biocontrol potential of *Mononchoides longicaudatus* on *Meloidogyne incognita* on tomato plants. *International Journal of Nematology* 8 : 89-91.

Garabedian, S. and Van Gundy, S.D. 1983. Use of avermectins for the control of *Meloidogyne incognita* on tomatoes. Journal of Nematology. 15 : 503-511.

Gautam, A., Siddiqui, Z.A. and Mahmood, I. 1995. Integrated management of *Meloidogyne incognita* on tomato. *Nematologia Mediterranea* 23 : 245-247.

Gintis, B.O., Morgan-Jones, G. and Rodriguez Kabana, R. 1983. Fungi associated with several developmental stages of *Heterodera glycines* from an Alabama Soybean field soil. *Nematropica* 13 : 181-200.

Giri, D., Banerjee, K., Laha, S.K. and Khatua, D.C., 1989. Some disease of horticultural and field crops. Environment and ecology. 7 : 821-825.

Gives, P.M. de, Davies, K.G., Clark, S.J., Behnke, J.M. 1999. Predatory behaviour of trapping fungi against srf mutants of *Caenorhabditis elegans*) and different plant and animal parasitic nematodes. *Parasitology* 119 : 95-104.

Gokte, N. and Swarup, G., 1988a. On the potential of some bacterial biocides against root knot and cyst nematodes. *Indian Journal of Nematology* 18 : 152-153.

Gokte, N. and Swarup G. 1988b. On the association of bacteria with larvae and galls of *Anguina tritici. Indian Journal of Nematology.* 18 : 313-318.

Gokte, N. and Swarup, G. 1988c. Efficacy of some bacterial isolates as seed treatment on penetration of *Anguina tritici* larvae to growing point of wheat seedlings and viability of larvae in the galls formed. *Indian Journal of Nematology.* 18 : 368-369.

Gopal Swarup, and Nandini Gokte 1986. Biological control. In : *Plant parasitic nematodes of India : Problems and Progress.* (eds. G. Swarup and D.R. Dasgupta). Nematological Society of India. I.A.R.I. New Delhi pp. 476-489.

Goswami, B.K. and Satyendra Singh, 1998. Efficacy of some soil mycoflora arising from the oil seed cakes of cottonseed, karanj, mahua and mustard amended soil against

M. incognita infecting okra. *Annals of Agricultural Research.* 1998, 19 : 158-161.

Goswami, B.K. and Uma Rao 1995. Fungi for biological control of Plant parasitic nematodes In : Gopal Swarup, Dasgupta, D.R. Gill, J.S. (Eds) Nematode pest management an appraisal of ecofriendly approaches, Nematological Society of India, p. 57-62.

Goswami, B.K. and Uma Rao, 1997. Studies on different isolates of *Paecilomyces lilacinus* collected from different agroclimatic regions in India. *India Journal of Nematology* 27 : 2, 238- 240.

Goswami, B.K., Uma Rao, and Satyendra Singh, 1998. Potentiality of some fungal bioagents against root knot nematode. *Meloidogyne incognita* infecting tomato, *Advances in IPM for horticultural crops, Proceedings of the First National Symposium on Pest Management in Horticultural Crops, environmental implications and thrusts, Bangalore, India* (Reddy, P.P. Kumar, N.K.K. and Verghese, A. eds.) 304-307.

Grewal, P.S., Sohi, H.S. Grabbe, K. (ed.) : Hilber, O. 1989. Effect of various fungal metabolites on *Apehlenchoides compositicola* Franklin and its multiplication on some fungi. *Mushroom Science. Proceedings of the twelfth International Congress on the Science and Cultivation of Edible Fungi*, Braunschweig, Germany, September 1987, Vol. 2. 1989, 813-820.

Gupta, M.C. and Satya kumar, 1998. Influence of certain organic amendments on the population dynamics of predatory nematodes. *Mononchus papillatus* and *Dorylaimus bastiani* in soil. *Annals of plant Protection Sciences* 6 : 48-52.

Hari, K. and Somasekhar, N. 1998. Utilisation of sugarcane wastes for the mass multiplication of fungal biocontrol agents. *Cooperative Sugar* 29 : 637-638.

Hazarika, K., Dutta, P.K. and Saikia 1998. Field efficacy of *Paecilomyces lilacinus* against root knot nematode in betelvine. *Journal of Phytological Research* 11 : 171-173.

Jain, R.K. and Sethi, C.L. 1988a. Influence of Endomycorrhizal fungi, *Glomus fasciculatum* and *G. epigaeus* on penetration and development of *Heterodera cajani* on cowpea. *Indian Journal of Nematology* 18 : 89-93.

Jain, R.K. and Sethi, C.L. 1988b. Interaction between vesicular arbuscular mycorrhiza and *Meloidogyne incognita, Heterodera cajani* on cowpea as influenced by time of inoculation. *Indian Journal of Nematology* 18 : 263-268.

Jairajpuri, M.S. and Bilgrami, A.L. 1990. Predatory nematodes. Aspects and Prospects. In : *Nematode Bio-control* (Jairajpuri, M.S., Alam, M.M., Ahmad, I. eds), 95-125.

Jatala, P., Kaltenbach, R. and Bocangel, M. 1979. Biological control of *Meloidogyne incognita acrita* and *Globodera rostochiensis* on potatoes. *Journal of Nematology* 11 : 303.

Jonathan, E.I., Padmanabhan, D. and Ayyamperumal. A. 1995. Biological control of root knot nematode on betelvine, *Piper betle*, by *Paecilomyces lilacinus Nematologia Mediterranea*, 23 : 191-193.

Kerry, B.R., Irving, F. and Horsney, J.C. 1986. Variation between strains of the nematophagous fungus, *Verticillium chlamydosporium* Goddard. I. Factors affecting growth *in vitro. Nematologica* 32 : 461-473.

Khan, Tabreiz, A., Hussain, S.I. and Azam, M.F. 1984. Effect of culture filtrates of eight species of *Aspergillus on the hatching and mortality of Meloidogyne incognita. Indian journal of Nematology* 14 : 51-54.

Khan, Tabreiz, A., Hussain, S.I. and Azam, M.F. 1984. Effect of culture filtrates of *Aspergillus niger* and *Rhizoctonia solani* on penetration and development of root knot nematodes and the plant growth of tomato var. marglobe. *Indian Journal of nematology* 14 :106-109.

Khan, M.R., and Goswami, B.K. 1995. Nematicidal efficacy of decomposed neem cake and fungal filtrates of *Paecilomyces lilacinus* and *Aspergillus flavus* against *Meloidogyne incognita. Proceedings of National symposium on Sustainable Agriculture in Sub-Humid Zone.* pp. 359-361.

Khan, M.R. and Tarannum, Z. 1999. Effect of field application of various micro organisms on *Meloidogyne incognita* on tomato. *Nematologia Mediterranea* 27 : 233-238.

Khan, M.W., Ashraf, S.S. and Haider, S.R. 1988. An analysis of biocontrol efficiency of *Paecilomyces lilacinus* in management of root knot nematodes. (Abstract), In Internation cofnerence on Research in Plant Sciences and Its Relevance to the Future, 7-11 March, 1988, Delhi, India, pp. 67.

Khan, S.T. and Khan, T.A. 1992. Effect of culture filtrates of soil fungi on the hatching and mortality of root knot nematode (*Meloidogyne incognita*). *Current Nematology* 3 : 53- 60.

Khan, T.A. and Husain, S.I. 1988. Studies on the efficacy of *Paecilomyces lilacinus* as biocontrol agent against a disease complex caused by the interaction of *Rotylenchulus reniformis. Meloidogyne incognita* and *Rhizoctonis solani* on cowpea. *Nematologia Mediterranea* 16 : 229-231.

Khan, T.A. and Husain, S.I. 1991. *In vitro* studies on the toxicity of culture filtrates of different fungi on the growth of *Rhizoctonia solani, New Agriculturist.* 1 : 107-110.

Khan, T.A. Khan, S.T., Fazal, M. and Siddiqui, Z.A. 1997. Biological control of *Meloidogyne incognita* and *Fusarium solani* disease complex in papaya using *Paecilomyces lilacinus* and *Trichoderma harzianum. International Journal of Nematology* 7 : 127-132.

Khan, T.A., and Saxena S.K. 1997. Effect of root dip treatment with culture filtrates of soil fungi on multiplication of *Meloidogyne javanica* and yield of tomato. *Tests of Agrochemicals and Cultivars* No. 18 : 50-51.

Khan, T.A. and Saxena, S.K. 1997b. Effect of root-dip treatment with fungal filtrates on root penetration, development and reproduction of *Meloidogyne javanica* on tomato, *International Journal of Nematology,* 7 : 85-88.

Khan, T.A. and Husain, S.I. 1990. Biological control of root knot and reniform nematodes and root rot fungus on cowpea. *Bioved.* 1 : 19-24.

Khan, T.A. and Saxena, S.K. 1996. Comparative efficacy of *Paecilomyces lilacinus* in the control of *Meloidogyne* spp. and *Rotylenchulus reniformis* on tomato. *Pakistan Journal of Nematology* 14 : 111-116.

Khanna, A.S., and Sharma, N.K. 1989. Host range of *Aphelenchoides agarici. Indian Journal of Nematology* 19 : 32-34.

Koshy, P.K. and Gulsar Banu, J. 1998. Use of fungal and bacterial antagonists for the biological control of nematodes in plantation crops. In : Biological Suppression of Plant diseases, Phytoparasitic nematodes and weeds of plant pathogens, Nematodes and Weeds. (Singh, S.P., Hussaini, S.S. eds.).

Kum Kum Walia and Gopal Swarup, 1985. Isolation of nematophagous and parasitic fungi from soil and other sources in India. *Indian Journal of Nematology* 15 :253.

Latha, T.K.S. and Sivakumar, C.V. 1998. Effect of culture filtrates of antagonistic organisms on cyst nematode, *Heterodera cajani* in blackgram. *Journal of Biological Control.* 12 : 143-145.

Leij, de F. A.A.M. and Kerry, B.R. 1991. The nematophagous fungus, *Verticillium chlamydosporium*, as a potential biological control agent for *Meloidogyne arenaria. Revue de Nematology.* 14 : 157-164.

Linford, M.B., Yap, F. and Oliveira, J.M. 1938. *Soil Science* 45 : 127-141.

Lisansky, S.G., Quinlan, R. and Tassni, G. 1993. In : *Bacillus thuringiensis* production Handbook, Laboratory Methods, Manufacturing, Formulation, Quality Control and Registration. C.P.L. press. C.P.L. Scientific Ltd. Newbury, p. 123.

Mahendra Singh, Samar, R., Sharma, G.L. and Singh, M. 1998. Effect of various levels of organic carbon on the development of *Pasteuria penetrans* on *Meloidogyne incognita. Indian Journal of Nematology* 28 : 72-76.

Mahendra Singh, and Sharma, G.L. 1995. Effect of soil moisture on development of Pasteuria penetrans on *Meloidogyne incognita* in brinjal crop. *Current Nematology* 6 : 161-165.

Maheswari, T.U. and Mani, A. 1988. Combined efficacy of *Pasteuriaa penetrans* and Paecilomyces lilacinus on the biocontrol on *Meloidogyne javanica* on tomato. *Intl. Nematol Network Newsl.* 5 (3) : 10-11.

Maheswari, T.U., Mani, A. and Rao, P.K. 1987. Combined efficacy of the bacterial spore parasite, *Pasteuria penetrans* (Thorne, 1940) and nematicides in the control of *Meloidogyne javanica* on tomato. *Journal of Biological Control.* 1 : 53-57.

Mahmood, I. and Siddiqui, Z.A. 1993. Integrated managment of *Rotylenchulus reniformis* by green manuring and *Paecilomyces lilacinus. Nematologia Mediterranea* 21 : 285-287.

Mani, A. 1988. Studies on the bacterial parasite *Pasteuria penetrans* : I. Spore viability after storage II, Culture on citrus nematode *Tylenchulus semipenetrans. Intn. Nematol. Network Newsl.* 5 (2) : 24-25.

Mani, A. and Anandam, R.J. 1989. Evaluation of plant leaves, oil cakes and agro industrial wastes as substrates for mass multiplication of the nematophagous fungus. *Paecilomyces lilacinus. Journal of Biological Control* 3 : 56-58.

Mani, A., Murthy, I.R., Rao, P.K. and Prasad, V.D. 1989. Growth of *Paecilomyces lilacinus* on natural substrates and its efficacy against citrus nematode. *Tylenchulus semipenetrans. Journal of Biological Control* 3 : 59-61.

Mani, M.p. Rajeswari, S. and Sivakumar, C.V. 1998. Managment of the potato cyst nematodes, *Globodera* spp. through plant rhizosphere bacterium *Pseudomonas fluorescens* Migula. *Journal of Biological Control* 12 : 131-134.

Mani, M.P. Sivakumar, C.V. and Ramakrishnan, S. 1999. Status of *Pasteuria penetrans* (Thorne) Sayre & Starr in root knot infested vineyards of Tamil Nadu, India. *Indian Journal of Nematology* 29 : 104-105.

Mankau, R., 1975. *Bacillus penetrans* n comb. Causing a virulent disease of plant parasitic nematodes. *Journal of Invertebrate Pathology* 26 : 333-339.

Mankau, R. 1980. Biocontrol : Fungi as nematode control agents. *Journal of Nematology* 12 : 244-252.

Mathew, M.J. Saju, K.A. Venugopal M.N. 1998. Efficacy of entomogenous fungi on biological suppression of *Pentalonia nigronervosa* f. caladii Van der Goot of cardmom (*Elettaria cardamomum* Maton). *Journal of Spices and Aromatic Crops* 7 : 43- 46.

Mittal, N. Saxena, G. and Mukerji, K.G. 1995. Integrated control of root knot disease in three crop plants using chitin and *Paecilomyces lilacinus. Crop protection.* 14 : 647-651.

Mohammad Akhtar, 1998. Effects of two compositae plant species and two types of fertilizer on nematodes in an alluvial soil, India. *Applied soil Ecology*, 10 : 21-25.

Mohilal, N. and Dhanachand, C. 1992. Observations on gut contents of the predatory nematodes of Manipur. *Current Nematology*, 3 : 89-94.

Mohinder Singh, Sharma, S.B. and Sharma, R. 1996. Parasitism of *Heterodera cajani* by *Fusarium solani* and other soil fungi. *Indian Journal of Nematology* 26 : 189-196.

Morgan-Jones, G. and Rodriguez Kabana, R. 1985. Phytonematode pathology : fungal modes of action - a perspective. *Nematropica* 15 : 107-114.

Morgan-Jones, G. Godoy, G. and Rodriguez Kabana R. 1981. *Verticillium chlamydosporium*, fungal parasite of *Meloidogyne arenaria* females. *Nematropica* 11 : 115-119.

Morgan Jones, G., White, J.F. and Rodriguez Kabana R. 1983. Phytonematode pathology : Ultrastructural studies. I. Parasitism of *Meloidogyne arenaria* eggs by *Verticillium chlamydosporium*. *Nematropica* 13 : 245-260.

Morgan-Jones, G., White, J.F. and Rodriguez-Kabana R. 1984. Phytonematode pathology: Ultrastructural Studies, 2. Parasitism of *Meloidogyne arenaria* eggs and larvae by *Paecilomyces lilacinus*. *Nematropica* 14 : 57-71.

Mujib. I. Azmi, 1983. Predatory behaviour of nematodes I. Biological control of *Helicotylenchus dihystera* through the predaceous nematodes. *Itonchus monohystera Indian Journal of Nematology 13 : 1-18.*

Nagesh, M. and Parvatha Reddy, P. 1995a. Comparative efficacy of *Paecilomyces lilacinus* (Thom.) Samson and *Verticillium lecanii* (A. Zimmerman) Viegas in combination with botanicals against, *Meloidogyne incognita* (Kofoid and White, 1919) Chitwood, 1949 infecting *Crossandra undulaefolia* L. *J. Biol. Control* 9 :109-112.

Nagesh, M. and Parvatha Reddy, P. 1995b. Eco friendly management of antagonistic potential for the control of *Meloidogyne incognita* on *Crossandra undulaefolia*. National Symp. on Nematode Problems of India - An Appraisal of the Nematode Management with Eco-friendly Approaches and Biocomponents, Indian Agri. Res. Inst. New Delhi, pp. 127-128.

Nagesh, M. and Parvatha Reddy, p. 1996a. Management of *Meloidogyne incognita* on *Crossandra undulaefolia* using vesicular arbuscular mycrorrhiza, *Glomus mosseae* and oil cakes. *Mycor. News*. 9 (1) : 12-14.

Nagesh, M. and Parvatha Reddy, P. 1996b. Management of root knot nematodes on carnation using antagonistic fungi and neem cake. *National Seminar on Integrated Crop Disease Management for Sustainable Agriculture*, Aurangabad, P. G105.

Nagesh, M. and Parvatha Reddy, P. 1998. A modified method for mass multiplication of *Pasteuria penetrans* (Thorne) Sayre & Starr. *Journal of Biological Control*. 12 : 63-66.

Nagesh, M., Parvatha Reddy, P., and Ramachandran, N. 1999. Integrated management of *Meloidogyne incognita* and *Fusarium oxysporum* f.sp. Gladioli in gladiolus using antagonistic fungi and neem cake. In : Nematology : Challenges and Opportunities in 21st Century (Ed. Usha K. Mehta). Proceedings of the 3rd International Symposium of Afro Asian Society of Nematologists, April 16-19, 1998, Coimbatore, India pp . 263-266.

Nagesh, M., Paravatha Reddy, P. and Rao, M.S. 1995. Management of *Meloidogyne incognita* on *Crossandra undulaefolia* using *Paecilomyces lilacinus* and *Pasteuria penetrans* formulations. National symp. on Perspectives in Eco- friendly Approaches to Plant Prot. Indian Agri. Res. Inst. New Delhi, pp. 47-48.

Nagesh, M., Parvatha Reddy, P. and Rao M.S. 1996. Integrated management of *Meloidogyne incognita* on tuberose using *Paecilomyces lilacinus* in combination with plant extracts. *Nematol. Medit.* 25 : 3-6.

Nagesh, M., Parvatha Reddy, P. and Rao, M.S. 1997a. Management of *Meloidogyne incognita* on *Crossandra undulaefolia* using arbuscular mycorrhiza, *Glomus mosseae* and oil cakes. *Mycorrhiza News* 9 : 12-14.

Nagesh, M. Parvatha Reddy, P. and Rao, M.S. 1997b. Integrated management of *Meloidogyne incognita* on tuberose using *Paecilomyces lilacinus* in combination with plant extracts. *Nematol. Medit.* 25 : 3-7.

Nagesh, M., Parvatha Reddy, P. and Rao, M.S. 1999. Comparative efficacy of VAM fungi in combination with neem cake against *Meloidogyne incognita* on *Crossandra undulaefolia. Mycorrhiza News.* 11 : 11-13.

Nagesh, M., Parvatha Reddy, P., Vijaykumar, M.V. and Nagaraju, B.M. 1998. Management of *Meloidogyne incognita* on *Polianthes tuberosa* using parasitic fungus, *Paecilomyces lilacinus* in combination with oil cakes. In : *Advances in IPM for horticultural crops. Proceedings of the First National Symposium on Pest management in Horticultural Crops : environmental implications and thrusts.* (Reddy PP; Kumar NKK and A. Verghese eds.) Navbharath Enterprises Bangalore, India, pp. 339-344.

Nakat, R.V., Acharya, A., Jonathan, E.I. Hazarika, K., Jha S., Singh , U.S., Wasnikar, A.R., Subba Rao, D.V. and Sitramaiah, K. 1995. Evaluation of *Paecilomyces lilacinus* for the control of root knot nematodes *Meloidogyne incognita* on betelvine. *National Symposium on Nematode Problems of India - An Appraisal of the Nematode management with Eco-friendly Approaches and Biocomponents*, Indian Agri. Res. Inst., New Delhi, p. 44.

Narula, N. 1999. Effect of rhizospheric bacteria on plant growth of wheat infected with *Heterodera avenae. Nematologia Mediterranea*, 27 : 311-314.

Nemec, S. and Meridith, F.I. 1981. Amino acid content of leaves in mycorrhizal and non mycorrhizal citrus rootstocks. *Ann. Bot.* 47 : 351-358.

Nordumeyer, D. and Dickson, D.W. 1981. Effect of oximecarbamates and organophosphates and one avermectin on the oxygen uptake of three *Meloidogyne* spp. *J. Nematol* 13 : 452-453.

Pandey, R. and Trivedi, P.C. 1992. Biological control of *Meloidogyne incognita* by *Paecilomyces lilacinus* in *Capsicum annuum. Indian Phytopathology.* 45 : 1, 134-135.

Parvatha Reddy, P. and Khan, R.M. 1988. Evaluation of *Paecilomyces lilacinus* for the biological control of *Rotylenchulus reniformis* infecting tomato as compared with carbofuran. *Nematol. Medit.* 16 : 113-116.

Parvatha Reddy, P. and Khan, R.M. 1989. Evaluation of biocontrol agent, *Paecilomyces lilacinus* and carbofuran for the management of *Rotylenchulus reniformis infecting brinjal. Pak. J. Nematol 7 : 55-59.*

Parvatha Reddy, P. and Khan R.M. 1991. Integrated management of the citrus nematode, *Tylenchulus semipenetrans* using oil cakes and *Paecilomyces lilacinus. Afro Asian J. Nematol* 1 : 221-222.

Parvatha Reddy, P., Khan, R.M. and Rao, M.S. 1996a. Integrated management of the citrus nematode *Tylenchulus semipenetrans* using pesticides and parasitic fungus, *Paecilomyces lilacinus. Pest Management in Hortil. Ecosystems* 2 : 61-63.

Parvatha Reddy, P. and Nagesh N. 1995. Nematode problems of ornamental plants in India. In 'Nematode Pest management - An Appraisal of Eco Friendly Approaches'. (G. Swarup D.R. Dasgupta & J.S. Gill, Eds.), pp. 244-251. Nematological Society of India, New Delhi.

Parvatha Reddy, P. and Nagesh M. 1998. Fungal and bacterial antagonists for biological suppression of plant parasitic nematodes on Horticultural crops. In : Biological Suppression of Plant diseases, phytoparasitic nematodes and weeds (Singh, S.P. Hussaini, S.S. eds.), Project Directorate of Biological (Control Bangalore, p. 138-153.

Parvatha Reddy, P., Nagesh M. and Devappa, V. 1997a. Integrated management of the burrowing nematode, *Radopholus similis*, using endomycorrhiza, *Glomus mosseae* and oil cakes, Pest Managmt. Hortil, Ecosystem. 3 : 25-29.

Parvatha Reddy, P., Nagesh, M. and Devappa, V. 1997b. Effect of integration of *Pasteuria penetrans, Paecilomyces lilacinus* and neem cake for the management of root knot nematode infecting tomato. *Pest managmt. Hortil. Ecosystme.* 3 : 100-104.

Parvatha Reddy P., Nagesh, M., Devappa, V. 1998. Management of *M. incognita* on tomato by integration of *Trichoderma harzianum, Glomus fasciculatum* and neem cake In : Advances in IPM for horticultural crops. Proceedings of the First National symposium on pest management in Horticultural Crops : environmental implications and thrusts (Parvatha Reddy, P., Kumar, N.K.K. and Verghese, A. eds.) Bangalore, India, pp. 349- 352.

Parvatha Reddy, P., Rao, M.S. and Nagesh, M. 1995. Integrated management of the citrus nematode, *Tylenchulus semipenetrans* Cobb. using VA mycorrhiza, *Glomus fasciculatum* (Thaxt.) Gerd. & Trappe and oil cakes. *Pest management in Hortil. Ecosystems* 1 : 37-41.

Parvatha Reddy, P., Rao, M.S. and Nagesh, M. 1996b. Management of the citrus nematode on acid lime by integration of parasitic fungi and oil cakes. *Pest Management in Hortil. Ecosystems* 2 : 15-18.

Parvatha Reddy, P., Rao, M.S. and Nagesh M. 1996c. Management of the citrus nematode, *Tylenchulus semipenetrans* by integration of *Trichoderma harzianum* with oil cakes. *Nematol. Medit.* 24 : 265-267.

Parvatha Reddy, P., Rao, M.S., Nagesh, M. 1996d. Integrated management of the citrus nematode. *Tylenchulus semipenetrans* using pesticides and parasitic fungus, *Paecilomyces lilacinus. Pest Management in Horticultural Ecosystems* 2 : 61-63.

Parvatha Reddy, P., Rao M.S. and Nagesh, M. 1998. Effect of bare root dip treatment of tomato seedlings in plant leaf extracts mixed with *Paecilomyces lilacinus* spores for the management of *Meloidogyne incognita.* In : *Advances in IPM for horticultural crops. Proceedings of the first National Symposium on Pest Management in Horticultural Crops : Environmental Implications and Thrust,* (Reddy, P.P., Kumar, N.K.K. and A. Verghese eds.) Bangalore, India pp. 334-338.

Parvatha Reddy, P., Rao, M.S. and Nagesh, M. 1999. Eco friendly management of *Meloidogyne incognita* on tomato by integration of *Verticillium chlamydosporium* with neem and calotripis leaves. *Nematologia Mediterranea.*

Parveen, G. and Alam, M.M. 1997. Effect of leaf on nematode antagonistic fungus, *Paecilomyces lilacinus, Annals of plant Protection Sciences* 5 : 175-178.

Patel, D.J., Vyas, R. V., Patel, B.A. and Patel, R.S. 1995. Bioefficacy of *Paecilomyces lilacinus* in controlling *Meloidogyne javanica* (Pathotype 2) on groundnut. International Arachis Newsletter. No. 15 : 46-47.

Patel, N.B., Vyas, R.V. and Patel D.J. 1998. Efficacy of *Paecilomyces lilacinus* for the management of *Meloidogyne javanica Pt.2 on groundnut. Proceedings of National Symposium on Rational Approaches in Nematode Management for sustainable Agriculture, Nov. 23-25, 1998 B.A. - college of Agriculture, Anand, India pp. 23-24.*

Patel, R.R. Patel, D.J. and Patel, B.A. 1991. Influence of organic amendments on growth and sporulation of nematophagous fungus. *Paecilomyces lilacinus* (Thorn.) Samson. *Current Nematology* 2 : 39-40.

Patel, R.R., Patel, D.J. and Patel, B.A. 1992. Growth and Sporulation of nematophagous fungus, *Paecilomyces lilacinus*(Thom). Samson as influenced by different cereal media. *Current Nematology* 3 : 99-100.

Pathak, J.J. and Saikia, B. 1999. Biological control of root knot nematode of betelvine. *Journal of Interacademicia.* 3 : 117-120.

Pathak, K.N. and Kumar B. 1995. Nematotoxic effects of *Trichoderma harzianum* culture filtrate on second stage juveniles of rice root knot nematode. *Indian Journal of Nematology* 25 : 223-224.

Perveen, S., Ehteshamul-Haque, S. and Ghaffar, A. 1998. Efficacy of *Pseudomonas areuginosa* and *Paecilomyces lilacinus* in the control of root rot- root knot disease complex on some vegetables. *Nematologia Mediterranea* 26 : 209-212.

Powell, K.A. 1993. The commercial exploitation of microorganisms in agriculture. In : *Exploitation of micro organisms*, ed. D.G. Jones, Chapman & Hall, London, pp. 441-459.

Prasad, S.S.V. Tilak, K.V.B.R. and Gollakota, K.G. 1972. *Journal of Invertebrate Pathology* 20 : 377.

Putter, J.G., Mac Connell, F.A. Preiser, F.A. Haidri, A.A. Rishich, S.S. and Dybas, R.A. 1981. Avermectins : Novel class of insecticides, acaricides and nematicides from a soil microorganisms. *Experientia* 37 : 963-964.

Rahman, P.F. and Ahmad, I. 1995. Community analysis of predatory nematode species from Aligarh soils. India, *Nematologia Mediterranea* 23 : 57-60.

Raj, M.A.J., Mani, A., Jaya Raj M.A. 1988. Biocontrol of *Meloidogyne javanica* with the bacterial spore parasite *Pasteuria penetrans. International Nematology Network Newsletter* 5 : 3-4.

Rajeshwari, S., Santhi, A. and Sivakumar, C.V. 1997. Effect of organic soil amendments, *Trichoderma viride* Pers. ex Fr. and carbofuran on the nematode trapping fungus *Arthrobotrys cladodes* var. *Macroides* (Drechsler, 1944) and plant parasitic nematodes. *Journal of Biological Control* 11 : 49-52.

Ramakrishnan, S., Sivakumar, C.V. and Poornima K. 1998. Management of rice root nematode. *Hirschmanniella gracilis* (de Man) Luc & Goodey with *Pseudomonas fluorescens* Migula. *Journal of Biological Control* 12 : 135-141.

Ramakrishnan, S. and Sivakumar, C.V. 1999. Biological control of rice root nematode with *Pseudomonas fluorescence.* In : *Proceedings of National Symposium on Rational Approaches In Nematode Management for Sustainable Agriculture*, Anand, India (Dhawan, S.C. and Kaushal, K.K.) pp. 43-46.

Rana, B.P. and Dalal, M.R. 1994. Effect of different inoculum levels of *Paecilomyces lilacinus* in controlling *Heterodera cajani* race A on mungbean (*Vigna radiata* (L) Wilczek). *Annals of Plant Protection Sciences* 2 : 43-45.

Rana, B.P. and Dalal, M.R. 1995. Management of *Heterodera cajani* in Mungbean with *Paecilomyces lilacinus* and carbofulfan. *Annals of plant Protection Sciences* 3 : 145-148.

Rangaswamy, G. and Bagyaraj, D.J. 1996. In : *Agricultural Microbiology*, Prentice Hall of India Pvt. Ltd. pp. 422.

Rangaswamy, S.D. Parvatha Reddy, P. and Nanje Gowda, D. 1999. Management of root knot nematode,*Meloidogyne incognita* in tomato by intercropping with marigold and mustard, *Pest Managmt. Hortil. Ecosystems* 5 : 118-121.

Rao, M.S. and Parvatha Reddy P. 1992a. Innovative approaches of utilization of bio control agents. *Paecilomyces lilacinus* and *Verticillium chlamydosporium* against root knot nematode on tomato. *First Afro Asian Nematol. Symp.* Aligarh Muslim Univ., Aligarh pp. 25-26.

Rao, M.S., and Parvatha Reddy P. 1992b. Prospects of management of root knot nematode on tomato through the integration of biocontrol agents and botanicals. *Indian Phytopath.* 46 : 337.

Rao, M.S. and Parvatha Reddy P. 1993. Interactive effect of *Verticillium chlamydosporium* and castor cake on the control of root knot nematode on egg plant. In "*Soil Organisms and Sustainability*" (D. Rajagopal, R.D. Kale and K. Bano, eds.), pp. 119-123. Indian Soc. of Soil Biol. and Ecol. Univ. of Agri. Sci. Bangalore.

Rao, M.S. and Parvatha Reddy, P. 1994. A method for conveying *Paecilomyces lilacinus* to soil for the management of root knot nematodes on egg plant. *Nematol. Medit.* 22 : 265-267.

Rao, M.S. and Parvatha Reddy, P. 1996. Effective utilization of wellgro and RD-9 Repelin for the management of root knot nematode on tomato. In "Neem and Environment" (R.P. Singh, M.S. Chari, A.K. Raheja and W. Kraus, eds.), pp. 651-656. Oxford & IBH Publishing Co. Pvt. Ltd., New Delhi.

Rao, M.S. Gowen, S.R. Pembroke, B. and parvatha Reddy P. 1996. Relationship of *Pasteuria penetrans* spore emncumberance on juveniles of *Meloidogyne incognita* and their infection in adults. *Nematol Medit* 25 : 129-131.

Rao, M.S. Kerry, B.R., Gowen, S.R., Bourne, J.M. and Parvatha Reddy, P. 1997a. Management of *Meloidogyne incognita* in tomato nurseries by integration of *Glomus deserticola* with *Verticillium chlamydosporium. J. Pl. Dis. Prot.* 104 : 419- 422.

Rao, M.S., Mohan Das, S. and Parvatha Reddy, P. 1993a. Integrated management of Root knot nematode on egg plant in nursery beds with the combinatios of *Glomus mosseae* and neem leaf. *Indian J. Nematol.* 23 : 14.

Rao, M.S., Mohan Das, S., Parvatha Reddy, P. and Khan R.M. 1993b. Synergistic effect of endomycorrhizal fungus (*Glomus fasciculatum*) and biocontrol fungus (*Paecilomyces lilacinus*) in the management of root knot nematode infecting tomato. In "*Soil Organisms and Sustainability*" (D. Rajagopal, R.D. kale and K. Bano , eds.), pp. 119-123. Indian Soc. of Soil Biol. and Ecol., Univ. of Agri. Sci. Bangalore.

Rao, M.S. Mohan Das, S., Parvatha Reddy, P. and Khan, R.M. 1993c. Management of reniform nematode on egg plant by integrating a botanical (castor cake), mycorrhiza (*Glomus fasciculatum*) and bio agent (*Paecilomyces lilacinus*). Golden Jubilee Symp. on Hortil Res. - Changing Scenario, Bangalore, pp. 259- 260.

Rao, M.S., Mohan Das, S. Parvatha Reddy, P. and Khan R.M. 1993c. Management of reniform nematode on egg plant by integrating a botanical (castor cake), mycorrhiza (*Glomus fasciculatum*) and bio agent (*Paecilomyces lilacinus*). Golden Jubilee Symp. on Hortil. Res. - Changing Scenario, Bangalore, pp. 259- 260.

Rao, M.S., Parvatha Reddy, P. and Mohan Das, S. 1993d. Integrated management of root knot nematodes on tomato using endomycorrhiza (*Glomus mosseae*) and biological control agent (*Paecilomyces lilacinus*). *Golden Jubilee Symp. on Hortil. Res. - Changing Scenario*, Bangalore, p. 258.

Rao, M.S., parvatha Reddy, P. and Mohan Das, S. 1995a. Studies on development of a bio rational management strategy against root knot nematode attacking tomato.

National Symp. on Nematode Problems of India - An Appraisal of the Nematode management with Eco-Friendly Appraoches and Biocomponents, Indian Agri. Res. Inst., New Delhi, P. 14.

Rao, M.S., Parvatha Reddy, P. and Mohan Das, S. 1995b. Effect of integration of endomycorrhiza (*Glomus mosseae*) and neem cake on the control of root knot nematode on tomato. *J. Plant Diseases and Prot.* 102 : 526-529.

Rao, M.S., Parvatha Reddy, P. and Mohan Das, S. 1996. Effect of integration of *Calotropis procera* leaf and *Glomus fasciculatum* on the management of *Meloidogyne incognita* infesting tomato. *Nematol. Medit.* 24 : 59-61.

Rao, M.S., Parvatha Reddy, P. and Nagesh, M. 1994. Use of biotic agent *Paecilomyces lilacinus* in combination with neem cake suspension treatment of okra seed for the management of root knot nematode under field conditions. *National Symp. on Emerging Trends in Pest Management*, Nauni pp. 121-122.

Rao, M.S., Parvatha Reddy, P. and Nagesh, M. 1997. Effective use of neem cake extract for the management of root knot nematodes infecting okra (*Abelmoschus esculentus*). *Pest Managmt. Hortil. Ecosystems* 3 : 95-99.

Rao, M.S., Parvatha Reddy, P. and Nagesh M. 1997b. Integrated management of *Meloidogyne incognita* on okra by castor cake suspension and *Paecilomyces lilacinus*. *Nematol. Medit.* 25 : 17-19.

Rao, M.S., Parvatha Reddy, P. and Nagesh M. 1997c. Integration of *Paecilomyces lilacinus* with neem leaf suspension for the management of root knot nematodes on egg plant. *Nematol. Medit.* 25 : 249-252.

Rao, M.S., Parvatha Reddy, P. and Nagesh, M. 1997d. Management of root knot nematode, *Meloidogyne incognita* on tomato by integration of *Trichoderma harzianum* with neem cake. *J. Pl. Dis. Prot.* 104 : 423-425.

Rao, M.S., Parvatha, Reddy, P., Nagesh, M., 1998a. Integrated management of *Meloidogyne incognita* on tomato using *Verticillium chlamydosporium* and *Pasteuria penetrans*. *Pest Management in Horticultural Ecosystems* 4 : 32- 35.

Rao, M.S., Parvatha Reddy, P., Nagesh M. 1998b. Integrated management of *Meloidogyne incognita* on tomato using oil cakes and a bioagent, *Verticillium chlamydosporium*. In Reddy-PP (ed); Kumar NKK (ed.); Verghese - A Advances in IPM for horticultural crops proceedings of the First national Symposium on Pest Management in Horticultural Crops : environmental implications and thrusts, Bangalore, India, pp. 345-348.

Rao, M.S., Reddy, P.P. and Nagesh M., 1998c. Evaluation of plant based formulations of *Trichoderma harzianum* for the management of *Meloidogyne incognita* on egg plant. *Nematologia Mediterranea* 26 : 59-62.

Rao, M.S., Parvatha Reddy, P. and Nagesh M. 1998d. Evaluation of *Paecilomyces lilacinus* Cultured on neem cake extract for the management of root knot nematode on egg plant. *Pest Managmt. Hortil. Ecosystems* 4 : 116-119.

Rao, M.S., Parvatha Reddy, P., and Nagesh, M. 1999. Bare root dip treatment of tomato seedling in calotropis or castor leaf extracts mixed with *Paecilomyces lilacinus* spores for the management of *Meloidogyne incognita*. *Nematologia Mediterranea* 27 : 323-326.

Rao, M.S., Parvatha Reddy, P., Somasekhar, N. and Nagesh M. 1997. Management of root knot nematodes, *Meloidogyne incognita* in tomato nursery by integration of endomycorrhiza, *Glomus fasciculatum* with castor cake. *Pest Managmt. Hortil. Ecosystems* 3 : 31-35.

Rekha, Arya, Saxena, S.K. Arya, R. 1998. *Trichothecium roseum* together with *Rhizoctonia solani* and *M. incongnita* on germination of seeds of tomato cv. 'Pusa Ruby'. Indian Journal of Nematology 28 : 2, 217-221.

Sankaranarayanan, C., Hussaini, S.S., Kumar, P.S., Prasad, R.D. 1998. Nematicidal effect of fungal filtrates against root knot nematodes. *Journal of Biological Control* 11 : 1-2, 37-41.

Sanknarayanan, C., Hussaini, S.S. Kumar, P.S. and Rangeshwaran, R., 1999. Antagonistic effect of *Trichoderma* and *Gliocladium* sp. against the root knot nematode (*M. incognita*) on sunflower. In : Dhawan - SC (ed.); Kaushal KK - *Proceedings of national symposium on rational approaches in nematode management fro sustainable agriculture*, Anand India 23-25, November, 1998. 1999, 25-27.

Santhi, A., Rajeshwari, S., Sivakumar, C.V. 1998. Soil application of *Pseudomonas fluorescens* (Migula) for the control of root knot nematode *Meloidogyne incognita* on grapevine. In : Usha K. Mehta (Ed.) Proceedings of the Third International Symposium of Afro-Asian Society of Nematologists held at Sugarcane Breeding Institute. Coimbatore, India, p. 199- 202.

Santhi, A., Rajeshwari Sundarababu, Sivakumar, C.V., Sundarbabu, R. 1999. Field evaluation of rhizobacterium, *Pseudomonas fluorescens* for the management of the citrus nematode, *Tylenchulus semipenetrans*. In : Dhawan, SC (ed.); Kausahl KK Proceedings of national syposium on rational approaches in nematode management for sustainable agriculture, Anand, India, 23-25, November, 1998, 38-42.

Santhi, A., Sivakumar, C.V. 1995. Biocontrol potential of *Pseudomonas fluorescens* (Migula) against root knot nematode, *Meloidogyne incognita* (Kofoid and White, 1919) Chitwood, 1949 on tomato, *Journal of Biological Control*, 9 : 2, 113-115.

Sathpathi, C.R. 1998. Bio efficacy, shelf life compatibility toxicity and residue of a biopesticide with inorganic pesticide, *Environment and Ecology* 16 : 4, 751-753.

Sayre, R.M. & Starr, M.P. 1985. Proceedings of Helminthological Society of Washington, 52 : 149-165.

Sekhar, N.S., Gill, J.S. 1990. Effect of temperature on attachment of *Pasteuria penetrans* spores to the second stage juveniles of *Meloidogyne incognita*. Indian Journal of nematology 20 : 1, 57-59.

Sekhar, N.S., Gill, J.S. 1990. Penetration and multiplication of *Meloidogyne incognita* as influenced by *Pasteuria penetrans. Indian Journal of Nematology* 20 : 2, 213- 218.

Sekhar, N.S., Gill, J.S. 1991. Efficacy of Pasteuria penetrans alone and in combination with carbofuran in controlling Meloidogyne incognita. *Indian Journal of Nematology* 21 : 1, 61-65.

Shalini Goyal, Trivedi, P.C. Goyal, S., 1998. Application of nematophagus funig in the management of root knot nematode infecting *Celosia argentea. Journal of Phytological Research* 11 : 1, 69-72.

Sharma, A., Trivedi, P.C. 1987. Screening of substrates suitable for the growth of *Paecilomyces lilacinus. International Nematology Network Newsletter.* 4 : 4, 24-26 : 4 ref.

Sharma, A., Trivedi P.C. 1989. Control of root knot nematodes on *Trigonella foenumgracum by Paecilomyces lilacinus. Nematologia Mediterranea* 17 : 2, 131-133: 5 ref.

Sharma, A. and Trivedi, P.C. 1989. Influence of inoculum levels of fungus *paecilomyces lilacinus* (Thom.) Samson on the biocontrol of root knot nematode, *Meloidogyne incognita* (Chitwood). *Intn. Nematol. Network Newsl.* 6 (2) : 27-29.

Sharma, W. and Trivedi P.C. 1994. Interaction of root knót nematode with VAM on *Abelmoschus esculentus* and their effect on net yield. In "*Vistas in Seed Biology*" (T. Singh and P.C. Trivedi, eds.) II : 61-81. Printwell, Jaipur.

Sharma, M., Saxena, S.K. 1992. Effect of culture filtrates of Rhizoctonia solani and *Trichoderma viridae* on hatching of larvae of root knot nematode (*M. incognita)* *Current Nematology* 3 :1, 61-64.

Sharma, R., Sharma, S.B. 1989. Sticky swarm disease of Heterodera cajani and *Meloidogyne javanica* caused by *Pasteuria penetrans*. International Pigeonpea - Newsletter, No. 10, 26- 27.

Siddiqui, Z.A. and Hussain, S.I. 1991. Studies on the biological control of root knot nematode. *Current Nematology* 2 : 1, 5-6.

Siddiqui, Z.A. Hussain - SI 1991. Control of *Meloidogyne incognita* and *Macrophomina phaseolina* on chickpea by fungal filtrates. Pakistan - Journal of Nematology. 1991, 9 : 2, 131-137; 12 ref.

Siddiqui, Z.A., Mahmood, I. 1992. Biological control of root rot disease complex of chickpea caused by *Meloidogyne incognita* race 3 and *Macrophomina phaseolina. Nematologia Mediterranea* 20 : 2, 199-202 : 9 ref.

Siddiqi, Z.A., Mahmood, I. 1993. Biological control of *Meloidogyne incognita race 3 and Macrophomina phaseolina by Paecilomyces lilacinus and Bacillus subtilis alone and in combination on chickpea. Fundamental and Applied Nematology 16 : 3, 215-218, 16 ref.*

Siddiqui, Z.A., Mahmood, I., 1993. Integrated control of a root rot disease complex of chickpea by fungal filtrates and green manuring. Nematologia Mediterranea. 21 : 161-164.

Siddiqui Z.A., Mahmood, I. 1994. Culture of *paecilomyces lilacinus* on leaf extracts and leaf residues for nematode control. *Bioresource Technology* 49 : 2, 187-189.

Siddiqui, Z.A., Mahmood, I., 1995. Some observations on the management of the wilt disease complex of pigeonpea by treatment with a vesicular arbuscular fungus and biocontrol agents for nematodes. *Bioresource Technology* 54 : 3, 227-230.

Siddiqui, Z.A., Mahmood, I. 1996. Biological control of *Heterodera cajani* and Fusarium udum on pigeonpea by *Glomus mosseae, Trichoderma harzianum*, and *Verticillium chlamydosporium. Israel Journal of Plant Sciences.* 44 : 1, 49 -

Siddiqui, Z.A., Mahmood, I. 1999. Role of bacteria in the management of plant parasitic nematodes : a review, *Bioresource Technology* 69 : 2, 167-179,: 4 pp.

Siddiqui, Z.A., Mahmood, I. 1999. Biological control of *Heterodera cajani* and *Fusarium udum* on pigeonpea by *Glomus mosseae, Trichoderma harzianum*, and *Verticillum chlamydosporium.* Zeitschrift fur Pflanzenkrankheiten und Pflanzenschutz 106 : 5, 530-533.

Siddiqui, Z.A. Irshad, Mahmood, Shamsul Hayat, Mahmood I; Hayat, S. 1998. Biocontrol of *Heterodera cajani* and *Fusarium udum* on pigeonpea using *Glomus mosseae, Paecilomyces lilacinus* and *Pseudomonas fluorescens.* Thai Journal of Agricultural Science, 31 : 3, 310-321.

Sikora, R.A. 1979. Predisposition to *Meloidogyne* infection by the endotrophic mycorrhizal fungus *Glomus mosseae.* In 'Root knot Nematodes (Meloidogyne species) Systematics, Biology and Control' (F. Lamberti & C.E. Taylor, Eds.). pp. 399-404. Academic Press, London.

Sikora, R.A. 1992. Annual Review of Phytopathology, 30 : 245-270.

Singh, B. and Dhawan, S.C. 1990. A new bacterial strain of *Pasteuria penetrans*, its host range and effect of temperature on spore attachment to second stage juveniles of pigeon pea cyst nematode, *Heterodera cajani, Indian J. Nematol.* 20 : 161-166.

Singh, B., Dhawan, S.C. 1992. Effect of soil texture on attachment of bacterial spores of *Pasteuria penetrans* to the second-stage juveniles of *Heterodera cajani.* Indian Journal of Nematology. 22 : 1, 72-74.

Singh, B. Dhawan, S.C. 1994. Effect of Pasteuria penetrans on the penetration and multiplication of *Heterodera cajani* in *Vigna unguiculata* roots. *Nematologia Mediterranea* 22 : 2, 159-161.

Singh, B; Dhawan, SC 1996. Suppression of Heterodera cajani by Pasteuria penetrans during three successive planting of cowpea. *Indian Journal of Nematology, 26 : 2, 216-221.*

Singh, M., Samar, R., Sharma, G.L. 1998. Effect of various soil types on the development of *Pasteuria penetrans* on *Meloidogyne incognita* in brinjal crop. *Indian Journal of Nematology* 28 : 1, 35-40.

Sitaramaiah, K. and Sikora, R.A. 1982. Effect of the mycorrhizal fungus, Glomus fasciculatum on the host parasite relationship of *Rotylenchulus reniformis* in tomato. Nematologica 28 : 412- 419.

Sivakumar, C.V. 1998. Bacterial antagonists for suppression of plant parasitic nematodes. In : Biological suppression of plant diseases, phytoparasitic nematodes and weeds. Singh S.P. Hussaini, S.S. (Eds.), Project Directorate of Biological Control, Bangalore p. 128-137.

Sivakumar C.V., Vidhyasekharan P. 1990. Control of Meloidogyne incognita on *Coleus forskohlii* with *Paecilomyces lilacinus* in farm yard manure amended and non amended soil. *Journal of Biological Control* 4 : 1, 68-69; 1 ref.

Sivakumar, M. 1999. Nin-chemical management of *Meloidogyne hapla* Chitwood on carrot (*Daucus carotae*). In : Usha K. Mehta (Ed.) Proceedings of the Third International symposium of Afro Asian Society of Nematologists held at Sugarcane Breeding Institute, Coimbatore, India p. 209-212.

Smith, G.S. 1987. Interactions of nematodes with mycorrhizal fungi, In : "*Vistas on Nematology*" (J.A. Veech and D.W. Dickson, eds.), pp. 292-306. Society of Nematologists, Maryland 500 pp.

Somasekhar, N. 1997. Host range and longevity of *Pasteuria penetrans* under different storage conditions. *Advances in IPM for horticultural crops. Proceedings of the first National Symposium on Pest management in Horticultural Crops : Environmental Implications and Thrusts. (Reddy PP (ed); Kumar - NKK (ed); Verghese - A) Bangalore, India 15-17 October 1997, 1998, 301-303.*

Somasekhar, N., Usha K. Mehta, Hari, K. 1998. Evaluation of sugarcane by products for mass multiplication of nematode antagonistic fungi. In : Usha K. Mehta (ed.) Proceedings of the Third International Symposium of Afro Asian Society of Nematologists held at Sugarcane Breeding Institute Coimbatore India p. 213.

Somasekhar, N., Usha K. Mehta, Hari K. 1998. Solubilization of insoluble phosphate by nematode antagonistic fungi. In : Usha K. Mehta (Ed.), Proceedings of the Third International Symposium of Afro - Asian Society of Nematologists held at Sugarcane Breeding Institute, Coimbatore, India, p. 213.

Sosamma, V.K. Geetha, S.M., Koshy, P.K. 1994. Effect of the fungus, *Paecilomyces lilacinus* on the burrowing nematode, *Radopholus similis* infesting betel vine *Indian Journal of Nematology* 24 : 1, 50-53 : 20 ref.

Sosamma, V.K. and Koshy, P.K. 1995. Effect of *Pasteuria penetrans* and *Paecilomyces lilacinus* on population build up of root knot nematode, *Meloidogyne incognita* on black pepper. *National Symposium on Nematode Problems of India - An Appraisal of the Nematode Management with Eco-Friendly Approaches and Biocomponents. Indian Agri. Res. Inst.*, New Delhi, p. 47.

Sosamma, VK., Koshy, P.K. 1997. Biological control of *Meloidogyne incognita* on black pepper by *Pasteuria penetrans* and *Paecilomyces lilacinus. Journal of Plantation Crops.* 25 : 1, 72-76.

Sreeja, T.P., Eapen, S.J, Ramana, K.V. 1996. Occurrence of *Verticillium chlamydosporium* Goddard in a black pepper (*Piper nigrum* L.) garden in Kerala, India. *Journal of Spices and Aromatic Crops*, 5 : 2, 143-147.

Srivastava, S.S. and Swarup, G. 1986. Biomanagement of root knot nematode (*Meloidogyne incognita*) through the predatory fungus, *Arthrobotrys conoides.* National Conference on plant Parasitic Nematodes of India - Problems and Progress, Indian Agri. Res. Inst. New Delhi, p. 85.

Stirling, G.R. 1991. Biological control of plant parasitic nematodes, CABI, Willingford, U.K., pp. 282.

Suhail Anver and Alam, M.M. 1997. Control of *Meloidogyne incognita* and *Rotylenchulus reniformis* singly and concomitantly on pigeonpea with *Paecilomyces lilacinus. Indian Journal of Nematology* 27 : 2, 209-213.

Suhail, Anver., Alam, M.M. and Anver, S., 1999. Control of *Meloidogyne incognita* and *Rotylenchulus reniformus* singly and concomitantly on chickpea and pigeonpea. *Archives of Phytopathology and Plantn Protection* 32 : 2, 161-172; 25 ref.

Suresh, C.K. and Bagyaraj, D.J. 1984. Interaction between a vesicular arbuscular mycorrhiza and a root knot nematode and its effect on growth and chemical composition of tomato. *Nematol Medit.* 12 : 31-39.

Swarup, G., Nandini Gokte, 1986. Biological Control. In : *Plant parasitic nematodes of India. Problems and Progress.* G. Swarup and D.R. dasgupta (eds.) I.A.R.I. New Delhi, pp. 476- 489.

Tabreiz Ahmad Khan and Syed Israr Hussain 1988. Studies on the efficacy of seed treatment with pesticides, oil cakes, neem leaf and culture filtrate of *Paecilomyces lilacinus* for the control of diseases caused by the presence of *Rotylenchulus reniformis, Meloidogyne incognita* and *Rhizoctonia solani* either individually or concomitantly on cowpea. *Indian Journal of Nematology* 18 : 192-198.

Tabreiz Ahmad Khan and Syed Israr Hussain 1989. Effect of culture filtrates of soil fungi on the mortality and hatching of reniform nematode (*Rotylenchulus reniformis*). *Indian Journal of Nematology* 19 : 35-40.

Timper, P. and Riggs, R.D. 1998. Variation in efficacy of isolates of the fungus ARF against the soybean cyst nematode *Heterodera glycines. Journal of Nematology* 30 : 461-467.

Trivedi, P.C. 1992. Evaluation of a fungus, *Paecilomyces lilacinus* for the biological control of root knot nematode, *Meloidogyne incognita* on *Solanum melongena.* In : Ooi-FAC (ed.) : Lim-GS (ed.); Teng-PS Proceedings of the 3rd international conference on plant protection in the tropics, Genting Highlands, Malaysia, 20-23 March 1990. 1992, Vol. 6, 29- 33; 13 ref.

Umesh, K.C., Krishnappa K. and Bagyaraj, D.J. 1988. Interaction of burrowing nematode, *Radopholus similis* and VA mycorrhiza, *Glomus fasciculatum* in banana. *Indian J. Nematol.* 18 : 6-11.

Verma, K.K., Gupta, D.C. and Paruthi, I.J. 1999. Preliminary trial on the efficacy of *Pseudomonas fluorescens* as seed treatment against *Meloidogyne incongita* in tomato. In : Dhawan - SC (ed.); Kaushal KK Proceedings of national symposium on rational approaches in nematode management for sustainable agriculture. Anand India 23-25 November, 1998, 79-81.

Vidya, K. and Reddy, B.M.R. 1998. Integrated management of *Radopholus similis* infecting banana using plant product, nematicide and biocontrol agents. *Mysore Journal of Agricultural Science*, 32 : 3, 186-190.

Vyas, R.V., Patel, B.A. Patel, D.J. and Patel R.S. 1996. Management of root knot nematodes in chickpea. *Pakistan Journal of Nematology* 14 : 2, 117-119; 5 ref.

Vyas, R.V., Patel, N.N. and Patel, D.J. 1999. Integrated management of root knot nematodes in cotton. In : Dhawan SC (ed); Kaushal KK Proceedings of national symposium on rational approaches in nematode management for sustainable agriculture. Anand, India, 23-25 November, 1998, 1999, 51-52; 4 ref.

Wajid Khan, M. 1998. Present status of biological suppression of plant parasitic nematodes. In : Singh, S.P., Hussaini S.S. (Eds.) *Biological suppression of Plant disease phytoparasitic nematodes and weeds*. Project Directorate of Biological Control Bangalore, p. 118-127.

Walia, R.K. and Bansal, R.K. 1992. Factors governing efficacy of nematode - oviparasitic fungus, *Paecilomyces lilacinus* : effect of fungus inoculum level. *Afro Asian Nematology Network* 1 : 9-11.

Walia, R.K. Bansal, R.K. and Bhatti, D.S. 1991. Effect of *Paecilomyces lilacinus* application time and method in controlling *Meloidogyne javanica* on okra. *Nematologia Mediterranea* 19 : 247-249.

Walia, R.K. and Dalal, M.R. 1994. Efficacy of bacterial parasite *Pasteuria penetrans* application as nursey soil treatment in controlling root knot nematode *Meloidogyne javanica* on tomato. *Pest Management and Economic Zoology*. 2 :1, 19-21 : Paper presented at the National Symposium on Emerging Trends in Pest Management, held during June 28-30, 1994 at the Dr. Y.S. Parmar University of Horticulture and Forestry, Solan, India.

Wang Bali and Gao Xuebiao, 1990. Progress in studies of nematophagous fungi. *Journal of South China Agricultural University* 19 : 113-117.

Yaqub Bhat, M., Hasamuddin and Munawar Fazal. 1999. Combined application of *Paecilomyces lilacinus* and oil cakes for protection of chickpea against *Meloidogyne incognita*. In : Nematology : Challenges and Opportunities in 21st Century (Ed. Usha K. Mehta). Proceedings of the 3rd International Symposium of Afro Asian Society of Nematologists, April 16-19, 1998, Coimbatore, India, pp. 258-262.

Zaki, A., Siddiqui and Mahmood, I. 1996. Biological Control of plant parasitic nematodes by fungi : A Review. *Bioresource Technology* 58 : 229-239.

Zaki, F.A. 1994a. Effect of culture filtrates of *Paecilomyces lilacinus* on *Meloidogyne javanica. Nematologia Mediterranea* 22 : 41-43.

Zaki, F.A. 1994b. Dose optimization of *Paecilomyces lilacinus* for the control of *Meloidogyne javanica* on tomato. *Nematologia Mediterranea*, 22 : 45-47.

Zaki, F.A. 1998. Biological control of *Meloidogune javanica* in tomato by *Paecilomyces lilacinus* and castor. *Indian Journal of Nematology* : 28 : 132 : 139.

Zaki, F.A. and Bhatti, D.S. 1988. Economical method for mass culturing of *Paecilomyces lilacinus* (Thom.) Samson. *Current Science* 57 : 153.

Zaki, F.A. and Bhatti, D.S. 1990. Effect of castor (*Ricinus communis*) and the biocontrol fungus *paecilomyces lilacinus* on *Meloidogyne javanica. Nematologica* 36 : 114-122.

Zaki, F.A. and Bhatti, D.S. 1991. Effect of culture media on sporulation of *Paecilomyces lilacinus* and its efficacy against *Meloidogyne javanica in tomato. Nematologia Mediterranea* 1991, 19 : 2, 211-212.

CHAPTER 13

AZOTOBACTER FOR THE CONTROL OF PLANT PARASITIC NEMATODES AND FOR OTHER BENEFITS

P.P.K. Chahal and V.P.S. Chahal

INTRODUCTION

Azotobacter is aerobic, non-symbiotic nitrogen fixing bacterium which occurs in most of our cultivated soils. Among all the nitrogen fixing bacteria, *Azotobacter* occupies an important place because of its multiple beneficial effects on plants. It is well documented fact that *Azotobacter* fixes nitrogen, produces growth promoting substances and antifungal compounds, excretes ammonia and more recently the authors have reported the ability of *Azotobacter* to control nematodes. This article summarises the beneficial effects of *Azotobacter* and results of some experiments conducted on the control of nematodes.

Azotobacter was first discovered by Beijerinck in 1901 who reported its ability to fix molecular nitrogen. According to the classification of Prokaryotes, *Azotobacter* has been placed in the seventh part of Bergey's Manual of Determinative Bacteriology.

1. IMPORTANT CHARACTERS

The genus *Azotobacter* has large ovoid cells occur singly, in pairs, do not form endospore but produce thick walled cyst and produces capsular slime. The cells are aerobic and gram negative. They may produce brown, black, cinnamon or green pigments which is the characteristic of species. Species of *Azotobacter* are *A. chroococcum*, *A. beijerinckii*, *A. vinelandii*, *A. paspali*, *A. insignis* and *A. macrocytogenes*. *A. chroococcum* is the most prevalent species in our soils. It produces brown/ black water soluble pigments, can utilize starch and mannitol. The cells are motile by peritrichous flagella and DNA base ratio GC (%) is 65-66.

2. DISTRIBUTION

There are several reports showing the presence of *Azotobacter* in most of the world soil (Mishustin and Shilnikova, 1969). The count of *Azotobacter* is generally x 10^3 cells/g of soil sample. Black soils have been found to contain more *Azotobacter* cells than the red soils.

Azotobacter chroococcum is the most occurring species in sandy loam soils of the Punjab state (Khullar and Chahal, 1977).

3. BENEFICIAL EFFECTS

Due to multiple beneficial effects of *Azotobacter* on crops, numerous workers have studied different aspects of this bacterium in detail :

Nitrogen fixation : *Azotobacter* not only fixes molecular nitrogen but assimilate inorganic form of nitrogen. The amount of nitrogen fixed by *Azotobacter* vary from strain to strain and may range from 2-20 mg/g of sugar oxidised (Sundara Rao and Iswaran, 1959). It has been reported that nitrogen assimilated by *Azotobacter* is released in the form of protein, amino acids and NH_4^+ (Sandrok, 1958). For fixation of 5-20 kg N/ha/year, *Azotobacter* needs 1000 kg of organic matter for oxidation. It has been pointed out by some workers (Moore, 1966; Dobereiner, 1966) that root exudates could support the active growth of *Azotobacter* for continuing the process of nitrogen fixation. Hardy *et al.* (1968) used acetylene reduction assay to determine the extent of nitrogen fixation by *Azotobacter* in the rhizosphere as well as in the culture medium. Chahal and Chahal (1999) isolated number of *Azotobacter* strains from the rhizosphere of different crops grown in diverse climatic conditions. All the strains of *Azotobacter* verified in nitrogen fixation from 1.18 to 16.0 mg/g sucrose utilized.

Number of field trials have been conducted on wheat, maize and cotton to evaluate the usefullness of *Azotobacter* alongwith different doses of inorganic nitrogen fertilizers/farm yard manure under Punjab conditions. The results have shown that there was no significant increase in the yield of any crop in any season due to *Azotobacter* inoculation (Ghai *et al.*, 1976; Dhingra *et al.*, 1979; Chahal *et al.*, 1979; Dhillon *et al.*, 1980).

The enhancement of germination of seeds and the growth rate of seedlings by *Azotobacter* can be attributed to the production of growth regulators (Brown and Burlingham, 1968; Brown, 1974; Vasisht *et al.*, 1979). Khullar *et al.*, (1978) reported that *Azotobacter* inoculation had shown increase in the chlorophyll contents of leaves of carrot, radish, brinjal and chillies. Shende *et al*, (1977) reported improvement in seed germination of rice and cotton.

Growth promoting substances : *Azotobacter* usually produces considerable amounts of biological active substances which includes vitamins of B group, nicotinic acid, pantothenic acid biotin, hetero auxin, gibberellins and cytokinins.

Resnitsina (1938) was the first to report the presence of auxin type substances in *Azotobacter* cultures by means of biological tests. The production of heteroauxin such as D- Indolyl acetic acid and gibberellins by *Azotobacter* has been reviewed by many workers (Rubinchih 1960). Brakel and Hilger (1965) showed that synthesis of auxin, B - indolyl acetic acid and B-indolyl lactic acid by *Azotobacter* was increased due to the presence of tryptophan. Vancura (1961) reported detection of gibberellic acid in the culture of *Azotobacter*.

Indole acetic acid and gibberellic acid have been the subject of controversies as to which is responsible for increasing the growth of plants inoculated with

Azotobacter. Vancura and Macura (1960) detected IAA in *Azotobacter* cultures. Vancura and Macura (1961) reported that gibberellic acid was the main growth promoting substance synthesized by *Azotobacter chroococcum* when grown in Ca Co_3 free medium. He reported the presence of 2-5 μg/ml of IAA. Such high figures were not found by Brown and Burlingham (1968). They reported only 0.03 μg gibberellic acid equivalent/ml.

Brown and Walker (1970) showed that small quantities of IAA could be synthesized in aerated cultures of *A. chroococcum*. They further reported that addition of L - tryptophan increased the synthesis of IAA. The accumulation of IAA (0.3 μg/ml) was highest in 7 days old cultures with or without tryptophan.

Barea and Brown (1974) detected IAA, atleast 3 gibberellins and 2 cytokinins. *Azotobacter* cultures could be initially screened for the production of plant growth substances by rapid yeast bioassay method developed by Barea *et al*., (1974).

Taller and Wong (1989) reported 3 cytokinins produced by *Azotobacter* after incubation of 10 days. The extract of the culture supernatent caused elongation of lettuce hypocotyls. Hypocotyls of rape (*Brassica campestris*), wheat (*Triticum turgidum*) and tomato (*Lycopersicon esculentum*) were increased with cultures of *A. paspali* (Abbas and Okon, 1993). Chahal and Chahal (1999) examined different strains of *A. chroococcum* and *A. vinelandii* for the production of gibberellins like substances and indole acetic acid. Most of the strains of *A. chroococcum* showed the production of gibberellins like substances. Out of 4 strains of *A. vinelandii* only 2 strains showed the production of gibberellins like substances. Gonzelez-Lopez *et al*., (1986) reported production of growth promoting substances by *A. vinelandii*.

Jackson *et al*. (1964) and Brown *et al*. (1968) observed beneficial effect on tomato due to *Azotobacter* inoculation and application of gibberellins. Barea and Brown (1974) treated tomato seedling with cultures of *A. paspali* and beneficial effect was attributed to plant growth regulating substances. Sharma and Chahal (1986) reported occurrence of growth regulators producing *Azotobacter* strains in Punjab soils.

(a) Excretion of ammonia

In nitrogen fixing bacteria, the immediate product of nitrogen fixation is ammonia. Fedrov (1952) demonstrated ammonia excretion by the young cells of *Azotobacter* when organic acids were used in sources of C. Mannital and glucose led to more rapid growth with ammonia release. It was reported that 7-13% of total amount of fixed nitrogen was excreted by the cells. Narula *et al*., (1980) studied 23 strains of *Azotobacter* and found that all the strains were capable of releasing ammonia but the amount of excretion varied with the strains. Chahal and Chahal (1999) tested 30 strains of *Azotobacter* and found that 26 strains of *A. chroococcum* varied in ammonia excretion from 2.5 to 40.7 μg/ml whereas 4 strains of *A. vinelandii* varied from 1.8 to 2.6 μg/ml.. Gordon and Brill (1972) obtained mutants of *A. vinelandii* in which nitrogenase system was derepressed by ammonia. In these mutants accumulation and release of ammonia occurred at faster rate than wild strains.

Chahal and Chahal (1999) obtained mutants derived from wild strain *A. chroococcum* 23 Ammonia excretion decreased in all the mutants and some of the mutants lost the ability to produce light brown pigments.

(b) Production of siderophores

Micro organisms solubilize Fe^{+++} by

producing extracellular low molecular weight iron transport co-factors known as siderophores (Neilands, 1981). These siderophores are broadly placed into 2 groups, phenolates and hydrovamates. Some bacteria like *Azotobacter* produce both types of siderophores. Siderophores can act as growth factors as well as phytopathogen suppressive agents.

A. vinelandii promotes plant growth by phytohormone production (Brown 1974; Brown, 1982). The siderophores produced by *A. vinelandii* are azotochelin and azotobactin which effectively release iron from insoluble ferriminerals commonly found in soil (Page and Van Tigerstrom, 1988) and transport the solubilised iron into the cells in manner typical of siderophore system (Knosp *et al.*; 1984). The yellow green fluorescent siderophores, azotobactin shares some structural similarities with pseudobactin, siderophores produced by *Pseudomonas* spp. which are involved in the suppression of diseases (Teintze *et al.*, 1981).

It has been reported that *A. vinelandii* enhanced the growth of *Agrobacterium tumefaciens* and *Erwinia carotovora* (Page and Dale, 1986). This stimulation was apparently siderophore mediated. Siderophores promoted by solubilizing mineral iron and ferrisiderophores so formed acted as iron sources for *Agrobacterium tumefaciens*. Catecholamine siderophores produced by iron - limited *A. vinelandii* was purified (Page and Von Tigerstom, 1988) and named as aminochelin (2, 3-dihydroxy benzoyl putrescine). Chahal and Chahal (1999) reported siderophore production by *A. chroococcum*.

(c) Biodegration

Biodegration of lindane and nitrogen fixation by *Azotobacter* was studied under liquid and soil conditions (Gupta and Narula, 1994). They found that degradation was significant under both the conditions at 20, 40 and 60 ppm concentration of lindane. Balajee and Mahadevan (1993) reported that *A. chroococcum* isolated from forest soils efficiently cleaned 2, 4-D. The cleanage was profoundly influenced by concentration of bacterial cells, moisture content of the soil and temperature at which the soil was incubated. The conditions which favoured the maximum degradation of 2, 4-D were 30°C (58% degradation) and 50% moisture content.

A bd-Alla (1994) reported utilization of 4-hydroxybenzoic acid, resorcinol, pyrocatechol and vanillic acid. He further concluded that *Azotobacter* isolated had good potential to utilize some phenolic compounds released during biodegradation of plant wastes.

Kinetic study of an anaerobic digestion of olive mill waste water that was previously fermented with *A. chroococcum* was carried out by Borja *et al.*, (1993). The specific rate of methane production was higher than that obtained in the anaerobic digestion of untreated olive mill waste water. Biotoxicity of the waste was reduced by 30% due to pre treatment.

Page (1992) observed production of polyhydroxy alkanoates by *A. vinelandii*. He reported that best molasses proved an excellent feedstock for the production of polyhydroxy alkanoate. The substrate cost for the production of polyhydroxy alkanoate from beet molasses was one third of that using glucose.

4. AZOTOBACTER AS BIOCONTROL AGENT

Biological control of phytopathogenic fungi: It has been reported that *Azotobacter* produced fungistatic compounds (Mishustin *et al.*, 1963; Mishustin, 1970). Pridachina *et al.*, (1982) have reported that *A. chroococcum* strain - 92 produced an antibiotic active against number of phytopathogenic fungi. *Azotobacter* not only inhibited the vegetative growth but also inhibited the spore germination of different fungi. Mellikarajunaiah and Bhide (1981) observed the inhibition of seed borne fungi and pearl millet by *Azotobacter*. Wildermuth (1982) gave the nature of antagonism among fungi and bacteria while indicating number of means by which bacteria can inhibit the fungi. Spores of fungi can be lysed in soil by number of means including lysis of fungal germ tube after and before germination, autolysis of spores action of antibiotics, metabolites and enzymes excreted by bacteria. Brown (1974) postulated that hormones like indol acetic acid, growth regulating substances and cytokinins alongwith antibiosis might be involved in control mechanism following bacterization. Chahal and Chahal (1998) reviewed the interactions between nematodes and fungi. Sharma and Chahal (1987) studied the antagonistic effect of *Azotobacter* strains on some plant pathogenic fungi (Table 1).

(a) Biological control of plant parasitic nematodes

For the first time we reported the ability of some strains of *Azotobacter* to cause inhibition in the emergence of juveniles from eggs of root-knot nematodes, *Meloidogyne incognita* and control of nematodes in tomato (Chahal and Chahal, 1985). In 1986, Chahal and Chahal reported antagonistic effect of *Azotobacter chroococcum* on the hatching of eggs and eggmasses of *M. incognita* (Table 2).

Table 1. Inhibition in spore germination of fungi by *Azotobacter* strains.

Fungi	Spore germination (%)		
	Control	*A. veinelandii-1*	*A. chroococcum-2*
Asperigillus flavus	73.4	72.1	70.2
Aspergillus phoenics	90.7	90.4	88.2
Fusarium equistii	85.7	73.5	66.5
Fusarium oxysporum	97.6	81.3	53.7
Penicillium oxalicum	96.5	96.0	93.4
Helminthosporium oryzae	80.9	79.3	71.4
Colletotrichum capsicum	74.5	61.3	42.3
Pythium aphanidermatum	83.1	79.5	69.1

C.D. at 5% Fungal Species = 2.41 *Azotobacter* species = 1.73
Fungal species x *Azotobacter* species = 3.37
More experiments are required to be conducted to evaluate *Azotobacter* strains for the control of fungal diseases.

Table 2. Hatching of eggmasses and eggs of *Meloidogyne incognita* in different cell concentrations of *Azotobacter chroococcum* (Data represent mean of 10 replications)

Azotobacter Cells/ml.	Hatching of egg masses in *Azotobacter* suspension (Days)			Total juveniles hatch in 9 days per 4 egg-masses	Hatching of eggs masses after transfer to sterile water (Days)				Total juveniles hatch egg masses in 21 days	Killing of egg masses over control (%)
	3	6	9		3	6	9	12		
Sterile water	1507	409	136	2052 (90)	47	29	4	1	2133	-
12×10^4	485	49	16	550 (32)	138	40	0	0	728	65.8 (75.5)
25×10^4	305	5	0	314 (12)	119	75	11	0	519	75.7 (86.6)
5×10^5	0	0	0	0(0)	106	150	15	0	271	87.3 (100)
1×10^6	0	0	0	0(0)	34	121	31	2	188	91.1 (100)

Data given in the brackets show killing/100 eggs and % killing over control.

Chahal and Chahal (1999) studied 30 strains of *Azotobacter* on the hatching of eggmasses and eggs of *M. incognita*. These strains were screened on the basis of nitrogen fixation, nitrogenase activity, excretion of ammonia, production of growth promoting substances and formation of siderophores. All the strains of *Azotobacter chroococcum* (cell concentration app. x 10^6 cells/ml) showed inhibition in emergence of juveniles from eggmasses of *M. incognita* which ranged from 12.5% to 65.2% during hatching period of 10 days. *A. vinelandii* strains showed inhibition in the range of 10.8 to 29.5%. The egg of *M. incognita* were more suceptible to the action of *Azotobacter* cells than eggs present in the eggmasses. Inhibitory effect of different strains of *A. chroococcum* varied from 17 to 82.4% during hatching period of 10 days in case of eggs. *A. vinelandii* strains showed inhibition in the range of 25 to 36.4%. Mutation in the selected strain of *A. chroococcum* did not improve inhibitory effect. Verma and Bansal (1996) showed inhibitory effect of *A. chroococcum* on hatching of *M. javanica*.

Chahal and Chahal (1986) after conducting *in vitro* experiments, studied the effect of *A. chroococcum* on the control of infection by *M. incognita* on brinjal (*Solanum melongena* L.) cultivar. Pusa purple long under pot culture conditions. They reported significant reduction in the gall formation, soil population and root population. The count of *A. chroococcum* decreased at the end of experiment.

Chahal and Chahal (1999) observed strain variations among different stains of *A. chroococcum* in their ability to control *M. incognita* under pot culture conditions (Table 3).

It has been found that different strains of *Azotobacter* varied in their ability to reduce number of galls/plant and total nematode population. Total chlorophyll contents of leaves and dry weight of shoot increased due to *Azotobacter* inoculation over control. *A. chroococcum* -23 proved better strain than other strains. Survival and establishment of different strains also varied.

The results of another pot culture experiment have shown that inhibitory effect of different strains of *Azotobacter* was dependent on the number of cells of *Azotobacter* in the rhizosphere of tomato plants. Mutants derived from A. chroococcum - 23 did not show more

Table 3. Effect of selected strains of *Azotobacter* on the control of *M. incognita* in tomato (*Lycopersicon esculentum* L.) variety Punjab Chhuhara in nematode infested soil (Average of 3 replications)

Treatments	Chlorophyll contents mg/g leaves	Dry weight of shoot (g)	Galls/plant	Nematode population	Azotobacter count (x 10^2 cells/g soil)
AC-23 + N	1.02	7.2	98	2660	3.4
AC-24 + N	1.05	7.6	109	2730	2.8
AC-28 + N	1.12	6.6	112	3250	1.3
AC-5 + N	1.10	7.8	195	3640	1.5
AC -3 + N	1.00	7.0	196	3750	2.1
AC-25 + N	1.13	6.9	202	2560	1.2
Control	0.98	6.5	75	3500	1.1
P N	0.82	6.1	210	3270	0.8

AC= *Azotobacter chroococcum* N = *Meloidogyne incognita*

Table 4. Effect of *Azotobacter chroococcum* -23 on the infection of *Meloidogyne incognita* and other growth characters of mungbean (*Vigna radiata* L.) variety ML-131 in steam sterilized soil (Average of 3 replications)

Characters	Treatments						
	C	N	R	R+N	A	A+R	A+R+N
Chlorophyll Contents (mg/g leaves)	1.67	1.62	1.78	1.69	1.69	1.76	1.77
Nodules / Plant	0	0	42.0	36	0	41	38
Leghaemoglobin (mg/g nodules)	0	0	4.39	2.25	0	4.42	4.02
Bacteroids ($2x\ 10^3$/g nodules)	0	0	5.4	3.7	0	5.2	3.8
Nitrogenase activity (mch/h/g dry weight nodules)	–	–	229.5	167.9	–	231.6	185.8
Dry weight of shoot (g)	6.4	6.1	8.2	7.50	6.5	7.7	7.9
Galls/plant	0	299	0	288	0	0	241
Azotobacter count ($x\ 10^3$ cells /g soil)	–	–	–	–	2.2	2.3	1.8
N content of shoot (%)	1.43	0.95	1.53	1.47	1.44	1.52	1.49

C= Control, N = Nematode R = *Rhizobium*, A = *Azotobacter*

reduction in the number of galls/plant and total nematode population as compared to control.

Chahal and Chahal (1999) reported inhibitory effect of *Azotobacter* strains on the emergence of juveniles from wheat galls. The results of pot culture experiment have shown significant reduction in the number of infected ears/plant. Due to inoculation with *Azotobacter* strains, grain yield/pot also increased.

As *M. incognita* has very wide host range and also infect pulses and adversely affect nodulation as well as nitrogen fixation (Chahal and Chahal, 1987; Chahal and Chahal, 1991) so pot culture experiment was conducted to study the effect of *Azotobacter chroococcum* -23 on infection of *M. incognita* in mungbean (*Vigna radiata* L.) variety ML-131 (Chahal and Chahal, 1999). The results of the pot culture experiment indicate beneficial effect of *Azotobacter* (Table 4).

M. incognita reduced all the growth characters of mungbean variety ML-131 such as total chlorophyll contents of leaves, number of nodules/plant, leghaemoglobin content of nodules, nitrogenase activity of nodules, bacteroid contents of nodules and N content of shoot. *Azotobacter* has been found to reduce the adverse effect of nematodes.

5. CONCLUSIONS

The favourable actions of *Azotobacter* on the plant growth may be associated with nitrogen fixation and production of growth promoting substances. Biodegration and recent identification of some new products produced by *Azotobacter* are few other interesting areas of research. *Azotobacter* is known to inhibit plant pathogenic fungi and more recently we have reported the ability of *Azotobacter* strains to control *M. incognita.* More research work is required to be conducted in order to exploit important aspects of *Azotobacter* with particular exphasis on mechanisms/mode of action for the control of plant pathogens and nematodes.

REFERENCES

Abbass, Z. and Okon, Y. 1993. Plant growth substances production by *Azotobacter paspali* in the rhizosphere. *Soil Biology and Biochem.*, 25 (8) : 1075-1083.

Abd-Alla, M.H. 1994. Utilization of some phenolic compounds by *Azotobacter chroococcum* and their effect on growth and nitrogenase activity. *Folia. Micro*, 39 : 57-60.

Balajee, S. and Mahadevan, A. 1993. Biodegradation of 2, 4-D in soil by *Azotobacter chroococcum. Toxicological and Environ. Chemistry*, 39 : 167-172.

Barea, J.M. and Brown, M.E. 1974. Effect of plant growth produced by *Azotobacter paspali* related to synthesis of plant growth regulating substances. *J. Appl. Bact.*, 37 : 583-589.

Beijerinck, M.W. 1901. Uber Oligonitrophile Mikroben. Zent. Bakt. Parasiti Infect Hyg. Abt., 2 (9) : 3.

Borja, D., Martin, A., Gomez, L., Ramos-Cromenzana, H. 1993. Anaerobic digestion of olive mill waste water pre-treated with *Azotobacter chroococcum. Resources Conservation and Recycling* 9 : 201-211.

Brakel, J. and Hilger, F. 1965. Elude quantitative et quantitative de la synthese de substances de nature auxinique par *Azotobacter chroococcum* in vitro. *Bull. Int. Agron. Stat. Res. Gembloux*, 33 : 469-472.

Brown, M.E. 1974. Seed and root bacterization. *Ann. Rev. Phytopatho.* 12 : 181-197.

Brown, M.E. 1982. Nitrogen fixation by free living bacteria associated with plants - fact or fiction?. In : Bacteria and Plant (Eds. M.E. Rhodes - Roberts and F.A. Skinner). pp. 25-41.

Brown, M.F and Burlingham, S.K. 1968. Production of growth substances by *Azotobacter chroococcum. J. Gen. Micro.* 53 : 135-142.

Brown, M.E., Jackson, R.M. and Burlingham, S.K. 1968. Effect produced on tomato plants by seed and root treatment with gibberellic acid and indol acetic acid. *J. Exp. Bot.*, 43 : 418- 423.

Brown, M.E. and Walker, M. 1970. Indol-3-acetic acid formation by *Azotobacter chroococcum. Plant Soil.* 32 : 250-255.

Chahal, V.P.S. 1984. *Azotobacter* in relation to crop productivity. Final Report of ICAR Adhoc Research Project. Submitted to ICAR, New Delhi.

Chahal, P.P.K. and Chahal, V.P.S. 1985. Control of root knot nematodes with antagonistic bacteria. Abstract in VI Int. Nitrogen Fixation Symp., held in Corvalis, Oregon, U.S.A. from August 4-10, 1985.

Chahal, P.P.K. and Chahal V.P.S. 1986. Effect of *Azotobacter chroococcum* on the hatching of eggmasses and eggs of *Meloidogyne incognita. Plant and Soil*, 95: 289-92.

Chahal, P.P.K. and Chahal, V.P.S. 1988. Biological control of root knot nematode with *Azotobacter chroococcum* on brinjal (*Solanum melanogena* L.). In : Advances in Plant Nematology (Eds. M.A. Maqbool; A. Morgan Golden; A. Ghaffar and L.R. Krusberg). pp. 257-263.

Chahal, P.P.K. and Chahal, V.P.S. 1992. Biocontrol of plant parasitic nematodes. In : Recent Advances in Nematology (Eds. B.K. Dwivedi), pp. 124-140.

Chahal, P.P.K. and Chahal, V.P.S. 1993. Nematodes of pulses and their control In : Integrated Pest Management (Eds. B.K. Dwivedi, M.M. Alam and G. Pandey). pp. 259-270.

Chahal, P.P.K. and Chahal, V.P.S. 1998. Interactions between nematodes and fungi. In : Plant Nematode Management : A Biocontrol Approach. (Ed. P.C. Trivedi). pp. 200-207.

Chahal, V.P.S. and Chahal, P.P.K. 1999. Final Technical Report, "Microbial Control of Plant Parasitic Nematodes" PL-480 Project submitted to USDA.

Dhillon, D.S., Kler, D.S. and Chahal, V.P.S. 1980. Effect of *Azotobacter inoculation* on yield of wheat. *Ind. J. Agri.*, 25 (3) : 533-535.

Dhingra, K.K., Gill, B.S., Singh Jagroop and Chahal, V.P.S. 1979. Effect of *Azotobacter inoculation*, farm yard manure, phosphorus and nitrogen levels on the yield of wheat. *Ind. J. Agron.* 24 (4) : 405-409.

Dobereiner, J. 1966. *Azotobacter paspoli* sp. umabacteria fixadora de nitrogenio na de Pespalum. *Pesq. agropec. Bras.*, 1 : 357-360.

Fedrov, M.V. 1952. Ammonia the primary product of nitrogen. *Mikrobiologiya.* 21 : 245-249.

Gonzelez Lopez, J., Salmeron, V., Maritinez- Tolado, M.V., Ballesteros, F. and Ramos Cormenzaka, A. 1986. Production of auxins, gibberellins and cytokinins by *Azotobacter vinelandii* ATCC - 12837 in chemically defined media and dialysed soil media. *Soil Biol. Biochem.* 18 : 119-120.

Ghai, B.S., Chahal, V.P.S., Singh, S. and Gupta, R.P. 1976. Effect of inoculation of wheat with *Azotobacter. Ann. Wheat Newsletter*, U.S.A. XXII : 15.

Gordon, J.K. and Brill, W.J. 1972. Mutants that produce nitrogenase in the presence of ammonia. *Proc. Nat. Acad. Sci.*, USA, 69 : 3501 - 3503.

Gupta, N. and Narula, N. 1994. Biodegradation of lindane by nitrogen fixing soil isolates of *Azotobacter chroococcum.* Proc. 34th Ann. Conf. AMI, held at P.A.U., Ludhiana from Feb. 11-14, 1994, pp. 116-119.

Hardy, R.D.W., Holstein, R.D., Jackson, E.K. and Burns, R.C. 1968. The acetylene-ethylene assay for nitrogen fixation. Laboratory and field evaluation. *Pl. Physiol.* 43 : 1185-1207.

Jackwon, R.M., Brown, M.E. and Burlingham, S.K. 1964. Similar effect on tomato plants of *Azotobacter* inoculation and application of gibberellins. *Nature*, 192 : 575-577.

Khullar, S. and Chahal, V.P.S. 1977. Distribution of *Azotobacter* in Punjab soil. *J. Res. P.A.U.* 19 : 123-125.

Khullar, S., Chahal, V.P.S. and Kaur, P.P. 1978. Effect of *Azotobacter* inoculation on the chlorophyll contents and other characters of carrot, radish, brinjal and chilli. *Ind. J. Micro.*, 18 (2) : 138-140.

Knosp, O., Von-Tigerstrom, M. and Page, W.J. 1984. Siderophores mediated uptake of iron in *Azotobacter vinelandii. J. Bacteroil.*, 159 : 341-347.

Mallikarjunaiah, R.R. and Bhide, V.P. 1981. Effect of *Azotobacter* on germination of spores of different fungi. *Zentl. Microbiol.*, 136 : 563-565.

Mishustin, E.N. 1970. Importance of non-symbiotic nitrogen fixing microorganisms in agriculture. *Plant and Soil*, 32 : 545-554.

Mishustin, E.N. and Shilnikova, V.K. 1969. The biological fixation of atmospheric nitrogen by free living bacteria. *Soil Biology*, UNESCO, PARIS.

Mishustin, E.N., Naumova, A.N. and Merinko, V.G. 1963. *Azotobacter* and its effectiveness. *IZV TSKI*, 4 : 42-54.

Moore, A.W. 1966. Non-symbiotic nitrogen fixation in soil and soil plant system. *Soil Fert.*, 29 : 227-235.

Narula, N., Lakshminarayan, K. and Tauro, P. 1980. Ammonia excretion by *Azotobacter chroococcum*. *Biotech, Bioeng.*, 23 : 467- 470.

Neiland, J.B. 1981. Microbial iron compounds. *Ann. Rev. Biochem.*, 50 : 715-731.

Page, W.J. 1992. Production of Polyhydroxy alkanoates by *Azotobacter vinelandii.* FEMS *Microbiology Rev.* 103 : 149-158.

Page, W.J. and Dale, P.L. 1986. Stimulation of *Agrobacterium tumefaciens* growth by *Azotobacter vinelandii* ferrisidirophores. *App. and Environ. Micro.* 51 (2) : 451-454.

Page, W.J. and Von Tigerstrom, M. 1988. Amino chalin, a cate cholamine siderophore produced by *Azotobacter vinelandii. J. Gen. Microbiol.*. 134 : 453-460.

Pridachina, E.D., Noriogeandskaya, E.B. and Krenlyk, E.G. 1982. *Azotobacter chroococcum* an organism producing a new antifungal antibiotic. *Bactakau Antibiotiki* 27 : 3-6.

Sandrak, N.A. 1958. Chemical composition and accessibility to plant of nitrogenous secretion. *Doki. Ts. Kh. A.*, 34 : 101-107.

Sharma, P.K. and Chahal, V.P.S. 1986. Occurrence of growth regulators producing *Azotobacter strains* in Punjab soils. *J. Res. P.A.U.* 23 (1) :95-97.

Sharma, P.K. and Chahal, V.P.S. 1987. Antagonistic effect of *Azotobacter* on some plant pathogenic fungi. *J. Res. P.A.U.* 24 (4) : 638-640.

Shende, S.T., Apte, R.G. and Singh, T. 1977. Influence of *Azotobacter* on germination of rice and cotton seeds. *Curr. Sci.* 46 : 675-676.

Sundara Rao, W.V.P. and Iswaran, v. 1959. Nitrogen fixation by *Azotobacter chroococcum* in culture. *J. Ind. Soc. Soil. Sci.*, 7 : 91-93.

Rasnitsina, E.A. 1938. Formation of growth substances of auxin type of bacteria. *Dokl. An. SSR.*, 18 : 353-356.

Rubenchik, L.I. 1960. *Azotobacter* and its use in Agriculture. Israel programme for scientific Translation. Old Bourne Press, London.

Taller, B.J. and Wong, T.Y. 1989. Cytokinins. In *Azotobacter vinelandii* culture medium. *Appl. Environ. Micro.*, 55 : 266-267.

Teintze, M., Hussain, M.B., Barnes, C.L. Leong, J. and Von der Helon, D. 1981. Structure of ferric pseudobactin, in siderophore from a plant growth promoting *Pseudomonas*. *Biochem;* 20 : 6446-6457.

Vancura, V. 1961. Detection of gibberellic acid in Azotobacter cultures. *Nature.* 192 : 88.

Vancura, V. and Macura, J. 1960. Indol derivatives in Azotobacter culture. *Nature,* 192 : 88-89.

Vancura, V. and Macura, J. 1961. The effect of root excretion on Azotobacter. *Folia Microbiol* 6 : 250 - 253.

Verma, V.K. and Bansal, R.K. 1996. Antagonistic effect of culture supernatant of Azotobacter chroococcum on larval hatching and mortality of *Meloidogyne javanica* under in vitro conditions. In : Natural Resources Management (Eds. R.C. Dogra, R.K. Behl and A.L. Khurana). pp. 158-163.

Visisht, K.D., Sharma, B.N. and Chahal, V.P.S. 1977. Influence of *Azotobacter* on the germination of okra seeds. *J. Res. P.A.U.*, 16 (4) : 487-489.

Wildermuth, A.H. 1982. Nature of antagonism of fungi by bacteria. *Soil Biochem. Biol.*, 14: 557-560.

CHAPTER - 14

Bacillus thuringiensis FOR THE CONTROL OF *Meloidogyne incognita*

V.P.S. Chahal and P.P.K. Chahal

INTRODUCTION

Nematicides have been valuable for the control of nematodes of crops. Their use has posed same problems because they tend to harm non target organisms such as man, animals and beneficial insects. In addition, they cause pollution of environement and underground water. Sometimes pests become resistant to pesticides which lead to the development of new races. Due to complexities, the use of microorganisms as agents to control pests of crops has become very significant in recent years. Number of microorganisms capable of reducing nematode population have been reported. The biological control of nematodes is entirely dependent on predators and parasites. Predators include turbellarians, tardigrades, enchytriads, insects, mites, fungi, amoeba and other bacteria. *Bacillus thuringiensis* a well known bacterium used the control of insects has been found by the authors a potential bio control agent for the nematodes.

1. IMPORTANT CHARACTERS

Bacillus thuringiensis is a spore forming aerobic, gram positive bacterium belonging to the genus *Bacillus*. More than 200 isolates of *B. thuringiensis* have been grouped into more than 12 serotypes. The classification of *B. thuringiensis* was done by the combination of H-antigens (serotypes) and biotypes particularly the esterase types (Norris, 1964). The classifications of *B. thuringiensis* have also been done by the flagellar agglutination, which shows the existence of flagellar antigen (H-antigen). New subspecies of *B. thuringiensis* has been proposed based on H-antigens. Subspecies of *B. thuringiensis* are *thuringiensis, finitimus, alestic, kurstaki, sotto, dendrolimus, kenyae, galleriae, canadensis, en-tomocidus, subtoxicus, aizawai, morrisoni, ostriniae, tolworthi, darmstadiensis, toumanoffii, kyushuensis, thompsoni, pakistani, israele-*

nsis, dakota, indiana, tohokuensis, kumanotoensis, tochigiensis and wuhanesis of these only, subspecies *thuringiensis, morrisoni, tolworthi, darmstadiensis, toumanoffii and Kumamotoensis* produce exotoxin.

The vegetative cells are rod-shaped with square ends measuring 1.0 to 1.2 by 3.0 to 5.0μ. The rods usually occur in short to long tangled chains. Spores are thin walled, ellipsoidal and measure 0.5 to 1.2 by 1.3 to 2.5μ. The spores generally lie obliquely in a sporangium. Crystalline, biopyramidal, proposal bodies as large as or sometimes larger than the spore are contained in the sporangium.

2. TOXINS PRODUCTION

The crystal produced by *B. thuringiensis* is toxic for many species of lepidopteron insects and mosquito larvae (Reeves and Garcia 1970), *B. Thuringiensis* produces severals toxins such as α -, β -, γ = exotoxin, δ- endotoxin (Heimpel, 1967) and louse factor (Gingrich *et al.*, 1974). McConnell and Richards (1959) reported β-toxin, which was composed of adenine, ribose, glucose and allaric acid with a phosphate group. The β - exotoxin is produced by some but not all strains of *B. thuringiensis* and the production is a strain specific property. β - exotoxin is thermostable. In general, one toxic crystal is formed in one sporangium and the shape is bipyramidal. The size and shape of toxic crystal depends on the strain. Spindle shaped inclusions can be oberseved which are either cubic or irregular.

The δ-endotoxin consist of a glycoprotein submit (Bulla *et al.* 1977). Lepidopterous insects show different responses to δ-endotoxin. Type I insects show paralysis after/ingestion of δ-endotoxin. Type II insects show gut paralysis after feeding on - endotoxin. Type III insects are killed due to ingestion of spores and δ-endotoxin. The mechanisms of action of *B. thuringiensis* toxins on plant-parasitic nematodes need investigations.

3. OCCURRENCE

B. thuringiensis occur in dead matter of insects, litters of sericulture forms and soils. The bacterium occur very widely in nature. *B. thuringiensis* var *israelensis* was isolated from samples collected from mosquito breeding site in Israel (Goldberg and Margalit, 1977). Protein crystals formed by bacteria during sporulation are highly effective stomach poison for mosquito larvae. Christoph (1985) studied germination and growth of *B. thuringiensis* var *israelensis* in the gut of *Aedes aegypti* and *A. vexams* larvae.

4. CONTROL OF ROOT KNOT NEMATODES

Root knot nematodes, *Mebidogyne incognita* (Kofoid and white) chit wood are plant parasitic nematodes of economically important crops. These cause significant reduction in the yield of different crops as well as in their quality. In view of health hazards and pollution of the envrionment by nematicides, many investigators have concentrated their efforts on biological control agents of plant parasitic nematodes (Mankau, 1975; Sayre, 1980, Mankauy 1980; Kerry and Crump, 1980; Wildermith, 1982; Stirling, 1984; Jatala, 1985; Jatala 1986; Chahal and Chahal 1986).

As mentioned above, *B. thuringiensis* has been exploited by many workers for the control of many insects. The authors perhaps for the first time investigated *B. thuringiensis* for the control of *M. incognita* (Chahal and Chahal, 1991).

Bacillus thuringiensis Berliner was multiplied on nutrient agar medium or nutrient both. The cells were centrigued and different dilutions were prepared for use in the experiment. *In Vitro* experiments were conducted to study the effect of *B. thuringiensis* on eggs and egg masses of *M. incognita*.

Table 1. Effect of different concentrations of *Bacillus thuringiensis* Berliner on the eggs and juveniles of *Meloidogyne incognita* after 10 days of hatching.

Cells/m.l.	Emergence of juveniles/20 eggs	Inhibition (%)	Emergence of juveniles /4 egg masses	Inhibition (%)	Viability of juveniles
Sterile Water	19	—	2124	-	+
2×10^4	2	89.50	374	82.4	-
5×10^5	0	100.00	138	93.5	-
1×10^6	0	100.00	27	98.7	-
2×10^6	0	100.00	18	99.6	-

Data represent mean value of 10 replications

The results summarized in the Table 1 shows that eggs of *M. incognita* were more susceptible to the action of *B. thuringiensis* Berliner. With increase in the concentrations of cells, there was linear increase in the per cent inhibition. Juveniles emerged from the eggs/egg masses could not remain viable in the cell suspension.

Chahal and Chahal (1993) kept 3 egg masses of *M. incognita* in vials containing 5 ml suspension of Thuricide (0, 0.25, 0.5, 0.75 and 1.0%) which is commercial product recommended for the control of many insect species. The result of *in vitro* experiment has shown that there was negligible emergence of juveniles from the egg masses kept in different concentration of thuricide. When egg masses were transferred to sterile distilled water no juvenile emerged from egg masses indicating permanent damage by *B. thuringiensis*.

Chahal and Chahal (1999) examined effect of different strains of *Bacillus thuringiensis* on wheat galls and on egg masses of *M. incognita*. The result of in vitro experiments is summarized in the Table 2.

It is obvious from the results that different strains of *B . thuringiensis* varied in causing inhibition in the emergence of Juveniles from egg masses of *M. incognita* and wheat galls. The juveniles if emerged from the egg masses or wheat galls became dead.

Chahal and Chahal (1991) studied the effect of supernatant, washed cells and whole broth of *B. thuringiensis* Berliner.

Table 2. Effect of different strains of *Bacillus thuringiensis* on wheat galls and on egg masses of *Meloidogyne incognita* (Data mean of 3 replications)

Strains (App. (X104 cells/ml)	Wheat galls		Egg Masses of *M. incognita*	
	Inhibition (%)	Survival of juveniles	Inhibition (%)	Survival of juveniles
Control (Sterile distilled water)	—	Active and Motile	—	Active and Motile
BT-868	98	Dead	97.7	Dead
BT-869	84	Dead	92.5	Dead
BT-1	86	Dead	93.5	Dead
BT-2	97	Dead	96.0	Dead
BT-3	98	Dead	98.0	Dead
BT-4	96	Dead	98.0	Dead
BT-5	95	Dead	98.5	Dead

BT = *Bacillus thuringiensis*

Table 3. Effect of supernantant, washed cells and whole broth of *B. thuringiensis* Berliner on the eggs and juveniles of *M. incognita* after 10 days of hatching (data mean of 10 replications).

Treatments	Emergence of juveniles/20 Eggs	Inhibition (%)	Emergence of juveniles /4 egg Masses	Inhibition (%)	Survival of Juveniles
Sterile distilled water	19	-	2124	—	Active and Motile
Supernatant	0	100	423	80.00	Dead
Washed cells (X.10^6 cells/ml)	0	100	32	98.50	Dead
Whole Cells (X.10^6 cells/ml)	0	100	27	100.00	Dead

All the eggs of *M. incognita* placed in bacterial supernatant washed cells or whole bacterial culture broth were killed since none of the juveniles could emerge from the eggs.

Eggs were susceptible to the action of *B. thuringiensis* as compared to egg masses. In egg masses, the eggs are surrounded by a gelatinous matrix, which protects the eggs from predators, dehydration and other adverse conditions (Bird and Soeffy, 1972). More over, it takes time. for metabolites to reach eggs situated in the interior of egg masses. *B. thuringiensis* killed eggs as well as eggs present in the egg masses but the level of inhibition varied from 89.5 to 100% and 82.4 to 99.6% respectively (Table 1). Freshly hatched juveniles emerging from egg masses were killed with in 24-48 hours irrespective of bacterial concentration. Eggs or egg masses kept in different concentrations of bacteria for 10 days did not show emergence of juveniles when washed several times and then transferred to sterile distilled water. This shows killing of eggs permanently.

Bacterial supernatent containing thermos table exotoxin killed eggs (100%) and egg masses (80%) as shown in the Table 3. Production of exotoxin by *B. thuringiensis* has been reported earlier. Exotoxin can withstand temperature of 120°C for 15 minutes, affects the polymerase system by acting as a competitive inhibitor of nucleotides.

The gelatinous matrix of egg masses was disintegrated due to bacterial action as revealed by microscopic studies. This may be due to the ability of *B. thuringiensis* to produce chitinase (Chigaleichik, 1976), an enzymc which hydrolyzes chitin (polymer of N-acctyl-glucosmine) present in the chitin protein layer of the eggs shell and gelatinous matrix of egg masses (Spiegel and Cohn, 1985) there by affecting the permeability. '(Bird and McClure, 1976; Godoy *et al.*, 1976). All the strains of *B. thuringiensis* were found to produce chitinase. Once permeability is altered, egg contents including developing larvae become pre disposed to possible physiological disorders brought about the entry of diffusible toxins. It is quite probably that chitinase activity along with exotoxin or some other metabolites may be responsible for killing of eggs.

Chahal and Chahal (1991) conducted pot culture experiment on pea (*Pisum sativum* L.) to study the effect of *B. thuringiensis* on the control of *M. incognita*. In case of sterilized soil freshly hatched juveniles @ 1000/pot were used for inoculations.

The results show (Table 4) that *M. incognita* reduced the dry weight of shoot and root both in naturally infested as well as in sterilized soils. Nematodes suppressed the root branching and root development, which resulted in reduced

Table 4. Effect of *B. thuringiensis* Berliner on the root infection of pea (*Pisum sativum* L.) variety Bonne ville by *M. incongita* in sterilized and naturally infested soil under pot culture conditions (Data represent mean value of 10 replications)

Treatment	Naturally infested soil				Sterilized soil			
	Dry weight of shoot (g)	Dry weight of root (g)	galls/ plant	inhibition %	Dry weight of shoot (g)	Dry weight of root (g)	galls/ plant	inhibition %
Control	2.3	234	85	—	3.4	285	95	—
B. thuringiensis	4.4	310	26	69.4	4.9	325	24	74.7

surface area for absorption of nutrients. Chahal *et. al.* (1979) reported reduction in the up take of nitrogen, iron and zinc due to nematode infection. Hunter (1958) opined that reduced uptake of nutrients suppress the growth of plants. Dry weight of shoot and root was reduced more in naturally infested soil than sterilized soil. This was due to synergistic effect of *Fusarium oxysporum* and *M. incognita* (Powel *et al.*, 1971) *F. oxysporum* was found associated with *M. incognita* in diseased plants and form disease complex.

B. thuringiensis reduced root knot infection both in naturally infested soil and sterilized soil by 69.4 and 74.7% respectively.

Chahal and Chahal (1999) reported the results of pot culture experiment on the control of *M. incognita* with different strains of *B. thuringiensis*.

The results (Table 5) show that *B. thuringiensis* have the ability to reduce number of galls/plant and total population of nematodes. Due to reduced nematodes infection, dry weight of shoot, chlorophyll contents and nitrate reductes activity of leaves increased. Minimum number of galls/plant was observed in case of *B. thuringiensis* - 868. Similarly pot culture experiments were conducted on tomato (*Lycopersicon esculantum* L.) variety Punjab chhuhara, in nematode infested soil. All the strains of *B. thuringiensis* decreased number of galls/plant and total population of nematodes. Minimum galls were recorded in case of *B. thuringiensis* - 868.

Table 5. Effect of different strains of *Bacillus thuringiensis* on the control of *M. incognita* in brinjal (*Solanum melangena* L) under pot culture conditions using naturally infested soil (Data represent mean value of 3 replications)

Treatments	Dry Weight of Shoot (g)	No. of Galls/plant	Total Population of nemetodes	Chlorophyll Contents of Leaves (mg/g leaves)	Nitrate Reductes Activity of Leaves (Nm no2/h/g)
Control	6.15	143	24,478	1.09	1.59
BT-868	8.87	72	12,581	1.21	1.83
BT-869	8.91	79	15,688	1.18	1.75
BT-1	7.57	75	13,495	1.11	1.69
BT-2	7.82	78	12,792	1.15	1.82
BT-3	8.40	92	16,297	1.20	1.72
BT-4	7.20	85	14,212	1.17	1.64
BT-5	7.00	98	15,782	1.21	1.71

BT = *Bacillus thuringiensis*

5. CONCLUSIONS

Bacillus thuringiensis being spore former and hetertrophic in nutrition have many advantages including survival in the soil under adverse conditions and easy to multiply as well as to apply under field conditions. *B. thuringiensis* strains differed in their ability to control nematode infection and inhibit emergence of juveniles from eggs of *Meloidogyne incognita*, *B. thuringiensis* not only kill insects and larvae of mosquito but have the potential to control *M. incognita*.

REFERENCES

Bird, A.F. and Mc Clure, M.A. 1976. The Tylenchoid (nematode) Eggshell, Structure, Composition and Premeability. *Parasitiology*, 72 : 19-28.

Bird, A.F. and Soeffky, A. 1972. Changes in the ultra structure of the gelatin matrix of *Meloidogyne incognita* during dehydration. *J. Nematol.* 4 : 166-168.

Bulla, L.A; Jr. Kramer, K.J. and Davidson, L.I. 1977. Characterization of the entomocidal parasporal crystal of *Bacillus thuringiensis. J. Bacterial*; 130 : 375-383.

Chahal, P.P.K. and Chahal V.P.S. 1986. Effect of *Azotobacter chroococcum* on the hatching of egg masses and eggs of *Meloidogyne incognita. Plant and Soil.* 95 : 289-291.

Chahal, V.P.S. and Chahal, P.P.K. 1991. Control of *Meloidogyne incognita* with *Bacillus thuringiensis.* In : Plant Soil interaction (Eds. : R.J. Wright *et al.*,) pp 677-680.

Chahal, P.P.K. and Chahal V.P.S. 1993. Effect of Thuricide on the hatching of eggs of Root knot Nematode (*Meloidogyne incognita*). *Current Nematole* 4(2) : 247-248.

Chahal, P.P.K. and Chahal V.P.S., N.P. Kaur and Arora, C.L. 1997. Effect of *Rhizobium* and *Meloidogyne incognita* on the up take of Nitrogen. Iron and Zinc in mungbean (*Vigna radiata* L.). *Nematode Medit.* 29 : 37-40.

Chahal, V.P.S. and Chahal P.P.K. 1999. Final Technical Report of PL-480 project "Microbial control of Plant Parasitic Nematodes" Submitted to USDA.

Chigaleichik, A.G. 1976. Chitinase of *Bacillus thuringiensis*, *Mikrobiologiya* 45 : 966-972.

Christoph, A. 1985. Germination of *Bacillus thuringiensis* var *israelensis* spores in the gut of Aeded Larval (Diptere : Culicidae). *J. invert. Pathology*, 45 : 1-8.

Goldberg, L. and Margalit, J. 1977. A bacterial spore demonstrating rapid larvaicidal activity against *Anopheles sergentii, Uranotaenia unguiculata, Culex univitattus, Aedes aegypti* and *Culex pipiens. Mosquito News* : 37 : 355- 358.

Heimpel, A.M. 1967. A critical review of *Bacillus thuringiensis* var. *thuringiensis* Berliner and other Crystalliferous bacteria *Am. Rev. of Entomology*, 12 : 282-322.

Hunter, A.M. 1958. Nutrient absorption and translocation of phosphorus as influenced by Root knot nematode (*Meloidogyne acrita*). *Soil Sci.*; 86 : 245-250.

Jatala, P. 1985. Biological control of nematodes. In : an advanced treatise on *Meloidogyne*. Vol. I (Eds Sasser, J.N. and carter, C.C.) pp 303-308.

Jatala, P. 1986. Biological control of plant parasitic nematodes. *Ann. Rev. phytopath.* 24 : 453-489.

Kerry, B.R. and Crump, D.H. 1980. Two fungi parasitic on females of cyst nematodes (*Heterodera* spp.) *Trans. Br. Mycol. Soc*; 74 : 119-125.

Mankau, R. 1975. *Bacillus penetrans* n. comb. Causing a virulent disease of plant parasitic nematodes. *J. Invert pathol.* 26 : 333- 339.

Mankau, R. 1980. Biological Control of nematode pests by natural enemies. *Ann. Rev. Phytopath* : 18 : 415-440.

Mc Connell, E. and Richards, A.G. 1959. The production by *Bacillus thuringiensis* Berliner of a heat stable substance toxic for insects. *Can. J. Microbiol.* 5 : 161-168.

Norris, J.R. 1964. The classification of *Bacillus thuringiensis J. App. Bacteriol.* 27 : 439-447.

Powell, N.T.; Melendez, P.L. and Batten, C.K. 1971. Disease complex in tobacco involving *Meloidogyne incognita* and certain soil borne fungi. *Phytopathol.* 61 : 1332-1337.

Spiegel Y and cohn, E. 1985. Chitin is present in gelatinous matrix of *Meloidogyne*. Revue, *Nematol*; 8 : 184-188.

Sterling, G.R. 1984. Biological control of *Meloidogyne javanice* with *Bacillus penetrans*. *Phytopathology*, 74 : 55-60.

Syre, R.M. 1980. Promising organism for bio control of nematodes. *Plant Disease*, 64 : 526-532.

Wildermuth, A.H. 1982. Nature of antagonism of fungi by bacteria. *Soil Biol Biochem*; 14 : 557-560.

CHAPTER -15

CURRENT OPTIONS FOR THE MANAGEMENT OF PLANT - PARASITIC NEMATODES

S.D. Mishra and S. Nageswari

INTRODUCTION

Plant-parasitic nematodes have long been recognized as constraints to crop production. However, studies on plant parasitic nematodes and their role in crop protection are relatively rare in comparison with the other crop pests such as insects, fungi and weeds. Nematodes are difficult soil pest problems to identify, demonstrate and control because of their secluded environment and inaccessibility. The degree of damage caused by nematodes depends upon the population density of nematodes, susceptibility of the crop, environmental condition such as soil fertility and moisture and also the presence of other pathogenic micro-organisms, which may interact with nematodes as they are known to cause complex plant diseases in association with fungi, bacteria and viruses. Since the diseases caused by nematodes are not sudden epidemic types, which spread gradually in extent, year after year and where the host is not out right killed due to highly specialised host parasitic relationship, the nematode problems are not appreciated by the farmers.

In recent years, there has been a marked decline in the work on nematicides and the recommendations for their use in nematode control. This has been partly because of the environmental lobby, costs and dangers of nematicides and the banning of many products on the world stage and partly because of the desire to produce a sustainable means of managing the pests. This has led to an increase in searching alternative means of nematode management. It indicates clear reversal of chemical trends to a re-evaluation and re-examination of the more traditional agricultural practices among other methods for improving crop yields in the presence of nematode pests. There has been a considerable shift in emphasis from straight control of the nematodes as isolated pests to wider or more integrated managment of the overall problems caused by pests and pathogens and other constraints to crop health. The appeal of Integrated Pest Management (IPM) is its emphasis on reducing the role of pest management in environmental degradation by using the safest tactics available in the context of economic and ecoenvironmental needs, and by invoking management only when it is determined to the necessary through bio monitoring and use of economic thresholds.

Elements of Nematode Management Programmes

The fundamental elements of nematode management programmes are :

- Sampling
- Prediction of nematode population change and crop loss.
- Management tactics.

1. SAMPLING

Many methods used to manage nematodes do not require sampling to achieve economic benefits. Nevertheless, for most management tactics, systems can be developed to optimize economic inputs and returns and minimize environmental risks, if accurate low cost methods of sampling are available to determine the species and in some cases the population density of nematodes in a field (Ferris, 1978; Noe, 1988).

Need for Sampling

- To evaluate the necessity of using nematodes.
- Where varieties of crops are available that are resistant to nematode pests, population monitoring is desirable if a cultivar is not the best one available in the absence of the pest (Roberts 1982) or if other nematode species may be present for which resistant is not available (Duncan and Ferris, 1983).
- Other cultural practices, such as the use of crop rotations or fallowing are employed to obtain best advantage in response to information about the species and number of nematodes in a field.

Direct and indirect methods are commonly used to estimate nematode levels in fields. With most annual crops, sampling should occur before sowing and may occur before or after an overwinter period that precedes sowing. In some systems it is useful to estimate nematode levels in fields after harvest and before over wintering (Belair and Boivin, 1988; Kinloch, 1982). In crops attacked by *Meloidogyne* spp. indirect estimation of nematode densities based on root knot galls, severity can be the most cost-effective sampling method if performed in the field at or prior to harvest. Root damage assessment requires less time and labour than processing soil and root samples for nematode counts and permits sequential sampling to optimize sample collection, because samples are analyzed individually, unlike samples for nematode counts that are usually bulked and often subsampled.

Another indirect sampling method is the use of seedlings planted in soil from samples as a bioassay based on symptoms of nematode infection (Brown, 1987; McSorely and Parrado, 1983). This method require effective planning by growers to obtain samples early enough to permit bioassay development before planting. It is useful only with nematodes that quickly cause distinctive root symptoms.

In most monitoring systems nematodes are extracted from samples of soil or plant material and identified and enumerated directly. Generally 10-25 samples from a defined area are composited and mixed to reduced processing expense. Nematodes may be recovered from the entire sample or from subsamples, depending on the extraction technique.

Studies of nematode spatial patterns have been used to optimize sampling either by computer simulation using actual counts or mathematical formulae descriptive of the spatial patterns (Francl, 1986; Noe and Campbell, 1985).

The mean density for which an optimum sample size is estimated depends on whether the objective is detection or qualification with reference to an economic threshold (Ferris 1984). It is useful to investigate the effect of mean population density on general sample requirements since economic thresholds vary greatly, based on host crops, adaptive and climatic conditions and crop price and management costs. Samples sufficiently large to reliably measure population density within 25-50% of the true mean have been suggested as acceptable for diagnostic use and are often the norm in field research (McSorley, 1982). Ranges of nematode population density between field are often large and unlike many pest organisms, fairly stable over short time intervals. Therefore, population estimates from samples are often sufficiently high or low to permit sound management decisions.

2. PREDICTION OF NEMATODE POPULATION CHANGE & CROP LOSS

Most applied prediction in Nematology is confined to qualitative descriptions of expected crop loss based on broad density classes of nematodes. Relationship between nematode population densities at two points in time have also been measured in a large number of systems, and to some extent have influenced general recommendations concerning cropping patterns. Forecasting crop losses due to nematodes has relied primarily on critical point damage functions derived by fitting experimental yield data to preplant population densities of nematodes under defined conditions. Data for these models are obtained from experiments in pots and microplots where nematode densities are estimated. The limited number of studies to develop simulation models of nematode host relationships is due partly to recognition that few options are available for nematode control during the course of a growing season.

Theoretical foundations for using damage functions to optimize management inputs based on probably crop response have been developed and will become increasingly useful as sample precision improves and as data bases are developed to support their use. Critical point damage functions are useful because most management options must be initiated prior to planting a crop and because nematode migration is relatively slow. Critical point models of the relationship of final population density (Pf) to initial density (Pi) are commonly used in Nematology for many of the same reasons as for the damage function. For management purposes, population models are used primarily to predict the influence of a given management that will remain to attack subsequent crops (Ferris *et al.*, 1986). Models have been used in temporal sequence to simulate the effects of management options on nematode populations and crop growth over more than one growing season and in perennial crops (Duncan 1983; Noe, 1988). Cumulative profits based on management costs and crop yield and value are one criterion to evaluate the relative value of large number of management options such as sequences of rotation crops, resistant varieties, fallow and nematicide treatments. The use of deterministic models to stimulate management sequences permits efficient organization of large amounts of information to help in developing management strategies.

Progress in forecasting depends heavily on improved methods to quantify nematode populations. Improved methods to quantify nematode populations would also provide incentive to develop the data bases which are necessary to extend models based on edaphic and climatic variables.

3. MANAGEMENT TACTICS

An objective of IPM is to reduce the use of pesticides by substituting other management tactics. Since nematodes do not have the migratory and reproductive capacity of many arthropod pests, careful scheduling of management tactics may reduce nematode populations to non-damaging levels for a period of time. Methods currently used to control losses due to plant parasitic nematodes have changed very little in a generation.

Exclusion : Exclusion is the principle behind regulatory methods of nematode management. Exclusion implies excluding the possibility of entry of a pest in a region where it is not known at the time of regulatory operation but is likely to establish in the absence of the same.

The spread of nematodes can be prevented, or at least, reduced, by use of nematode - free seed or planting material and production of clean seedlings from nematode - free seed beds or soil.

Cultural and physical methods of Nematode Management

Cultural practices play a vital role in nematode management strategies for subsistence farming, since low value crops rarely receive any major input of fertilizer and pesticides. Various cultural practices that can be used for effective management of nematode pests are being discussed here:

Table 1. Plant parasitic nematodes of current or future regulatory significance

	Plant parasitic nematodes present in India
Anguina agrostis	*Aphelenchoides besseyi*
A. graminis	*A. fragariae*
Belonolaimus longicaudatus	*A. ritzemabosi*
Bursaphelenchus xylophilus	*Cactodera cacti*
Cactodera cacti	*Globodera rostochiensis*
Dolichodorus heterocephalus	*Heterodera avenae*
Heterodera goettingiana	*H. zeae*
H. leuceilyma	*Hirschmanniella oryzae*
H. trifolii	*Meloidogyne graminis*
Hopoloaimus columbus	*Pratylenchus coffeae*
Longidorus sylphus	*Radopholus similis*
Meloidogyne chitwoodi	The Plant parasitic nematodes not present in India but may be disastrous if established
M. naasi	*Ditylenchus destructor*
Nacobbus aberrans	*D. dipsaci*
Rotylenchulus macrodoratus	*Heterodera glycines*
Trichodorus viruliferus	*H. schachtii*
Xiphinema diversicaudatum	*Rhadinaphlenchus cocophilus*

(a) Application of fertilizers and organic amendments

Application of inorganic fertilizers to soil and incorporation of various organic amendments influence nematode populations. Their application may increase the growth and yield of nematode infected plants by improving the soil structure and increasing the water holding capacity, supplying mineral nutrients to deficient soils, and providing substrate on which parasites of nematodes may propagate.

In India, application of fertilizers improved the growth of rice and increases the number of nematodes, although popuplations of *Hirshmanniella oryzae* are lowered in fields by application of urea (Mathur and Prasad, 1971). In Japan, control of rice root nematode is obtained by application of calcium silicate and compost. The yield of irrigated wheat infected with *Pratylenchus thornei* is improved by the application of nitrogen, when the nematode population is near threshold level. At the population level of nematodes, greater than threshold levels, application of nitrogen does not compensate for nematode damage and yield is reduced.

Application of lime and ash leads to an increase in the bacterial feeding nematode populations, probably due to an increase in soil pH, populations of *Acrobeloides* sp., *Protorhabditis* sp., *Rhabditilis* sp. and other bacterial feeding nematodes increase in limed soils and the effect is visible months after the application of lime. These nematodes may play a role in organic matter decomposition.

From time immemorial, organic amendments which are mainly bioproducts and wastes from agricultural and other activities include oilseed cakes, plant crop residues, plant composts, green manure, agroindustrial wastes, ash, animal, wastes and human wastes. These are added to soil to improve the crop yield. The benefits of adding organic material or residues to the soil is mainly due to a decrease in the levels of soil structure and ecology and/or by the action of waste-derived chemicals on the soil fauna and flora.

Most of the work on nematode control using organic amendments has been done in India. Mathur and Prasad, (1973) observed a reduction in the population of *Hirichmanniella oryzae* and improved growth of rice on application of mustard and neem oilcakes before planting. Oil cakes of linseed, mustard, groundnut, castor, mahua and neem are all capable of reducing the number of root knot galls on the roots of the plants when they are incorporated into soil infested with the root knot nematodes.

Mishra (1988) reported that application of neem and karanj cakes @ 1000 kg/ha were very effective in reducing nematode populations infecting chickpea and lentil. While karanj cake provided better yield, the neem cake showed more nematicidal value. Neem seed kernel and neem seed coat applied in a decomposed form were also found to be nematicidal (Mishra and Mojumder, 1993). Das (2000) found that the soil application of neem seed powder was highly effective against *Meloidogyne incognita*, *Heterodera cajani* and *Rotylenchulus reniformis* infecting pigeonpea.

The maximum benefit of the use of organic amendments will be gained by very small farmers who have limited financial resources but do have access to locally produced wastes or one sort or another.

(b) Time of Planting

Nematode population levels fluctuate both with season of the year and with cropping cycles. Control of some nematodes can be obtained by manipulating dates to avoid periods of peak nematode activity. Late planted deep water rice crop escapes ufra disease caused by *Ditylenchus angustus* (McGeachie and Rahaman, 1983). Tomato, brinjal and other susceptible vegetables are planted in the cooler winter (rabi) months to avoid root knot nematode damage (Sethi and Gaur, 1986). Gaur *et al.* (1979) were of the opinion that adjusting the date of sowing in November might enable chickpea to escape excessive root knot nematode injury.

(c) Tillage practices

Cultivation of land (deep ploughing and other tillage operations) modifies both macro and micro-climates of the soil and affects the activity of soil biota. There is an increasing interest in minimum tillage practices to preserve soil structure, control soil erosion, reduce the turn around time and reduce non productive costs. Wheat sown by direct drilling is less severally damaged cyst nematode than when it is traditionally sown.

(d) Solarization

Solar heating of moist soil by means of polythylene mulching during the summer months is effective in controlling soilborne nematodes. The soil is mulched or covered with clear polyethylene for several weeks before the crop is planted. Irrigation before covering soil with polyethylene improves the effect of solarization in reducing the nematode densities. Soil solarization reduces the population densities of nematodes parasitic to chickpea and pigeonpea (Sharma and Nene, 1990). It is an environment friendly approach without any toxic residual effects.

(e) Trap cropping and antagonistic plants

Many plants belonging to 57 families possess nematicidal properties (Sukul, 1992). Plants with nematicidal properties when intercropped with nematode susceptible crops or grown as cover crop, can effectively reduce populations of nematodes and improve yields. Their acceptability and usefulness depends on their economic or other value to the farmers. Crops such as *Tagetes* spp., sesame, mustard and asparagus which have local use or market can be readily recommended. Some weeds that have nematicidal root exudates can be recommended as cover crops. *Sphenoclea zeylanica*, a succulent weed common in lowland rice fields produces nematicidal exudates that can give a 99% kill of the rice root nematode, *Hirschmanniella* spp. (Mohandas *et al.*, 1981). It decomposes quickly when incorporated into the soil and can be used as a green manure crop before rice.

(f) Flooding

Flooding routinely controls several species of *Meloidogyne* capable of attacking rice. *M. graminicola*, which can survive in paddy, can be managed by growing seedlings in flooded conditions that prevent nematode penetration (Bridge and Page, 1982). Vegetables (Thamas and Stoner, 1953) and tobacco (deGurian, 1970) can often be grown successfully following rice without damage from *Meloidogyne* spp.

(g) Cropping systems

A cropping system can be defined as the sequence of growing crops in a given field along with the required technologies for their production. The sequence would be of two types : spatial and temporal. The spatial sequence referes to the arrangement of different crop species in a piece of land. Mixed cropping, intercropping and strip cropping are examples of this system. Temporal sequence referes to distribution of crops over a period of time. Crop rotation, relay cropping, discontinuous planting and fallow rotation are examples of this system.

The damage caused by root knot nematodes is reduced when susceptible hosts are grown as a mixed crop with nematode antagonistic crops. Cultivation of tomato with castor results in considerable reduction in the number of root knot galls on tomato (Hackney and Dickerson, 1975). The number of root knot galls on the susceptible brinjal is reduced when it is planted in association with margiold (Choudhury, 1981). Intercropping of onion and maize reduce the galling by *M. incognita* on potato roots.

Crop rotation is the best known method of temporal sequence management to reduce damage by nematodes. It is the most widely practiced method of disease control in agriculture. The principle of nematode control by crop rotation is not to grow susceptible crops at long intervals so that nematode population levels fall below damage threshold level between the crops, and to prevent buildup of nematode populations to damaging levels (Johnson *et al.*, 1975; Rao *et al.*, 1984). Rotation of wheat with various legume species reduces the population density of cereal nematode (Brown 1987).

(h) Resistant cultivars

Resistant varieties offer many of the advantages for nematode management as rotation of crops, with the additional feature of permitting production of crops best suited to the needs of the grower. With notable exceptions, resistant varieties of most major agronomic and horticultural crops are available for use against some of the important nematode pathogens.

(i) Destruction of infected crop residues

Uprooting and burning of infected root systems kill nematodes present in the soil and reduces the carry over populations. Destroying crop residues including roots or lifting roots onto the soil surface is traditional in many parts of the country (Sethi and Gaur, 1986).

(j) Biological control

A number of fungal and bacterial antagonists of nematodes have good potential for biological control (Kerry, 1990; McClure, 1990). Large scale use of commercially available biocontrol agents in the field level is still at infant stage.

(k) Chemical control

In the era of eco-friendly management, chemicals have a very limited role. IPM enables restricted or no use of chemicals. The means by which chemicals are used to control nematodes has changed considerably and there is a sense of urgency in the recognized need for additionally changes to suit the guidelines of environmental safety. The use of many synthetic nematicides is legally restricted based on soil texture, organic matter content, depth to water table and proximity to drinking water wells.

4. CONCLUSION

Plant parasitic nematodes, the hidden enemies of farmers, cause extensive and insidious damage to agricultural crops in our country. These nematodes, though extensive in occurrence, cause greater crop damage when they get favourable soil conditions, cultivations of susceptible crops continuously one after the other and year after year in the same field. The lack of general awareness about these pathogens may be one of the main reasons of the damage. The physioligcal damages to plants are expressed in the form of yellowing, stunting, premature drying, poor growth, shedding of leaves and finally reduction in yield. As farmers have little perception of these invisible pests due to the microscopic size and subterranean habitat of the nematodes, the damage to crop productivity continues unchecked. Scientists have not developed low cost options for the management of important nematodes for low input agriculture. The need for development of programmes to educate farmers about the nematode problems and to develop inexpensive, environmentally safe and effective management of the nematode disease is acute. Any nematode population management planning programme may permit nematodes to subsist in different cropping systems by not in damaging proportions.

Studies on crop diversity, nematode inter relationships and quantification of production constraints to a crop are essential in developing integrated plant protection programmes. The physical and biological environment of a given region as well as the accompanying socio-economic conditions are of utmost importance in designing a meaningful research program for management of nematodes. There is a great need for inter disciplinary research since what looks good to a nematologist may not be appealing to an agronomist or economist. Recommendations for nematode management need to be product which are practical, reliable, sustainable in the long term and adaptable to different farming systems, farmers and circumstances.

REFERENCES

Barker, K.R. 1989. Yield relationships and population dynamics of *Meloidogyne* spp. on flue-cured tobacco. *Ann. Appl. Nematol.* 21 : 597-603.

Belair G. and Bovin, G. 1988. Spatial pattern and sequential sampling plan for *Meloidogyne hapla* in muck grown carrots. *Phytopathology* 78 (5) : 604-7.

Bridge, J. and Page, S.L.J. 1982. The rice root knot nematode, *Meloidogyne graminicola*, on deep water rice. *Rev. Nematol*, 5 : 225-32.

Brown, R.h. 1987. Control strategies in low value crops. In *Principles and practices of nematode control in crops* (Brown, R.H. and Kerry, B.R., eds.) Merrickville, Australia : Academic Press, pp. 351-382.

Choudhury, B.C. 1981. Root knot nematode problem in various crop plants in Bangladesh. In *Proceedings of the III International Meloidogyne Project Research and Planning Conference on Root knot Nematodes, Meloidogyne spp. Region VI*, Jakarta, Indonesia, pp. 142-145.

Das, D. 2000. Interaction studies of pigeonpea cyst nematode (*Heterodera cajani*) reniform nematode (*Rotylenchulus reniformis*) and root knot nematode (*Meloidogyne incognita*) infecting pigeonpea and their management through neem formulations. Ph.D. thesis, I.A.R.I. New Delhi -12.

de Gurian, G. 1970. Le Proleme *Meloidogyne* sur tabc a Madagascar. *Cah. ORSTOM, Ser. Biol.* 11 : 187-208.

Duncan, L.W. 1983. Predicting effects of plant parasitic nematode communities on crop growth. Ph.D. thesis, Univ. Calif. Riverside.

Duncan, L.W. and Ferris, H. 1983. Effects of *Meloidogyne incognita* on cotton and cowpea in rotation. *Proc. Beltwide Cotton Prod. Res. Conf.*., pp. 22-26.

Ferris, H. 1978. Nematode economic thresholds : derivation, requirements and theoretical considerations. *J. Nematol.* 10 : 341-50.

Ferris, H. 1984. Probability range in damage prediction as related to sampling decisions. *J. Nematol.* 16 : 246-51.

Ferris, H; Ball, D.A.; Been, L.W. and Gudmunson, L.A. 1986. Using nematode count data in crop management decisions. *Calif. Agric.* 40 : 12-14.

Francl, L.J. 1986. Improving the accuracy of sampling field plots for plant parasitic nematodes. *J. Nematol.* 18 (2) : 190- 95.

Gaur, H.S.; Mishra, S.D. and Sud, U.C. 1979. Effect of date of sowing on the relations between the population density of the root knot nematode, *Meloidogyne incognita* and the growth of three varieties of chickpea, *Cicer arietinum. Ind. J. Nematol.* 9 : 152-59.

Hackney, R.W. and Dickerson, O.J. 1975. Marigold, castor bean and chrysanthemum as controls of *Meloidogyne incognita* and *Pratylenchus alleni. J. Nematol.*, 7: 84-90.

Johnson, A.W.; Dowler, C.C. and Hauser, E.W. 1975. Crop rotation and herbicide effects on population densities of plant parasitic nematodes. *J. Nematol,* 7; 158-168.

Kerry, B.A. 1990. An assessment of progress toward microbial control of plant parasitic nematodes. *J. Nematol.* 22 (4S) : 621-31.

Kinloch, R.A. 1982. The relationship between soil population of *Meloidogyne incognita* and yield reduction of soybean in the coastal plain. *J. Nematol.* 14 : 162-68.

Mathur, V.K. and Prasad, S.K. 1971. Occurrence and distribution of *Hirschmanniella oryzae* in the Indian Union with description of *H. mangaloriensis n.* spp. *Ind. J. Nematol.*, 1 : 220-226.

Mathur, V.K. and Prasad, S.K. 1973. Control of *Hirschmanniella oryzae* associated with paddy. *Ind. J. nematol.*, 1 : 220-226.

McGeachie, I. and Rahaman, L. 1983. Ufra disease : a review and a new approach to control. *Tropical Pest Management,* 29 : 325-332.

McClure, M.A. 1990. Biological control of nematodes. In *New Directions in Biological Control Alternatives for Suppressing Agricultural Pests and Diseases*, Liss, New York, pp. 255-69.

McSorley, R. 1982. Simulated sampling strategies for nematodes distributed according to a negative binomial model. *J. Nematol,* 14 : 517-22.

McSorley, R. and Parrado, J.L. 1983. A bioassay sampling plan for *Meloidogyne incognita Plant Dis.*, 67 : 182-84.

Mishra, S.D. 1988. Management of plant parasitic nematodes infecting chickpea and lentil by neem and Karanj seed cakes. *Int. Soil Zool. Colloquium,* 88-89.

Mishra, S.D. and Mojumder, V. 1993. Nematicidal efficacy of decomposed neem products on *Meloidogyne incognita* infecting mungbean and their residual effect in chickpea. *World Neem Conference,* India. 2 : 657-662.

Mohandas, C; Rao, Y.S. and Sahu, S.C. 1981. Cultural control of rice root nematode with *Sphenoclea zeylanica. Proc. Ind. Acad. Sci.* 90 : 373-376.

Noe, J.P. 1988. Theory and practice of the cropping systems approach to reducing nematode problems in the tropics. *J. Nematol,* 20 (2) : 204-13.

Noe, J.P. and Barker, K.R. 1985. Overestimation of yield loss of tobacco caused by the aggregated spatial pattern of *Meloidogyne incongita. J. Nematol.* 17 : 245-51.

Noe, J.P. and Campbell, C.L. 1985. Spatial pattern analysis of plant parasitic nematodes. *J. Nematol,* 17 : 86-93.

Rao, Y.S.; Prasad, J.S.; Yadav, C.P. and Padalia, C.R. 1984. The influence of rotation crops in rice soils on the dynamics of parasitic nematodes. *Biol. Agric. Horti*, 2 : 69-78.

Roberts, P.A. 1982. Plant resistance in nematode pest management. *J. Nematol*, 14 : 24-33.

Sethi, C.L. and Gaur, H.S. 1986. Nematode management : An overview. In Plant Parasitic Nematodes of India - Problems and progress, pp. 424-45.

Sharma, S.B. and Nene, Y.L. 1990. Effects of soil solarization on nematodes parasitic to chickpea and pigeonpea. *J. Nematol* 22 : 658-64.

Sukul, N.C. 1992. Plants antagonistic to plant parasitic nematodes. *Ind. Rev. Life Sci.*, 12 : 23-25.

Thames, W.H. and Stoner, W.N. 1953. A preliminary trial of lowland culture rice in rotation with vegetable crop as a means of reducing root knot nematode infestations in the Everglades. *Plant Dis. Reptr.*, 37 : 187-92.

CHAPTER - 16

CURRENT TRENDS IN INTEGRATED NEMATODE MANAGEMENT IN FRUIT CROPS

P. Parvatha Reddy and M.S. Rao

INTRODUCTION

Fruits are the most important priceless blessings that nature has bestowed upon mankind. They are not only rich in Vitamins 'A' and 'C' and minerals like Calcium and Iron but also have low calorific values and are low in fats. The present area under fruit crops is estimated to be about 2.9 million hectares with an estimated production of 25 million tonnes.

The destructive plant parasitic nematodes are one of the major limiting factors in fruit production throughout India. Fruit crops being perennial in nature, harbor and encourage the build up of nematode population. For centuries, man has been plagued by these microscopic organisms feeding on the roots of crop plants essential for his survival and well being. Roots damaged by the nematodes lose efficiency in the utilization of available soil moisture and nutrients resulting in reduced functional metabolism. Furthermore, the weakened and damaged roots become easy prey to many fungi and bacteria which accelerate root decay. Visible symptoms of nematode attack often include reduced growth, varying degrees of chlorosis, wilting and sometimes death of plants. These deleterious effects result in reduced yields and poor fruit quality in several fruits such as citrus, banana, grapevine, pineapple and papaya. Nematode management is therefore important for high yields and quality that are required by the high cost of modern crop production.

Realization that many environmental and health hazards are linked with the use of chemicals. Researchers all over the world are engaged in standardizing the pest management strategies by following non-chemical and eco-friendly approaches such as cropping systems, botanicals and biological control agents. Standardization of methods for effective utilization of biocontrol agents is very important for evolving ecologically sound integrated nematode management strategies. This concept of Integrated Nematode Management is targeted at prevention, population reduction of plant parasitic nematodes and development of resistant varieties of fruit plants. INM seeks to stabilize populations of target nematodes at acceptable levels resulting in favourable long term socio-economic and environmental consequences.

A. CITRUS

Nematodes are one of the important limiting factors in citrus production throughout India. Citrus crop being perennial in nature harbour and encourage the build up of nematode population throughout the year. The first record of an association between a nematode and citrus appears to be that of Rao (1956), who found *Meloidogyne* sp., parasitizing citrus roots in Andhra Pradesh. But it was not until the discovery in 1961 of *Tylenchulus semipenetrans*, on the roots of citrus trees in Uttar Pradesh (Siddiqui, 1961) that a nematode was found to cause a disease condition of citrus, later caller "slow decline". At present 122 species within 57 genera of plant parasitic nematodes have been reported in association with citrus roots from many parts of the country. However, most of these nematodes are not known pathogens of citrus and their true relationship with their host plant still remains to be established.

Of the 122 species of plant parasitic nematodes reported in association with citrus roots from India, proof of pathogenicity to citrus is available for only six species viz., *Tylenchulus semipenetrans* (Siddiqui, 1961), *Pratylenchus coffeae* (Siddiqui, 1964), *Hoplolaimus indicus* (Gupta and Atwal, 1972), *Meloidogyne africana* (Chitwood and Toung, 1960), *M. indica* (Whitehead, 1968) and *M. javanica* (Mani, 1986). Except for the citrus nematode (*T. semipenetrans*), other nematode species that are pathogenic on citrus have limited geographic distribution. This review will be devoted to these six cases of nematode-citrus relationships that have been described in some detail.

The citrus nematode — *Tylenchulus semipenetrans* Cobb, 1913

Siddiqui (1961) reported the citrus nematode for the first time from India in a paper presented before the Indian Science Congress and observed that about 80 per cent of the citrus trees at Aligarh, Uttar Pradesh, were infested with this nematode. The nematode causes "slow decline" and is considered to be one of the factors responsible for die-back of citrus trees in India (Chona *et al.*, 1965).

Yield increases of 40 to 200 per cent have been obtained following the control of the nematode over a range of growing conditions (Mukhopadhyaya and Dalal, 1971; Mukhopadhyaya and Suryanarayana, 1969; Swarup and Seshadri, 1974).

OCCURRENCE AND DISTRIBUTION

The citrus nematode is known to be widely distributed in major citrus growing regions of the country. It has been reported from Uttar Pradesh (Siddiqui, 1961); Delhi, Punjab, Rajasthan, Maharashtra, West Bengal, Assam (Chona *et al.*, 1965); Himachal Pradesh (Mukhopadhyaya, 1970a), Sikkim (Edward and Rai, 1970); Orissa (Khuntia and Das, 1969); Haryana, (Mukhopadhyaya and Suryanarayana 1969); Bihar (Prasad and Chawla, 1965); Karnataka (Swamy, *et al.*, 1973); Kerala (Nair, 1965); Tamil Nadu (Muthukrishnan and Sivakumar, 1973) and Andhra Pradesh (Krishnamurthy Rao and Thammiraju, 1975).

(a) Symptoms

'Slow decline' of citrus is a diseased condition of trees with symptoms similar to those caused by drought and malnutrition. Affected trees exhibit reduced vigour, chlorosis and falling of leaves, twig dieback and consequently, reduced fruit production (Prasad and Chawla, 1965). This decline of the tree is gradual and persists until the crop is so small that tree maintenance may become uneconomical.

As the nematode feeds on roots and reproduces, a large proportion of the feeder roots of citrus trees, particularly in the upper soil layers, is inactivated or destroyed, the uptake of water and minerals from the soil is reduced and the symptoms appear in the above ground tree parts. Heavily infested roots are darker in color with branch rootlets shortened, swollen and irregular in appearance than in normal ones (Chona *et al.*, 1965).

Soil particles usually cling tightly, even after washing, to the gelatinous egg masses which cover the protruding part of the nematode body. In heavily infested roots, the cortex separates readily from the vesicular stele.

(b) Hosts range

Mainly host range is confined to Rutaceae family. A total of 75 Rutaceous species mainly citrus and citrus hybrid are the hosts of this nematode. Other non Rutaceous hosts reported from India are grapes (Parvatha Reddy and Singh, 1978a), *Clerodendron inerme, Jasminus sambac* (Bajaj *et al.*, 1988), *Garcina mangostana* (Chawla *et al.*, 1980) and loquats (Chona *et al.*, 1965).

(c) Economic threshold level

Citrus is a perennial crop for which to find out economic threshold level is slightly difficult. This also varies from field to field depending upon the agropractices followed. In India, economic threshold level has not been estimated under field conditions. However, it has been estimated in California that soil stages (juveniles/100 g soil) below 800 represents a non damaging population level. The orchards with more than 1600 juveniles/100 g soil may respond economical to nematicide treatments and at level more than 3600 during peak population growth period, treatment may improve yield substantially. The females per gram root are also used to define the damage level, with counts of 300, 700 and 1400 representing low, moderate and high ranges, respectively. In Florida, it was estimated that yield was not measurably reduced and populations were below 2000 juveniles/100 cc soil during peak period of the soil population development (Duncan and Cohn, 1986).

(d) Interaction with other micro-organisms

The role of root rot fungi in the disease syndrome caused by *T. semipenetrans* is significant. Misra and Edward (1977) reported severe root damage to citrus when the nematode was accompanied by secondary invaders like *Fusarium oxysporum* and *Macrophomina phaseoli*. The nematode helps weakly pathogens fungus like *Fusarium* to cause secondary infection of citrus roots (Chhabra, 1972).

In the presence of *T. semipenetrans*, significantly more citrus plants become infected by *Fusarium oxysporum* and *F. solani* and the combined activity of the nematode and *F. solani* is thought to be a factor in the debilitation of citrus. *Fusarium* spp. seem to cause more damage in combination with the citrus nematode in coll, wet soils and with *Phytophthora* spp. in warm, wet soil.

MANAGEMENT

(i) Physical methods

The nematode can be controlled by dipping the bare roots of citrus seedling in hot water at 45°C for 25 min or 46.7°C for 10 min (Bindra *et al.*, 1967). During summer exposing the soil to hot sun can control the nematodes substantially (Sehgal *et al.*, 1990).

(ii) Cultural methods

Singh and Sitaramaiah (1973) reported that castor cake was highly effective against the citrus nematode. Mobin and Khan (1969) have also observed that neem, mahua, groundnut and mustard oil cakes were effective in reducing the nematode population in citrus rhizosphere. Application of neem cake at 20 kg per tree at an interval of 4 months can effectively reduce the nematode population (Mani and Dakshinamurthy, 1986).

The citrus nematode is easily transported to new orchards through infested seedlings and budded plants which are usually carried 'balled' with infested soil. Therefore, nurseries should never be established on or near old citrus orchards and nursery soil should be sterilized before planting. Care should also be taken not to spread infection through tools, machinery and irrigation water used in infested groves.

Irrigation of citrus orchards with sewage water reduces the citrus nematode population to a very low level.

A number of plants posses nematicidal properties. Different plant parts and their extract have been tested against *T. semipenetrans* by different workers. Aqueous and methanol garlic extracts were toxic to *T. semipenetrans in vitro* (Nath *et al.* 1982). The extracts of leaves, stem and buds of *Datura stramonium, Ipomoea carnea, Tagetes patula* and *Lowsonia alba* caused 50-100% mortality at 4 mg/ml or 1:5 dilution after 48 hrs, 1986). Similarly, water extract of *Calotropis procera* and *Nerium oleander* and essential oil of *Cymbopogon* grasses have been found toxic to *T. semipenetrans* (Sangwan *et al.*, 1985; Verma *et al.* 1989). Castor (*Ricinus communis*) oak (*Calotropis procera*) and Datura (*Datura stramonium*) chopped leaves applied @ 20 and 40 g/kg soil in 9 sq meter area around tree trunks can result in reduction of nematode population and increase in yield (Baghel, 1995).

(iii) Biological control

Several nematophagous fungi such as *Arthrobotrys cladodes, Dactylaria* sp., *Monacrosporium gephyrophagum, Laginidium* sp. etc., can trap the *T. semipenetrans* and could be exploited for nematode control (Gowda *et al.* 1982). Endoparasitic fungus, *Paecilomyces lilacinus* is a potential biocontrol agent for controlling citrus nematode (Mani *et al.*, 1989). Vesicular arbuscular mycorrhizal fungus, *Glomus mosseae* when added to the soil with *T. semipenetrans* in *Citrus jambhiri*, limited the nematode development and partially neutralized the adversed effect of the nematode (Baghel *et al.*, 1990).

Since the parasitic fungus, *Paecilomyces lilacinus* has many attributes of a successful biological control agent, experiments were conducted in nematode infested citrus grove to determine its efficacy in controlling *T. semipenetrans*. It was observed that fruit diameter of the orange Cv. 'Valencia' was significantly affected by application of 3 nematicides and *P. lilacinus*. The fruit diameter from fungus treated tree was greater than that from trees treated with temik, vydate and mocap. Similarly, the number of nematodes in roots and soil around citrus trees inoculated with *P. lilacinus* was significantly lower than that in the non-treated or the nematicide treated trees (Jatala, 1986).

(iv) Chemical control

In early seventies, fumigant nematicides such as DBCP, DD etc. tested against citrus nematode were very effective (Chhabra and Bidra, 1971; Mukhopadhyaya and Dalal, 1971). Due to health and environmental reasons these fumigants are banned now.

Other non fumigant compounds like carbofuran, ethoprophos, profos, phorate, fensulfothion, dimethoate etc., have also been tested against citrus nematode under field conditions. Dichlofenthion @ 45 l/ha reduced the nematode population by 80% (Mukhopadhyaya, 1970b). Mukhopadhyaya and Dalal 1971) achieved 97.6% control by ethoprophos when applied @ 4.0 kg ai/ha and 39.9% increase in yield of sweet lime during second year. Chhabra *et al.* (1977) found fensulfothion very effective @ 3.0 kg ai/ha in controlling the nematodes on pumello and increased the yield by 68-76%.

Significant reduction in the citrus nematode population (76 to 84 per cent) in sweet orange and increase in yield (61 to 244 per cent) was observed with soil application of aldicarb and carbofuran both at 6 kg a.i. per ha (Baghel and Bhatti, 1983). The effectiveness of aldicarb and carbofuran was also demonstrated for the control of *T. semipenetrans* infecting acid lime (parvatha Reddy, 1986).

(v) Plant resistance

Citrus rootstocks which are resistant or highly tolerant to the citrus nematode would offer practical and effective means of control. Trifoliate orange (*Poncirus trifoliata*) and its hybrids (Citrumello) were highly resistant to the citrus nematode (Parvatha Reddy and Singh, 1978; Chhabra and Bindra, 1974; Parvatha Reddy and Agarwal, 1987; Mani and Reddy, 1986). The citrus rootstock hybrids evolved at the Indian Institute of Horticultural Research Bangalore, by crossing Rangpur lime (*Citrus limonia*) with *P. trifoliata* viz., CRH-3, CRH-5 and CRH-41 were highly resistant to the citrus nematode (Parvatha Reddy *et al.* 1987).

(vi) Integrated management

Different oil cakes such as neem, castor, karanj when integrated with parasitic fungi *Paecilomyces lilacinus* (Parvatha Reddy, *et al.*, 1991), *Verticillium chlamydosporium*, *V. lecanii* (Parvatha Reddy *et al.* 1996a), antagonistic fungus, *Trichoderma harzianum* (Parvatha Reddy *et al.*, 1996b) and VA mycorrhiza, *Glomus fasciculatum* (Parvatha reddy *et al.*, 1995) gave better plant growth and more nematode reduction in acid lime under glasshouse conditions. Neem cake @ 1 kg/plant combined with carbofuran at 2 kg/ha reduced 33.0% nematode population and increased the yield by 33.8% (Baghel 1995). Integration of *P. lilacinus* with carbofuran was found effective in increasing the plant growth parameters of acid lime and in reducing the citrus nematode population both in soil and roots (Parvatha Reddy *et al.*, 1996c).

The lesion nematode — *Pratylenchus coffeae* (Zimmerman, 1898) G. Goodey, 1951

P. coffeae appears to be the most pathogenic species and is responsible for the "Citrus slump disease" in florida. The nematode root rot of citrus in Uttar Pradesh is principally caused by *P. coffeae* (Siddiqui, 1964). This nematode constitutes the dominant factor in the nematode complex of *citrus* in Aligarh. O'Bannon and Tomerlin (1973) reported 49 to 80 per cent growth reduction in young trees up to 4 years depending upon different rootstocks.

P. coffeae is a pest of citrus in India, Japan and the U.S.A. In India, it has been reported from uttar Pradesh, Punjab and South Western India on citrus (Sethi & Swarup, 1971; D'Souza *et al.*, 1970).

(a) Symptoms

P. coffeae infected citrus plants exhibit poor growth, general dieback and produce undersized fruits.

P. coffeae is a migratory endoparasite of the root cortex where it feeds and multiplies. The favored place of entrance is slightly back the elongating zone in the piliferous region. A colony or 'nest' is established showing brownish lesions on the root. When decay sets in due to secondary invasion by bacteria and fungi, the nematodes quit and migrate through the soil to attack fresh roots. The first molt takes place within the egg and three molts occur outside. *P. coffeae* requires about 45 to 48 days to complete the entire life cycle. Of this time, 15 to 17 days are necessary for eggs to hatch, 15 to 16 days are spent as larvae, and it takes 15 days from maturity to egg production.

(b) Host range

Citrus limon, C. sinensis, C. reticulata and banana have been found to be hosts of *P. coffeae* in Uttar Pradesh (Siddiqui, 1964). The nematode is transferable from banana to citrus plants and from lemon to orange plants and vice versa.

MANAGEMENT

(i) Plant resistance

Four selections of a *Microcitrus australis x M. australasica* hybrid and Rubidoux 70-A5 trifoliate orange (*Poncirus trifoliata*) show some resistance to *P. coffeae.*

(ii) Chemical methods

Infestation in citrus seedlings of *P. coffeae* was eliminated by treating with fensulfothion and fenamiphos at 4.4 kg a.i. per ha and the stem diameter of the treated seedlings increased by 44 and 46 per cent, respectively after 9 months. The population of *P. coffeae* can be effectively controlled by applying aldicarb or carbofuran at 4 kg a.i. per ha (Baghel and Bhatti, 1983).

The lance nematode — *Hoplolaimus indicus* Sher, 1963

Khan, *et al.*, (1964) reported *H. indicus* from the soil around the roots of *C. sinensis, C. aurantifolia* and *C. reticulata* from India. This nematode is only known from India where it is widely distributed.

OCCURRENCE AND DISTRIBUTION

H. indicus has been reported on citrus from Punjab, Delhi, Rajasthan (Sharma *et al.*, 1969) and Uttar Pradesh (Rashid *et al.* 1973).

(a) Symptoms

H. indicus at 1,000 numbers per 500 g of soil caused some growth reduction of citrus, but a marked reduction occurred at or above an initial population of 2,000 (1 male; 4 female). None of rough lemon, karna khatta, sweet orange, mandarin, lemon or trifoliate orange was completely resistant to nematode attack at population levels above 4,000 per 500 g of the soil. Lemon proved somewhat more tolerant than other species. The greatest nematode populations were found in roots and soil of trees 10 to 20 years old and in an intermediate stage of decline; smallest populations were associated with trees one to five years old at initial stages of decline and 10 to 20 years old in the final stages of decline.

(b) Host range

H. indicus is reported on sugarcane, banana, pea, guava, tomato, rice, brinjal, ragi, amaranthus, black gram, chillies, cowpea, cucumber, groundnut, okra, maize, mustard, paragrass, cauliflower, harayali grass, mango, peach, wheat bajra, sorghum and cabbage besides citrus. In Citrus, it has been reported on *C jambhiri, C. karna, C. reticulata, C. sinensis, C. limon* and *Poncirus trifoliata* (Khan *et al.*, 1964).

Root knot nematodes – *Meloidogyne* spp.

In India, Rao (1956) reported that *Meloidogyne* sp. caused considerable damage to citrus in Andhra pradesh when a susceptible crop like tobacco or bhendi is grown as an intercrop. This is the first report of nematode damage to citrus in India. Chitwood and Toung (1960) observed the root knot nematode resembling *M africana* infecting *C. sinensis* from Delhi and proved its pathogenicity to citrus. Whitehead (1968) reported *M. indica* on citrus from India. Mani (1986) reported that *M. javanica* caused severe crop loss of acid lime in Andhra Pradesh and it was pathogenic to acid lime and sweet orange. It is confined to only coastal Andhra Pradesh.

(a) Symptoms

M. javanica infected trees are poor in vigour, unthrifty in appearance and show severe stunted growth. They fail to flower and produce fruits even after several years of planting. The roots have conspicuous galls on pioneer and fibrous roots. In advanced stage, large cavities can be observed in place of galls. Eggs masses can be seen as thin films spread over the root surface (Mani, 1986).

(b) Host range

M. javanica has been reported on *C. aurantifolia* and *C. sinensis* (Mani, 1986), whereas *M. indica* infects *C. aurantium, C. sinensis* (Whitehead, 1968) and *C. aurantifolia* (Parvatha Reddy *et al.*, 1981).

MANAGEMENT

(i) Cultural Method

It is essential to select root knot nematode free planting material to prevent the introduction of nematode inoculum into orchards. It is preferable to avoid growing vegetables as intercrops or as rotational crops in citrus orchards and nurseries.

Crotalaria roots exert a toxic effect on *Meloidogyne* sp. population. *Meloidogyne* larvae freely entered roots of *Crotalaria* but failed to survive. The possibility of growing *Crotalaria* in tree basin for the control of root knot nematodes infecting citrus needs further investigation.

Field experiments carried out in lime orchards indicated that application of neem cake at 20 kg per tree basin reduced the population of *M. javanica* by 25 per cent after three months of application. Repeated applications once in four months significantly reduced the root knot nematode population.

(ii) Plant resistance

It was observed that *C. reticulata*, Palestine and Rangpur limes, Rough lemon, Sour orange, Sexton, Thentriton tangelo and Volkamer lemon were not infected by *M. javanica*. Sour orange, Lime, Rough lemon, Grape fruit, Troyer citrane, Cleopatra mandarin and Trifoliate orange were highly resistant to *M. incognita* (Race 1 and 3) as well as *M. javanica*.

(iii) Biological control

A. fungal parasitic, *Paecilomyces lilacinus* consistently and effectively controlled the population of *Meloidogyne* spp.

(iv) Chemical control

Aldicarb and carbofuran at 4 kg a.i. per ha reduced the larval population of *M. javanica* on lime in soil by 77.8 and 62.8 per cent, respectively.

B. BANANA

Nematodes constitute one of the major limiting factors to banana production. They cause extensive root necrosis resulting in serious economic consequences. The burrowing nematode, *Radopholus similis* is the most important pest on banana. *Helicotylenchus multicinctus*, *Pratylenchus coffeae* and *Heterodera oryzicola* are the other nematodes that infect banana plantations. At present 71 species belonging to 33 genera of plant parasitic nematodes have been reported in association with banana roots from India.

Rodopholus similis Burrowing nematode

This nematode in India was first reported on banana from the Palghat district of Kerala by Nair *et al.* (1966). It causes rhizome rot of banana.

OCCURRENCE AND DISTRIBUTION

It has been reported from Tamil Nadu (Rajagopalan and Chinnarajan 1976), Karnataka (Venkitesan, 1976), Maharashtra (Parvatha Reddy and Singh 1980), Andhra Pradesh (Parvatha Reddy and Khan, 1987), Gujarat (Sethi *et al.* 1981), Madhya Pradesh (Tiwari and Dave, 1985), Bihar and Orissa.

The nematode does not seem to have any special adaptations for survival in the absence of a susceptible host. Free living stages and eggs do not survive for more than 12 weeks. After a plantation is removed, the nematodes in the corms may probably survive for as long as these remain succulent.

The burrowing nematode is spread from one locality to another through planting material (Corms). Trimming sets, which is often done before they are planted, is usually not sufficiently rigorous to remove infections that extend deeply into a set. *R. similis* may be disseminated when water that drains from infested areas gets recycled into irrigation system. Soil that adheres to implements, tyres of motor vehicles and shoes of plantation workers may also spread *R. similis*.

Biotypes

Two biotypes of *R. similis* are known, one that attacks citrus and the other that attacks bananas. The citrus biotype attacks many plant species belonging to diverse families including banana. The banana biotype seems to be restricted to *Musa* spp., Sweet potato, *Pueraria phaseoloides javanica* and possibly to a few other plants but not citrus. Venkitesen (1976) reported that the population of *R. similis* isolated from pepper roots did not infect *Citrus sinensis*, *C. reticulata*, *C. aurantifolia*, *Coffea arabica*, coconut and arecanut; but infected banana only. The population isolated from banana infected pepper, coconut and arecanut; while those from arecanut infected pepper and banana but not coconut. And thus he concluded that these four isolates were different "host types".

(a) Symptoms

Wounding of banana roots by the burrowing nematode usually induce reddish brown cortical lesions which are diagnostic of the disease. These lesions are clearly seen when an affected root is split longitudinally and examined immediately. Root and rhizome necrosis is manifested by varying degrees of retarded growth, leaf yellowing and falling of mature plants.

With the increase in nematode population, feeding roots are invaded and destroyed as fast as they are formed. The resulting setback in the uptake of plant nutrients leads to debility of the plant and production of smaller fruits. The lesioning of the primary roots together with the girdling and death of these anchor roots makes the plant prone to 'tip over' by wind action. It causes retarded growth and extensive root and rhizome necrosis resulting in serious economic consequences viz., fertilizers are not effectively utilized, the period from planting to harvesting is lengthened, maximum bunch weights are not attained, the quality of fruit is impoverished and fields have to be replanted every two to three years because of drastic reduction in plant numbers. Root population of *R. similis* is indirectly correlated with the yield (Charles *et al.*, 1985).

(b) Host range

The banana biotype of *R. similis* is known to infect *Musa* spp., *Ipomoea batatas, Pueraria phaseoloides javanica*, pepper arecanut and coconut.

(c) Economic threshold level

The burrowing nematode is responsible for 30.76 to 41 per cent yield loss in banana (Rajagopalan and Naganathan, 1977a; Nair, 1979; Parvatha Reddy *et al.*, 1996d; Vadively *et al.*, 1987). Root population of *R. similis* is indirectly correlated with the yield. The burrowing nematodc, *R. similis* was responsible for 31.61 per cent loss in banana fruit yield as determined by paired plot treatment experiment (Parvatha Reddy, Rao and Nagesh, 1996d).

(d) Interaction with other pathogens

The incidence of Panama wilt caused by *Fusarium oxysporum f. cubense* was doubled in the presence of *R. similis* during an experimental period of 3 months. When Gros Michel bananas infected with *R. similis* were inoculated with *F. oxysporum f. cubense*, the onset of wilt was considerably shortened. The nematodes caused the breakdown of resistance to Panama wilt in Lacatan bananas.

MANAGEMENT

(i) Physical methods

To avoid introducing inoculum into a new plantation, sets may be disinfected by heat therapy. The pared sets can de disinifected by dipping them in a hot water bath at 55°C for 25 minutes without causing significant losses.

Dipping of banana suckers in 0.1 per cent aldicarb solution increased plant height, bunch weight, fingers per bunch, induced early flowering and effectively reduced the burrowing nematode population (Venkitesan and Charles, 1983).

Paring and pralinage with carbofuran at 1.2 g a.i. per sucker recorded low incidence of *R. similis* and highest yield (Rajendran and Naganathan 1977a. Soil application of fenamiphos at 3 kg a.i. per ha was effective in increasing plant height, plant girth number of leaves and yield and in reducing the nematode population (Parvatha Reddy, *et al.*, 1991b).

(ii) Cultural methods

The number of nematodes in soil may be decreased either by fallowing, crop rotation or intercropping. The banana biotype of *R. similis* has a narrow host range and limited survival period in soil in the absence of host. Therefore, destruction of the infected area either mechanically with a bulldozer or by injection or by injection of a herbicide into the pseudostem and replanting the area with nematode free sets seems a practical approach to rehabilitate the infected plantations. After destroying in infested plantation, planting of cover crops like *Panicum maximum* var. *Trichoglume* and *Phaseolus atropurpureus* did not support *R. similis* population. Planting sugarcane in soil following the destruction of bananas eliminated *R. similis* after 10 weeks (Loos, 1961).

Rotation of banana with paddy, sugarcane, green gram, cotton or turmeric suppressed the nematode population and increased the yield (Rajendran, *et al.*, 1979). Mulching with black polythene at 20 per cent moisture depletion recorded highest yield with low population of *R. similis* (Vadivelu *et al.*, 1987). Intercropping with sunnhemp, coriander, marigold or lucerne reduced nematode population and increased yield (Vadivelu *et al.*, 1987). Banana intercropped with *Crotalaria* reduced *R. similis* population in root and increased plant growth and yield (Charles *et al.*, 1985a).

Fallowing for a period of 3 months after banana effectively suppressed the burrowing nematode population. Flood fallowing for about 5 months destroyed not only the fusarium but also the burrowing nematodes (Rajendran *et al.*, 1979).

Application of neem cake at 800 per plant (two split applications) or neem coated urea at 110 g per plant reduced the population of the burrowing nematode and also increased the yield of banana (Vadivelu *et al.*, 1987).

(iii) Biological methods

In an interaction study of vesicular arbuscular mycorrhizal (VAM) fungus, *Glomus fasciculatum* and *R. similis*, the multiplication of nematodes in the root was suppressed and soil population was significantly lower. The banana plants inoculated with mycorrhiza also recorded higher N, P, K, Ca, Mg, Sugars, Phenols and Amino acids in their root system suggesting VAM as a potential biocontrol agent in the management of *R. similis*.

(iv) Host resistance

Banana Cvs./hybrids viz., Kadali, Octoman, Then Kunnan, Kunnan, Dulmal Bhog, Malbhug, Tongat, Pedalimoongil, Pisang Lilin, Anil Vazhai, Velatothan, Dudhsagar, Chinachampa, Chinia Kurlan, Benid, Ayiramkapoovan, Pisang Seriby, Vennettukunnan, Beecha, Brilechinia, Local Dwarf, Manick Champa, Singhlal, Pey Kunnan, Elakkibale Pakea, Annaikomban, Bodles Altafort and Hybrid-74 (Matti x Pisang Lilin) were found tolerant/resistant to *R. similis* (Parvatha Reddy *et al.*, 1989; Ravichandra and Krishnappa, 1985; Charles *et al.*, 1983; Vadivelu *et al.*, 1987).

(v) Integrated management

Integration of neem cake with *Glomus mosseae* was most effective in reducing the burrowing nematode population in soil and roots, while karanj cake with *G. mosseae* gave maximum increase in plant growth parameters and fruit yield of banana. Mycorrhizal root colonization and number of chlamydospores of *G. mosseae* were maximum in neem cake amended soil (Parvatha Reddy *et al.*, 1997).

Integration of neem cake at 400 g per plant with *P. penetrans* at 100 g soil (300 spores) was found effective in reducing *R. similis* population both in soil and roots of banana by more than 50 per cent and increased plant growth parameters.

Integration of karanj cake with *G. mosseae* gave maximum banana fruit yield, while neem cake + *G. mosseae* was most effective in reducing *Radopholus similis* population both in soil and roots (Parvatha Reddy *et al.*, 1997).

The spiral nematode — *Helicotylenchus multicinctus* (Cobb, 1893) Golden, 1956

H. multicinctus in an important pest of banana throughout India and is responsible for 20 per cent loss in fruit yield. It causes extensive root necrosis, dieback and disfunction leading eventually to debility of the entire plant.

(a) Symptoms

The spiral nematode incite discrete, relatively shallow, necrotic lesions on banana roots. *H. multicinctus* causes extensive root necrosis, die-back and disfunction leading eventually to debility of the entire plant.

Groups of 8-26 eggs were observed in discolored cortical tissues. Eggs hatch in 48-51 hr in tap water at 30°C and the first molt is believed to take place outside the egg shortly after hatching. The second stage larvae are recognized by their ventral digitate tail process. The female gonad primordium is 2 and 6 celled in the early third stage larvae and at the third molt, respectively. During the fourth molt, the male and the female gonads complete their development and the vulva and vagina are seen in female which perhaps complete entire life cycle within the roots.

The adults and larvae, penetrate the epidermis of the root within 36 hr of inoculation. The females and the males feed directly on parenchyma cells and after 4 days, the nematodes get usually wholly within the cortex to a depth of 4-6 cells. The head of each nematode is usually oriented parallel to the long axis of the root, but the posterior portion is curved and occupies several contiguous cells. The nematode withdraws cytoplasm from surrounding cells, the cell walls are distorted or ruptured, and the nucleous if intact, is enlarged. Evacuated cells together with those near 'nests' of eggs, soon become discolored and necrotic. The characteristic macroscopic brown flecks or discrete necrotic lesions then appear.

(b) Host range

This nematode is known to infect banana, cocoa, sweet potato, citrus sugarcane, cassava, coffee, maize, mango, oil palm, rice, rubber, tea, yam, avocado, Bermuda grass and grapevine.

(c) Economic threshold level

The spiral nematode causes serious decline of banana and is responsible for 33.83 per cent loss in yield, 55.88 per cent loss in number of fruits per bunch and delayed fruiting by 134 days (Vadivelu *et al.*, 1987).

MANAGEMENT

(i) Plant resistance

Banana Cvs./hybrids viz., Neyvannan, Rasthali, Wather, Peyan, Kullan, Krishna Vazhai, Karpurvalli, Sirumalai, Co-1, kunnan, Robusta, Gross Michel, Vennettu Kunnan, Hybrid-94, 100, 106 and 109 (all Matti x Tongat) and Hybrid-74 (Matti x Pisang Lilin) were found tolerant/resistant to *H. multicinctus* (Vadivelu *et al.* 1987).

The lesion nematode — *Pratylenchus coffeae* (Zimmerman, 1898) T. Goodey, 1951

P. coffeae has been reported to cause decline of banana plantations. The lesion nematode causes most damage to banana in association with *F.o.f. cubense* (Vadivelu *et al.*, 1987).

(a) Symptoms

The symptoms caused by *P. coffeae* are very similar to those caused by *R. similis*. Both *R. similis* and *P. coffeae* are associated with the blackhead toppling diseases of bananas, *P. coffeae* enter the roots and feeds on cortical cells resulting in the production of dark lesions in the cortex. The corm is similarly infected resulting in necrosis of the outer tissues.

The life cycle of *P. coffeae* is similar to that *R. similis* except that *P. coffeae* males penetrate and feed on the underground plant tissues and cause decay. The egg to egg cycle is completed in about 27 days at 26-32°C. The optimum temperature for reproduction is 29.5°C.

MANAGEMENT

(i) Plant resistance

Banana Cvs./hybrids viz., Kunnan, Vennettukunnan and Hybrid-74 (Matti x Pisang Lilin) were found tolerant/resistant to *P. coffeae* (Vadivelu *et al.* 1987).

(ii) Integrated management

Management of the burrowing nematode, *Radopholus similis* infecting banana was achieved by integrating oil cakes (castor, karanj, neem) with endomycorrhiza, *Glomus mosseae* (Parvatha Reddy, *et al.*, 1997).

The cyst nematode — *Heterodera oryzicola* Rao & Jayaprakash, 1978

The infestation of a cyst nematode, *Heterodera oryzicola* on banana variety Nendran was observed for the first time in India at Banana Research Station, Kannara, Kerala during July 1983 (Charles and Venkitesan, 1984). These nematodes feed on fibrous roots and cause necrosis. In a subsequent survey, this nematode was recorded in several locations in Trichur, Ernakulam, Quilon and Calicut districts of Kerala especially on banana cultivated in paddy land. *H. oryzicola* also infects paddy, *Cynodon dactylon*. This creates practical difficulties in adoption of crop rotation with banana in rice fields. The infested plants are progressively debilitated leading to substantial yield losses. *H. oryzicola* was also reported on banana cultivars Saldathi and Njalippnan from Goa (Koshy *et al.* 1987).

(a) Symptoms

H. oryzicola feed on fibrous roots of banana and causes necrosis. The infected banana plants are progressively debilitated leading to substantial yield losses in Kerala.

H. oryzicola attacks the minute tertiary feeder roots and not observed to attack the fleshy roots and rhizome portions. The affected feeder roots turn black in color and detach easily (Charles, 1989).

(b) Economic importance and losses

H. oryzicola has been reported to cause considerable damage to banana in certain pockets of Kerala. An initial inoculum level of 100 and 1000 viable cysts per plant could reduce bunch weight by 20.5 to 56.6 per cent. The nematode infection was observed to affect the quality of fruits by increasing acidity and reducing total sugars (Charles, 1989).

MANAGEMENT

(i) Culture practices

Fallowing for a period of three months after banana effectively suppressed the burrowing nematode population (Rajendran *et al.*, 1979).

(ii) Chemical control

The nematode can be controlled by treatment of suckers (by hot water or chemicals) before planting.

Root knot nematodes — *Meloidogyne* spp.

The root knot nematodes, *Meloidogyne incognita* and *M. javanica* attack bananas. The infected plants were stunted, having small leaves and galled roots. Root knot nematodes have not been reported to cause yield reductions in bananas, but may interact with other nematodes or soil pests.

The reniform nematode — *Rotylenchulus reniformis* Linford & Oliveira, 1940

The reniform nematode, *Rotylenchulus reniformis* attack bananas. The nematode feed on secondary or tertiary roots and cause necrosis and destruction of feeder roots. In field plots where *R. reniformis* was the predominant parasitic nematode, yield reduction of 19.3 tonnes per ha was observed.

C. PAPAYA (*Carica papaya* L.)

Nematodes are one of the important limiting factors in production of papaya throughout the country. Papaya being perennial in nature, harbors and encourages the build up of nematode population round the year. The first record of an association between a nematode and papaya appears to be that of Prasad (1960) who reported *Meloidogyne* spp. parasitizing papaya roots at New Delhi. At present, 32 nematode species withing 20 genera have been found associated with papaya roots from various parts of the country (Khan and Khan 1998). However, most of these nematodes are not known pathogens of papaya and their true relationship with the host plant is yet to be established.

Root knot nematodes (*Meloidogyne incognita* and *M. javanica*), the reniform nematode (*Rotylenchulus reniformis*) and the dagger nematodes (*Xiphinema basiri*) have been identified as proved pathogens on papaya.

Root knot nematodes — *Meloidogyne* spp.

M. incognita and *M. javanica* have been reported to be the major nematode pests of papaya in India. Ponte (1980) and Taylor *et al.* (1982) reported 10 to 20 per cent reduction in papaya fruit yield due to root knot nematodes. The root knot nematodes on papaya have been recorded from Delhi, Bihar, Haryana, Karnataka, Madhya Pradesh, Maharashtra, Orissa and Uttar Pradesh.

(a) Symptoms

The symptoms noticed due to nematode infection in papaya plants are leaf yellowing dropping of leaves, reduction in leaf production and premature dropping of fruits. Studies have shown that *R. reniformis* is mainly a phloem feeder (Sivakumar and Seshadri, 1972). The infected tissue at the site of feeding show hypertrophy, hyperplasia, thickening of cell walls, granular and enlarged nuclei and nucleoli, Infection with *Meloidogyne* sp. caused localization of total lipids.

MANAGEMENT

(i) Physical method

Treatment of *M. incognita* infected papaya seedling roots at 50°C for 10 min. gave satisfactory control provided the main field was not infested.

(ii) Chemical methods

Nursery treatment : Highest germination (94%) was obtained with carbofuran at 4 kg a.i. per ha. Maximum plant height was recorded in fensulfothion at 2 kg a.i. per ha.

Main field treatment : Carbofuran and aldicarb at 2 kg a.i. per ha were effective against nematodes and gave highest yield. Nayak (1986) reported that aldicarb followed by carbofuran and ethoprop all at 6 kg a.i. per ha were effective in controlling *M. incognita* on papaya.

Neem cake at 2.5 tons per ha and carbofuran at 6 kg a.i. per ha gave reduction in root knot nematode population in both soil and roots. Highest yield of 11 kg per plant was recorded in carbofuran followed by neem cake (10.4 kg), whereas control plants yielded 5.5 kg (Nayak, 1986).

(iii) Biological methods

Simultaneous application of *Paecilomyces lilacinus* and *Trichoderma harzianum* both at 1 g per plant gave maximum increase in plant growth parameters and highest reduction in reproduction factor and root galling in papaya (Khan, 1991).

(iv) Host resistance

The papaya Cv. Pusa 22-3 was resistant, while Thailand, 'Waimanalo, Mukund Farm, Pusa 1-15 and Pusa 1-45 were moderately resistant (Parvatha Reddy *et al.* 1988). Naik (1986) reported that cvs. CO 2 and CO 3 were found to inhibit the development and reproduction of *M. incognita*. Khan *et al.*, (1992) found that Cv. Co 2 was resistant, while Peradeneya, Selection No. 7 and Washington were moderately resistant against *M. incognita* Race 1.

The reniform nematode — *Rotylenchulus reniformis* Linford & Oliveira, 1940.

Prasad *et al.* 1964) first reported this nematode on papaya from Northern India. The reniform nematode has been found in the states of Gujarat, Madhya Pradesh, Orissa, Punjab, Tamil Nadu and Uttar Pradesh (Khan and Khan, 1998).

(a) Symptoms

Nematode affected plants are stunted, foliage is slightly yellow in color with reduced leaf size. Under severe nematode infection, roots show discoloration with dark brown lesions which are prone to attack by soil pathogens. Egg masses on root are covered with soil particles.

Sivakumar and Seshadri (1972) studied histopathological changes in papaya due to the infection of *R. reniformis*. Nematode damaged cortical cells and reached up to vascular tissue and fed on phloem. Histological studies have shown that the reniform nematode, *R. reniform* is mainly a phloem feeder in papaya (Sivakumar and Seshadri, 1972). The infected tissue at the site of feeding shows hypertrophy, hyperplasia, thickening of cell walls, granular protoplasm and enlarged nuclei and nucleoli.

MANAGEMENT

(i) Chemical methods

Foliar application of Nemacur and Vydate on papaya was effective for the management of reniform nematode (Ayala *et al.*, 1971).

Rajendran and Naganathan (1981) reported that carbofuran at 2 kg a.i. per ha gave significant reduction in soil population of *R. reniformis* and increased papaya fruit yield by 38.4 per cent over control.

(ii) Host resistance

Papaya cultivars CO-1, CO-2, CO-3, CO-4, CO-5, Thailand and Giant were tolerant/resistant to *R. reniformis*. CO-2 and CO-3 were found to inhibit the development and reproduction of *M.*

incognita (Nayak, 1986). It has been reported that papaya Cvs. Solo and Washington were resistant, while Coorg Honey Dew moderately resistant. Papaya Cv. Mammoth was found resistant, while CO1, Peradeneya, Sel. 7 and Washington were moderately resistant (Khan and Khan, 1998).

The Dagger nematode — *Xiphinema basiri* Siddiqui, 1959

The pathogenic capability of *X. basiri* on papaya has been established. Inoculation of 2000 nematodes per 2 kg soil caused significant reduction in plant height (30.4%) and weight (38.6%) on papaya.

(a) Symptoms

The nematode feeds on the roots of papaya and induced stubbiness of roots, thickness at root tips resulting in galling and fish hook formation.

(b) Host range

X. basiri is associated with banana, citrus, fig, guava, litchi, mango, mulberry, plum, rose, tomato, okra, and castor in India.

D. GRAPEVINE (*Vitis vinifera* L.)

Several nematode species are known to be associated with grapes but the major ones belong to genera *Meloidogyne, Pratylenchus* and *Xiphinema.* Other nematodes found associated with this crop are *Tylenchulus semipenetrans, Rotylenchulus reniformis, Paratrichodorus minor* and *Longidorus attentuatus.* Besides, several new species of nematodes have been recorded on grapes from different parts of India.

All species of root knot nematode have almost the same life cycle except for duration of different stages which depend considerably on soil condition, environmental factors and plant host. However, detailed information on this aspect with regard to grapevine is meagre.

Root knot Nematodes - *Meloidogyne* spp.

Three species of root knot nematodes, *M. javanica, M. incognita* and *M. arenaria* are recognized as major pests of grapes causing economic damage. Other species found associated with this crop are - *M. hapla, M. thamesi* and *M. nataliei.*

(a) Symptoms

The root knot nematode infestation is not manifested by any typical above ground symptoms. Patches of poorly branched vines with scant foliage, pale and small leaves, and poor bearing are the indications of root knot nematode damage. In young plants, premature decline, weak vegetative growth are commonly associated with nematode attack. The visibly unthrifty growth is generally attributed to moisture stress, low fertility, nutritional deficiency and other adverse conditions. However, the confirmation of nematode attack is possible by assaying soil sample and roots. The root system shows typical localized swellings particularly on feeder roots and young secondary roots and females may be found on internodal trunk just below the ground level. *Meloidogyne incognita* has been reported to stimulate the production of many new fine rootlets above the site of nematode infection resulting in "hairy root" condition. Depending upon the variety of grape, *M. javanica* forms galls of varying size and shape and distorts the normal appearance of roots.

MANAGEMENT

(i) Integrated management

Concomitant infection of *M. arenaria* and *Glomus fasciculatum* on grapevine cv.

Thompson Seedless negated the growth stimulatory effect of VAM fungus and reduced the growth suppressive effect of nematode on the host. The fungus stimulated multiplication of nematodes whereas, nematodes had no effect on fungus spore production. In combined inoculation of *Glomus mosseae* and *M. javanica* on grapevine cv. Perlette, Mycorrhizal fungus adversely affected nematode multiplication. Mycorrhizal colonization was not affected by nematodes. Growth suppressive effect of nematode was neutralized by fungus.

Root lesion Nematode — *Pratylenchus* spp.

Pratylenchus vulnus commonly found on grape is economically most important among lesion nematodes and its occurrence and severity is more in heavier soils.

(a) Symptoms

Plants infected with *P. vulnus* show loss of vigor and reduction in fruit production. Infected young vines remain very weak, often fail to establish root system and eventually die. Below ground symptoms on root distinctly shows lesions, which are initially brown and later turn black. In severe infection, black lesions combine and girdle the roots. Infected plant have generally reduced root system. Nematode penetration has been found restricted to few layers of cortical cells only. Adult female and eggs were not found in meristematic and vascular tissues. Inoculation of 1000 *P. vulnus* per vine has reported to be pathogenic and resulted in suppression of root and shoot growth, *P. vulnus* infection resulted in reduction in potassium and zinc uptake, which was measured in leaf blade and petiole of vines.

Simultaneous inoculation of *P. vulnus* and *Xiphinema index* at 1000 and 500 nematodes, respectively per vine (cv. Thompson Seedless) enhanced the stunting of shoot and root as compared to individual inoculation of either nematode alone.

Dagger Nematodes — *Xiphinema* spp.

Dagger nematodes are ectoparasites and feed on succulent tissues of young roots. *X. index* and *X. americanum* have been found pathogenic on grapes. Other species found associated with vines are *X. divericaudatum* and *X. mediterraneum, X. aurcicum, X. italiae, X. vuitteneri, X. brevicoides, X. elongatus* and *X. insigne.*

(a) Symptoms

Dagger nematodes feeding on grape roots result in terminal swelling, cessation of root elongation and distortion due to malformation of rootlets. Discoloration, decay of roots and death of growing points of feeder roots occur at later stage. Epidermal and outer cortical cells collapse at feeding site and show necrosis. Multinucleate enlarged cells have been found beneath the layer of necrotic cells. *X. index* rarely preferred feeding on root tip. *X. index* inoculated at 500 nematodes per vine altered auxin relationship and reduced vine growth. Concomitant inoculation of grape cv. Thompson Seedless with 500 *X. index* or *P. vulnus* caused greater stunting of vine shoot and root than did inoculation of either nematodes alone, Individually, *P. vulnus* caused greater stunting than *X. index, X. index* Leftover roots of dead and old plants in vineyards act as source of viral inoculum of dagger nematode year after year. *X. americanum* have been found transmitting tobacco mosaid virus, grape yellow vine virus, peach rosetter mosaic virus and tomato ring spot virus.

MANAGEMENT

A number of methods like cultural, chemical, plant material, intercropping and use of resistant root stocks have been adopted to bring nematode population at levels that do not cause economic losses.

(i) Cultural methods

Use of nematode free planting material is one of the most important cultural practices adopted to avoid nematode infection. *Tagetes patula* used as intercrop, reduced 39- 44% nematode population and significantly increased vine yield.

(ii) Physical methods

Hot water treatment has been used to disinfect planting material but accurate timings and suitable uniform temperature (125°F) are very important for obtaining good results.

(iii) Resistance root stocks

Use of nematode resistant root stocks has an added advantage over other methods of control. The cost factor is reduced to minimum. In grapevine, Dogridge (*Vitis champini*), Salt creek (*V. champini*), 1613 (*V. solanis* x Othello), Harmony (1613 x Dogridge), St. George (*M. rupestris*), A x R1 (Aroman x Ganzin1), are well known nematode resistant root stocks. Lake Emerald and Tompa are recently recognized as resistant root stocks. Banquabad Cardinal, Early Muscat, Jasbeli, Loose Perlette and Reisling (all commercial varieties) have been recognized as resistant to *M. javanica*.

(iv) Chemical methods

Soil fumigants 1,3 dichloropropene (1,3 D) at 60-170 gallons per acre, and methyl bromide (MBr) at 60-170 gallons per acre and 450-560 kg/ha at 60-100 cm depth have been used to clean field soil before planting vines. Single application of aldicarb at 5 kg/ha improved both mean yield (42.2%) and mean cane mass. Aldicarb at 5 kg/ha or Oxamyl at 10 kg/ha applied in two equal splits to control nematodes of vineyard gave significant yield improvement of 58% and 97% respectively and reduced nematode population. Aldicarb at 25 kg/ha applied once annually at bud breaking stage to control nematode in vine, gave best results in terms of nematode reduction and crop yield. Application of carbofuran, benfurocarb or phorate at 6 kg ai/ha, in *M. javanica* infested vineyard gave 63-68% nematode reduction and improved yield (120-133%). Carbofuran at 13 g/sq m at bud breaking stage is recommended for controlling *M. javanica*.

E. PINEAPPLE

At present, 26 species belonging to 19 genera of plant parasitic nematodes have been reported in association with pineapple from India. The most common nematode found to be detrimental to crop production are *Meloidogyne* species and *Rotylenchulus reniformis*.

The root knot nematode — *Meloidogyne incognita* (Kofoid and White, 1919) Chitwood 1949, *M. javanica* (Treub, 1885) Chitwood, 1949)

M. incognita has been found on pineapple plants in various parts of the country. Losses for the first crop were greater than 40% due to the attacks of root knot nematodes.

Population of *M. incognita* showed two seasonal peaks, once during April-May and another in November. Soil temperature range of 21.2°C-27.5°C and soil moisture range 14.0 - 18.6% were found to favor the population development.

(a) Symptoms

The root growth is greatly slowed down. The nematode also reduces plant weight, leaf weight and fruit weight by 47.5%. The root infested simultaneously by several larvae stops in growth and may produce many secondary roots.

MANAGEMENT

(i) Crop rotation : Yield increases of pineapple was observed after rotating with pigeon pea due to decrease in infestation by *M. incognita*. Crop rotation with pangola grass (*Digitaria decumbens*) results in the elimination of *M. incognita* after one year.

(ii) Resistant varieties : *Annas ananasoides* and three of its hybrids are highly resistant to attacks by *M. incognita*.

The Reniform nematode — *Rotylenchulus reniformis* Linford & Oliveira, 1940

(a) Economic importance and losses

The reniform nematode can reduce the pineapple plant populations by more than 83% in the fields.

MANAGEMENT

(i) Culture practices

Crop rotation : Rotation of pineapple with sugarcane or pangola grass is effective in controlling *R. reniformis*.

Irrigated Fallow : Irrigated fallow would reduce reniform nematode population as nematode prefers low moisture regimes in pineapple crop.

(ii) Resistant varieties

Pineapple varieties/species, Venezolana, *Ananas ananasoides* and several of its hybrids are highly resistant to *R. reniformis* and do not allow the initial infestation to be maintained.

(iii) Chemical control

Application of carbofuran at 2.25 kg a.i. per ha, oxamyl at 1 kg a.i. per ha and Fenamiphos at 4.5 kg a.i. per ha were effective in controlling *R. reniformis* and in increasing fruit yields.

REFERENCES

Ayala, A., Acosta, N. and Adsuar, J.A. 1971. A preliminary report on the response of *Carica papaya* to foliar applications of two systemic nematicides. *Nematropica* 1 : 10.

Baghel, P.P.S. 1995. Nematode problems in citrus, pp. 203-210. In *Nematode Pest Management - An Appraisal of Eco-friendly Approach* eds. Swarup, G. Dasgupta D.R. and Gill, J.S., Nematological Society of India, New Delhi.

Baghel, P.P.S. and Bhatti, D.S. 1983. Evaluation of pesticides for the control of phytonematodes on citrus. *Third Nematol. Symp.*, Solan pp. 38-39.

Baghel, P.P.S., Bhatti, D.S. and Jalali B.L. 1990. *Proceedings of National Conference on Mycorrhiza at Haryana Agricultural University* Hisar, P. 118.

Bajaj, H.K. Kanwar, R.S. Baghel, P.P.S. and Bhatti, D.S. 1988. The Motia (*Jasminum sambac*) - a new host record of citrus nematode, *Tylenchulus semipenetrans* Cobb, 1913, H.A.U.J. Res. 18 : 146-147.

Bindra, O.S., Chhabra, H.K. Chadha, K.L. and Mehrotra, N.K. 1967. A study on the correlation of citrus nematode population with decline of citrus. *J. Res. PAU* 7 : 197-202.

Charles, J.S.K. 1989. Bioecology of the Cyst nematode infecting banana. Ph.D. thesis, Kerala Agricultural University, Trichur, 122 pp.

Charles, J.S.K., Venkitesan T.S. and Thomas Y. 1983. Field screening of banana against the burrowing nematode, *Radopholus similis* (Cobb) Thorne, Third Nematol Symp., Solan p. 45.

Charles, J.S.K., Venkitesan, T.S. and Thomas Y. 1985. Comparative efficacy of antagonistic intercrops with carbofuran in control of burrowing nematode, *Radopholus similis* in the banana cultivar nendran. *Indian J. Nematol* 15 : 241-242.

Chawla, M.L., Samathanam, G.J. and Sharma, S.B. 1980. Occurrence of citrus nematode (*Tylenchulus semipenetrans* Cobb, 1913) on roots of mangosteen (*Garcinia mangostana*). *Indian J. Nematol.* 10 : 240-242.

Chhabra, H.K. 1972. Citrus nematode and its control. *Pesticides* 6 (2) : 88-89.

Chhabra, H.K. 1978. Distribution pattern of *Tylenchulus semipenetrans* and determination of suitable distance and depth for soil sampling. *Indian J. Nematol.* 7 : 66-68.

Chhabra, H.K. and Bindra, O.S. 1971. Screening of nematicides against citrus nematode, *Tylenchulus semipenetrans* Cobb. *Indian J. Entomol* 33 : 472-473.

Chhabra, H.K. and Bindra, O.S. 1974. Screening of citrus rootscoks against citrus nematode. *Tylenchulus semipenetrans* Cobb, in Punjab. *Indian J. Hort.* 31 : 194-195.

Chhabra, H.K., Bindra, O.S. and Singh, I. 1977. Further studies on the control of the citrus nematode, *Tylenchulus semipenetrans* Cobb, 1913 infecting *Citrus maxima* Marrill., In *Proceedings of the International Symposium on Citricuulture*, Bangalore, p. 38.

Chidwood, B.G. and Toung, M.C. 1960. Host parasite interactions of the asiatic pyroid citrus nematode. *Plant Dis. Reptr.*, 48 : 848-854.

Chona, B.L., Sethi, C.L., Swarup, G. and Gill, J.S. 1965. Citrus nematode (*Tylenchulus semipenetrans* (Cobb, 1913) associated with die-back disease of citrus. *Indian J. Hort.* 22 : 370-373.

D'Douza, G.I. Kumar, A.C., Viswanathan, P.R.K. and Shamanna, H.V. 1970. Relative distribution and prevalence of plant parasitic nematodes in the coffee tracts of South-Western India. *Indian Coffee* 34 : 329-331.

Duncan, L.W. and Cohn, E. 1986. Nematode parasites of citrus. pp. 321-346. In "Plant Parasitic Nematodes in Subtropical and Tropical Agriculture" eds. Luc. M., Sikora, R.A. and Bridge, J., CAB International Wallingford, UK.

Edward, J.C. and RAI, B.B. 1970. Plant parasitic nematodes associated with hill orange (*Citrus reticulata Blaco*) in Sikkim. *Allahabad Farmer* 44 : 251-254.

Gowda, D.A., Narayana Swamy, B.C. and Setty, K.G.H. 1982. Nematophagous fungi from cultivated soils in Karnataka. *J. Soil Biol. Ecol.* 2 (2) : 95-96.

Gupta, J.C. and Atwal, A.S. 1972. Biology and ecology of *Hoplolaimus indicus* (Hoplolaiminae : Nematoda). Influence of population density on tomato and citrus plants (Abstract). In *Proceedings of the Eleventh International Nematology Symposium* p. 25.

Jatala, P. 1986. Biological control of plant parasitic nematodes. *Ann. Rev. Phytopathol.* 24 : 453-489.

Khan, A.M., Siddiqui, M.R. Khan, E., Hussain, S.I. and Saxena, S.K. 1964. List of stylet bearing nematodes reported from India - I. Nematology publication 1, Aligarh Muslim University, Aligarh, U.P., India,19 pp.

Khan T.A. 1991. Studies on a diseases complex of papaya caused by *Meloidogyne incognita* and *Fusarium solani*, Final Tech. Rept. D.S.T. Project, Aligarh Muslim

Univ., Aligarh, 47 pp.

Khan, T.A. and Khan, S.T. 1998. Nematode diseases of papaya, pp. 344-360. In : "Nematode Diseases in Plants" ed. Trivedi, P.C., CBS publishers and Distributors, New Delhi.

Khan, T.A. Khan, S.T. and Saxena, S.K. 1992. Effect of interaction of *Meloidogyne incognita* and *Fusarium solani* on the relative response of papaya cultivars, First Afro Asian Nemato. Symp., Aligarh, p. 14.

Khuntia, N. and Das, S.N. 1969. Plant parasitic nematodes associated with fruit trees in Orissa (Abstract). *Proceedings of the All India Nematology symposium*, pp. 22-23.

Koshy, P.K., Sosamma, V.K. and Faleiro, J.R. 1987. Occurrence of *Heterodera oryzicola* on banana in Goa. *Indian J. Nematol.* 17 : 334.

Krishnamurthy, Rao, B.H. and Thammiraju N.B. 1975. Nematodes associated with citrus, pp. 55-58. In "*Citrus Decline in Andhra Pradesh*" eds. Reddy, G.S. and Murthy, V.D., Andhra Pradesh Agricultural Univeristy, Hyderabad.

Loos, C.A. 1961. Eradication of the burrowing nematode, *Radopholus similis* from banana. *Pl. Dis. Reptr.* 45 : 457- 461.

Mani, A. 1986. Occurrence of *Meloidogyne javanica* on citrus in Andhra Pradesh. *Int. Nematol. Network Newsl.* 3 (2) : 11-13.

Mani, A. and Dakshinamurthi, V. 1986. Citrus nematode and root knot nematodes : Distribution, crop loss symptoms and management, pp. 249-260. In "*Plant parasitic Nematodes Problems and Progress*" eds. Swarup, G. and Dasgupta, D.R., Indian Agric. Res. Int. New Delhi.

Mani, A. Murthi, I.R., Rao, P.K. and Prasad, V.D. 1989. Growth of *Paecilomyces lilacinus* on natural substrates and its efficacy against citrus nematode, *Tylenchulus semipenetrans. J. Biol. Control*, 3 (1) : 59-61.

Mani, A. and Reddy, G.S. 1986. Studies on the varietal reaction of some species of *Citrus* and *Poncirus to Tylenchulus semipenetrans* Cobb. *Indian J. Nematol.* 16 : 267-268.

Misra, S.L. and Edward J.C. 1977. Observations on histopathology of some of the more important rootstocks for commericial citrus affected by citrus nematodes. *Allahabad farmer* 53 : 7-18.

Mobin, M. and Khan, A.M. 1969. Effect of organic amendments on the population of rhizosphere fungi and nematodes around the roots of some fruit trees (Abstract), *Proc. All India Nematol. Symp.*, New Delhi, p. 68.

Mukhopadhyaya, M.C. 1970a. Nematodes associated with fruit trees and vegetable crops in Himachal Pradesh. *J. Res. Punjab. Agric. Univ.* 7 : 656-658.

Mukhopadhyaya, M.C. and Dalal, M.R. 1971. Effect of two nematicides on *Tylenchulus semipenetrans* and on sweet lime yield. *Indian J. Nematol* 1 : 95-97.

Mukhopadhyaya, M.C. and Suryanarayana, D. 1969. Citrus decline in Haryana : Role of *Tylenchulus semipenetrans* and its control. *Indian phytopath.* 22 : 495-497.

Muthukrishnan, T.S. and Sivakumar, C.V. 1973. Some observations on nematodes of citrus, banana and papaya in Tamil Nadu. *Proc. Workshop on Fruit Research*, Coimbatore.

Nair, K.K.R. 1979. Studies on the chemical control of banana nematodes. *Agri. Res. J. Kerala* 17 : 232-235.

Nair, M.R.G.K. 1965. On the occurrence of the citrus nematode. *Tylenchulus semipenetrans* Cobb, in Kerala. *Agri. Res. J. Kerala*, 3 : 105.

Nair, M.R.G.K., Das, N.M. and Menon, M.R. 1966. On the occurrence of the burrowing nematode, *Radopholus similis* (Cobb, 1893) Thorne, 1949 on banana in Kerala, *Indian J. Ent.* 28 : 553-554.

Nath, A., Sharma, N.K. Bhargawa, S. and Thapa, C.P. 1982. Nematicidal properties of garlic. *Nematologica* 28 : 253- 255.

Nayak, G.M. 1986. *Studies on Papaya in Relation to Certain Plant Diseases*. M.Sc. (Agri.) Thesis, Univ., of Agri. Sci., Bangalore, 106 pp.

O'Bannon, J.H. and Tomerlin, A.T. 1973. Citrus tree decline caused by *Pratylenchus coffeae. J. Nematol* 5 : 311- 316.

Paravatha Reddy, P. 1986. Chemical control of plant parasitic nematodes infecting citrus and grapevine. *Indian J Hort* 43 : 303- 305.

Parvatha Reddy, P. and Agarwal P.K. 1987. Resistance in citrus rootstock to the citrus nematode, *Tylenchulus semipenetrans. Indian J. Hort.* 44 : 111-114.

Parvatha Reddy, P. and Khan, R.M. 1987. First report on the occurrence of *Radopholus similis* (Cobb, 1893) Thorne, 1949 in Andhra Pradesh, India. *Intn. Nematol. Network Newsl.* 4 (1) : 11.

Parvatha Reddy, P., Khan, R.M. and Agarwal, P.K. 1987. Selection of citrus rootstocks and hybrids resistant to the citrus nematode, *Tylenchulus semipenetrans. Pakistan J. Nematol*, 5 : 69-72.

Parvatha, Reddy, P., Khan, R.M. and Agarwal, P.K. 1989. Screening of banana germplasm against the burrowing nematode, *Radopholus similis. Indian J. Hort.*, 46 : 276-278.

Parvatha Reddy, P., Khan, R.M. and Rao, M.S. 1991a. Integrated management of the citrus nematode, *Tylenchulus semipenetrans* using oil cakes and *Paecilomyces lilacinus. Afro Asian, J. Nematol.* 1 : 221-222.

Parvatha Reddy, P., Khan, R.M. and Rao, M.S. 1991b. Evaluation of chemicals for the management of the burrowing nematode, *Radopholus similis. Curr Nematol* 2 : 1-2.

Parvatha Reddy, P., Nagesh, M., Rao, M.S. and Devappa, V. 1997. Integrated management of the burrowing nematode, *Radopholus similis* using endomycorrhiza, *Glomus mosseae* and oil cakes. *Pest Management in Hortil. Ecosystems* 3 ; 25-30.

Parvatha Reddy, P., Rao, M.S., Mohandas, S. and Nagesh, M. 1995. Integrated management of the citrus nematode, *Tylenchulus semipenetrans* Cobb. using VA mycorrhiza, *Glomus fasciculatum* (Thaxt.) Gerd & Trappe and oil cakes. *Pest Management in Hortil. Ecosystems* 1 : 37-41.

Parvatha Reddy, P., Rao, M.S. and Nagesh, M. 1996a. Management of the citrus nematode, *Tylenchulus semipenetrans* by integration of *Trichoderma harzianum* with oil cakes. *Nematol. Medit.*, 24 : 265-267.

Parvatha Reddy, P., Rao, M.S. and Nagesh, M. 1996b. Management of the citrus nematode on acid lime by integration of parasitic fungi and oil cakes. *Pest Management in Hortil. Ecosystems* 2 : 15-18.

Parvatha Reddy, P., Rao, M.S. and Nagesh, M. 1996c. Integrated management of the citrus nematode, *Tylenchulus semipenetrans* using pesticides and parasitic fungus, *Paecilomyces lilacinus. Pest Mangmt. in Hortil. Ecosystems* 2 : 61-63.

Parvatha Reddy, P., Rao, M.S. and Nagesh, M. 1996d. Crop loss estimation in banana due to the burrowing nematode, *Radopholus similis, Pest Mangmt. in Hortil. Ecosystems* 2 : 85-86.

Parvatha Reddy, P. and Singh, D.B. 1978a. The association of the citrus nematode with grape roots in a commercial orchard. *Curr. Sci.* 47 :640-641.

Parvatha Reddy, P. and Singh, D.B. 1978b. Evaluating the reaction of some species and varieties of *Citrus* and *Poncirus* to the citrus nematode. *Indian J. Nematol.* 8 : 82-84.

Parvatha Reddy, P. and Singh D.B. 1980. Plant parasitic nematodes associated with banana in South India, Proc. Nat. Seminar Banana Prod. Technol., Coimbatore, pp. 161-163.

Parvatha Reddy, P., Singh D.B. and Nagalingam, B. 1981. Fighting citrus nematodes in Andhra Pradesh. *Indian Hort.* 26 (3) : 24-25.

Ponte, J.J.D.A. 1980. *Meloidogyne* - importanica and control no nordeste. *Sociedade Brasileira Nematologia* 4 : 1-14.

Prasad, S.K. 1960. The problem of plant parasitic nematodes in the Indian Union. *Indian J. Ent.* 22 : 301-304.

Prasad, S.K. and Chawla, M.L. 1965. Observations on the population fluctuations of citrus nematode, *Tylenchulus semipenetrans* Cobb. 1913. *Indian J. Ent.* 27 :450-455.

Prasad, S.K. Dasgupta, D.R. and Mukhopadhyaya M.C. 1964. Nematodes associated with commercial crops in Northern India. I. Uttar Pradesh. *Indian J. Nematol.* 3 : 8-23.

Rajagopalan,P. and Chinnarajan, A.M. 1976. Plant Parasitic nematodes associated with banana in Tamil Nadu. South Indian Hort. 24 : 69-72.

Rajagopalan, P. and Naganathan, T.G. 1977. Scheme for studies on the nematode parasite of banana. *Tamil Nadu Agric. Univ. Ann. Rep.* 6 : 129.

Rajendran, G. and Naganathan, T.G. 1981. Evaluation of systemic chemicals for the control *Rotylenchulus reniformis* on papaya. *Indian J. Nematol* 11 : 130.

Rajendran, G., Naganathan, T.G. and Sivagami, V. 1979. Studies on banana nematodes. *Indian J. Nematol.* 9 : 54.

Rao, V.T. 1956. Stray notes on some crops pest out breaks of south India. *Indian J. Ent.* 18 : 123-126.

Rashid, A., Khan, F.A. and Khan, A.M. 1973. Plant parasitic nematodes associated with vegetables, fruits, cereals and other crops in North India. I. Uttar Pradesh. *Indian J. Nematol.* 3 : 8-23.

Ravichandra, N.G. and Krishnappa, K. 1985. Reaction of banana cultivars popularly grown in Karnataka to the burrowing nematode, *Radopholus similis*. *Indian J. Nematol* 15 : 240.

Sethi, C.L., Nand, S. and Srivastava, A.N. 1981. Occurrence of *Radopholus similis* (Cobb, 1893) Thorne, 1949 in Gujarat, India, *Indian J. Nematol* 11 : 116.

Sethi, C.L. and Swarup G. 1971. Plant parasitic nematodes of North Western India III. The genus, *Pratylenchus*. *Indian Phytopath* 24 : 410-412.

Sharma, B.B., Sharma, H.C. and Khan, E. 1968. Nematodes associated with sweet orange and mandarin plants in relation to die back (Abstract), *Proc. All India Nematol, Symp.*, New Delhi, p. 16.

Sangwan, N.K. Verma, K.K. Verma, B.S., Mali M.S. and Dhindsa, K.S. 1985. Nematicidal activity of essential oils of *Cymbopogon grasses*. *Nematologica* 31 (1) : 93-99.

Siddiqui, M.R. 1961. Occurrence of the citrus nematode, *Tylenchulus semipenetrans* Cobb, 1913 and the reniform nematode, *Rotylenchulus reniformis* in India. *Proc. 48th Indian Sci. Cong.* 3 : 504.

Siddiqui, M.R. 1964. Studies on nematode root rot of citrus in Uttar Pradesh, India. *Proc. Zool. Soc.* Calcutta 17 : 67- 75.

Singh, R.S. and Sitaramaiah, K. 1973. *Control of plant parasitic nematodes with organic amendments of soil.* Research Bulletin 6, G.B. Pant University of Agriculture and Technology, Pantnagar, U.P., 289 pp.

Sehgal, M., Chawla, M.L. Singh, S. and Balakrishnan, K. 1990. Studies on interactive effect of moisture and temperature on *Tylenchulus semipenetrans. Curr. Nematol* 1 : 35-36.

Sivakumar, C.V. and Seshadri, A.R. 1972. Histopathology of infection by the reniform nematode, *Rotylenchulus reniforms* Linford and Oliveira, 1940 on castor, papaya and tomato. *Indian J. Nematol.* 2 : 173-181.

Swamy, B.C.N., Rajagopal B.K. and Anahosur, K.H. 1973. Citrus nematode, a new record from Mysore state. *Mysore, J. Agric. Sci.* 7 : 138-139.

Swarup, G. and Seshadri, A.R. 1974. *Nematology in India - Problems and Progress.* pp. 303-311. In "*Current Trends in Plant Pathology*" eds. Raychaudhuri, S.P. and Verma, J.P., Dept. of Botany, Univ. of Lucknow, Lucknow.

Taylor, A.L., Sasser, J.N. and Nelson, L.A. 1982. *Relationship of climate and soil characteristics to geographical distribution of Meloidogyne species in agricultural soil.* Co-op. publ., Dept. Plant Path., North Carolina State Univ. & U.S. Agency, Int. Devp., Raleigh, N.C., 65 pp.

Tiwari, S.P. and Dave, G.S. 1985. Burrowing nematode, *Radopholus similis* (Cobb) Thorne 1949 on banana in Madhya Pradesh. *Curr. Sci.* 54 : 1286.

Vadivelu, R.S., Rajendran, G. Naganathan, T.G. and Jayaraj S. 1987. Studies on the nematode pests of banana and their management - A review. *Proc. Third Group Discussions on the Nematological problems of plantation crops*, Sugarcane Breeding Inst., Coimbatore, pp. 32-33.

Venkitesan, T.S. 1976. Studies on the burrowing nematode *Radopholus similis* (Cobb, 1893) Thorne, 1949 on pepper *(Piper nigrum* L) and its role in slow wilt disease. Ph.D. Thesis, Univ. Agric. Sci., Bangalore, 122 pp.

Venkitesan, T.S. and Charles, J.S.K. 1983. Sucker dip treatment for banana nematodes control, *Third Nematol. Symp.*, Solan, p. 45.

Verma, K.K., Verma, B.S. Sangwan, N.K. and Dhindsa, K.S. 1989. Toxicity of some indigenous plant extracts to root knot, seed gall and citrus nematode, *Pesticides* 23 (11): 25-27.

Whitehead, A.G. 1968. Taxonomy of *Meloidogyne* (Nematodea : Heteroderidae) with descriptions of four new species. *Trans. Zool. Soc. London* 31 : 263-401.

CHAPTER - 17

INTEGRATED PEST MANAGEMENT FOR PLANT PARASITIC NEMATODES

R.K. Jain

INTRODUCTION

During 1950, the world population was around 2.8 billion and it rose to about 5.5 billion by mid of 1992, of which 80 per cent is the share of developing countries and is growing @ 1.7 per cent a year. Among developing countries, the largest share is for India i.e. 2.1 per cent. Considering the fact that Indian population is increasing @ 2.2 per cent per annum, the cropped land per capita and per capita food availability would continue to decrease progressively. Even to hold the per capita food supply in the year 2000 at 1975 level, we would require 66, 75 and 100 per cent increase in legumes, cereals and vegetable production respectively. The problem is further compounded due to the vageries of pest and diseases which destroy about 35 per cent of all food crops worldwide in the fields. Post harvest losses are upto 20 per cent due to insects, microorganisms, rodents and birds etc. Thus, approximately 50 per cent of potential world food supply is being destroyed annually by pests and diseases.

In order to fulfil the food requirements of increasing population, the production technology and protection technology in the field of agriculture should be given equal weightage. The fruits of success in production technology can not be harvested without progress in protection technology.

Pest control includes everything that makes life hard for the pest species and tends to kill them or to prevent their increase or spread over large areas. The control can be accomplished by natural control (e.g. unfavourable climatic factors, physical factors, natural occurrence of biocontrol agents etc.) and applied control (e.g. cultural methods, mechanical, physical, biological, chemical, genetical and regulatory methods etc.). However, any of the above mentioned methods used unilaterally may not solve the pest problem in long term. So, it was suggested that pest control must be extended to a system based as approach based on the principles of applied ecology. Such a system has been called as Integrated Pest Control, which was later on designated as Integrated Pest Management (IPM). Hence, the concept of IPM was first articulated by Entomologists.

Hence, "IPM is defined as utilization of all suitable techniques to reduce and maintain the pest populations at levels below those causing injury of economic importance to agriculture or bringing together two or more methods of control into a harmonised system designed to maintain pests at levels below those at which they cause harm".

1. NECESSITY OF PEST CONTROL

The success of synthetic organic insecticides, such as DDT and BHC following the conclusion of second world war, began a new era of pest control. The demonstration of highly effectiveness of these chemicals resulted into the formulation of hundreds of effective synthetic pesticides (acaricides, fungicides, herbicides, insecticides, nematicides, and rodenticides) but with their continuous use, the failure of synthetic organic insecticides to control all pests and adverse consequences of their widespread use during the last 4 decades have changed this attitude to a more rational one. Development of pesticide resistant populations, resurgence of treated populations, elevation of secondary pests to the status of primary pest and environmental and ground water pollution hazards have posed many critical problems associated with the use of synthetic organic insecticides.

The pattern of crop protection for an agricultural ecosystem has been classified into five phases, which may be applicable in many crop situations.

(i) Subsistence phase : Here the crops are usually grown under non-irrigated conditions, crop yields are low, crop protection is dependent on natural control.

(ii) Exploitation phase : The crop protection programmes were developed, mainly through the use of chemicals. The crop yields are high thereby leading to excessive use of chemicals for managing pest populations and obtaining higher yields.

(iii) Crisis phase : The overdependence and excessive use of chemicals led to development of resistance in pest populations. Moreover, the occasional pests became serious primary pests. The combination of pesticide resistance, pest resurgence etc. increased the cost of crop production.

(iv) Disaster phase : The production cost increases so high due to the excessive and higher dose of chemicals, that crop production becomes profitless. The environmental pollution and residue problems occur at high levels.

(v) Integrated Pest control phase : Inte- grated pest control programmes include acceptance and utilization of ecological factors and compatibility, of other control of measures that result in favourable economic, ecologic and sociologic consequences.

In India, most of the pest control programmes are either in the exploitation phase or crisis phase and the integrated pest management concept needs to be adopted at an early stage to avoid the crisis and disaster phase.

Integrated pest management strategies include :

(1) Pest Avoidance

(2) Pest Eradication

(3) Pest Population Manipulation

(4) No action, which may be considered, because of the low population build-up of the pest population.

Among various pests and diseases, which damages the crops, plant parasitic nematodes present a formidable pest problems for the successful cultivation of economically important crops. The estimated overall annual yield loss of the world's major crops due to damage by phytoparasitic nematodes (Table 1) is 12.3 per cent (Sasser and Freckman, 1987). Further, it has been observed that losses in developing countries were relatively higher (14.6 per cent) than the developed countries (8.8 per cent).

The monetary losses due to nematodes on 21 crops, 15 of which are life sustaining were estimated at 77 million dollars. This figure may well exceed 100 million dollars, if all the crops are taken into account.

In India, extensive studies on crop loss assessment has not been conducted. Nevertheless, some experimental estimates as reported by various workers is presented in Table 2.

2. NEMATODE MANAGEMENT STRATEGIES

In order to reduce and minimise the adverse effect of increasing nematode populations on crop plants and for economical crop production different control options has been tried. Once the nematode populations are established in a field, eradication is neither practical nor

Table 1. Estimated annual yield losses due to damage by phytoparasitic nematodes (World basis)

Life sustaining crops	Loss (%)	Economically Important crops	Loss (%)
Banana	19.7	Cacao	10.5
Barley	6.3	Citrus	14.2
Cassava	8.4	Coffee	15.0
Chickpea	13.7	Cotton	10.7
Coconut	17.1	Cowpea	15.1
Corn	10.2	Egg plant	16.9
Fieldbean	10.9	Forages	8.2
Millet	11.8	Grapes	12.5
Oat	4.2	Guava	10.8
Peanut	12.0	Melons	13.8
pigeonpea	13.2	Misc. others	17.3
Potato	12.2	Okra	20.4
Rice	10.0	Ornamentals	11.1
Rye	3.3	Papaya	15.1
Sorghum	6.9	Pepper	12.2
Soybean	10.6	Pineapple	14.9
Sugarbeet	10.9	Tea	8.2
Sugarcane	15.3	Tobacco	14.7
Sweetpotato	10.2	Tomato	20.6
Wheat	7.0	Yam	17.7
Average	10.7%	Average	14.0%
	Overall average 12.3%		

Sasser and Freckman, 1987.

Table 2. Estimated crop losses due to important phytoparasitic nematodes in India

Crop	Nematode	Estimated loss (%) (Rs.) / avoidable yield loss	Remarks
Wheat*	*Heterodera avenae*	Rs. 40,000,000	Seshadri and Das gupta, 1980
Barley*	*Heterodera avenae*	Rs. 30,000,000	
Wheat	*Anguina tritici*	Rs. 75,000,000	Purohit and Bhat 1981
Coffee	*Pratylenchus coffeae*	Rs. 20,000,000	Seshadri and Das gupta, 1980
Okra	*Meloidogyne incognita*	28.1 per cent	Reddy and Singh, 19
Brinjal	*Meloidogyne incognita*	33.7 per cent	Reddy and Singh, 19
Frechbean	*Meloidogyne incognita*	43.5 per cent	Reddy and Singh, 19
Cowpea	*Meloidogyne incognita*	28.6 per cent	Reddy and Singh, 19
Pea	*Meloidogyne incognita*	20.0 per cent	Reddy and Singh, 19
Potato	*Globodera rostochiensis*	Total failure of crop	Reddy and Singh, 19
Rice	*Hirschmanniella oryzae*	30-87 per cent	Mathur and Prasad 1997
	Hirschmanniella mucronata	43 per cent	Prasad and Rao 1984
Cotton	*Meloidogyne incognita*	17.7 - 19.9 per cent	Jain et al., 1999
Cotton	*Rotylenchulus reniformis*	14.9	Palaniswamy and Balasubramanian, 1981
Maize	*Rotylenchulus reniformis*	6.0	Palaniswamy and Balasubramanian, 1981
Fingermillet	*Rotylenchulus reniformis*	4.8	Palaniswamy and Balasubramanian, 1981
Cowpea	*Rotylenchulus reniformis*	13.2	Palaniswamy and Balasubramanian, 1981
Blackgram	*Rotylenchulus reniformis*	8.7	Palaniswamy and Balasubramanian, 1981
Citrus	*Tylenchulus semipenetrans*	15	Anon, 1971
Tobacco	*Meloidogyne incognita*	50	Patel et al., 1988
Groundnut	*Meloidogyne arenaria*	51	Anon, 1993
Maize	*Rotylenchulus reniformis*	6	Palaniswamy and Balasubramian 1981
Pigeonpea	*Heterodera cajani*	14.2	Ramila Saxena and Reddy 1987
Rice	*Aphelenchoides besseyi*	12.2	Sivakumar, 1988

desired. So, we have to manage the nematode populations and learn to live with them. The direct and indirect benefits of nematode control include increased quantity and quality of produce, improved health of plants thereby reducing their chances of sufferings from nematode diseases and increasing the plants, ability to withstand adverse growing conditions. The selection of different nematode management options is dependent upon the biology of nematode species, the host crop, means of nematode spread, survival and ecological relationships.

(a) Chemical Methods : The chemicals used for controlling nematodes are designated as nematicides. The use of soil fumigants, which are applied to the soil and diffuse through the soil as gas had great influence in the development of science of Nematology. Kuhn (1881) used CS_2 against *Heterodera schachtii* while chloropicrin a broad spectrum biocide was used in 1919 for the control of nematodes. The discovery of DD soil fumigant in 1943 and EDB in 1945 played remarkable role in demonstrating the nematode damage and crop losses in field. DBCP was developed in 1954, which had greater advantage over earlier nematicides as it was available as liquid formulation and could be applied with irrigation water. The use of methylisothiocyanate precursor compounds like dazomet, methamsodium, methylisothiocynates mixture like vorlex etc. also helped in controlling nematodes besides other soil borne fungi, insects and weeds etc.

The era of soil fumigants was followed by the advent of non- volatile nematicides such as fensulfothion, aldicarb, carbofuran, ethoprop etc. Several workers have reviewed the work on chemical control of nematodes (Castro and Thompson, 1971; Evans, 1973; VanGundy and McKenry (1977); Wright (1981); Sethi and Gaur, 1986, and Bhatti and Jain (1984). The use of chemicals happens to be a costly proposition especially for Indian farmers but these methods can not be totally dispensed away with. They are normally used when nematode population is too high to permit successful cultivation of crops. However, the concept of nematode control changed with more emphasis on nematode management practices, which are less expensive, eco-friendly, less hazarduous and easy to apply. Hence, the importance of living with the diseases was advocated with emphasis on biological control, use of cultural methods, host plant resistance and integrated nematode management etc. In recent past, many of the effective chemical nematicides like DBCP, MBr, aldicarb etc. has been banned for use because of toxicity hazards and environmental pollution. Application of nematicides for practical nematode control in India at the Farmer's level is not very popular except in a few cases like in potato in Nilgiris, tobacco in Gujarat and Karnataka. The judicious use of chemicals as nursery treatment/seed treatment in vegetables and pulses against root-knot nematode has also been advocated. The cultural and biological methods are alternative to chemicals that needs systematic development and greater emphasis in future.

(b) Cultural Methods : Cultural methods of nematode control are normal practices carried out at little or no extra cost to reduce nematode population and crop losses. Cultural methods of nematode control are the attempts to adopt crop husbandry practices so as to minimise the crop losses due to nematodes. This can be achieved by following practices such as : (a) crop rotation (b) summer solarization and deep summer ploughing (c) flooding (d) phytotherapy and use of organic amendments (e) use of nematode free transplants (f) growing trap crops and antagonistic crops (g) destruction of infected plant residue (h) time of planting (i) manuring and irrigation etc.

These practices has been in use for the management of mineral nutrition, weeds and insects much before the knowledge of their effects on nematode population. In order to achieve successful nematode management, the thorough knowledge of following aspects is needed.

(i) Racial composition of nematode community.

(ii) Biology, host range, population dynamics, rate of decline in the absence of host.

(iii) Host status of various adaptable crops.

(iv) Prevailing agroclimatic conditions and

(v) Economics

The efficacy of most cultural practices in reducing nematode population varies with

(i) Soil type

(ii) Local climatic conditions

(iii) Crop husbandry practices

(iv) Target nematode species or race involved

(c) Crop rotation and cropping sequence : It is well recognised that soil becomes sick if monoculture is practised. The control of phytoparasitic nematodes by adopting cropping sequences is based on the fact that some nematode species are able to feed and multiply on certain crop plants but not on others. The success of crop rotation depends on proper selection of crops in the sequence. The length of rotation depends on the initial nematode population level and rate of population decline during rotation. Hence information on the nematode species, its biotypes, host range, biology, survival, multiplication, relationship between populations and yield losses must be obtained prior to adopting any crop rotation/cropping sequence for nematode management.

The use of trap crops is expensive without any return. However, use of antagonistic crops like (*Poa pratensis* and *Poa trivialis* against pigeonpea cyst nematode, *Tagetes erecta* against *Pratylenchus penetrans* and root-knot nematode etc. have been reported to be effective (Gommers, 1981). Flooding can be employed to reduce the incidence of phytoparasitic nematodes if sufficient amount of water is available. Nematodes are killed due to aphysixation and production of chemicals lethal to nermatodes such as butyric acid, propionic acid and production of H_2S gas because of low pH as a result of decomposition of organic matter. This practice does not seem to be practical as it adversely affects soil structure, fertility status and pH of soil.

Under tropical climatic conditions, harnessing of solar energy by deep summer ploughings has been recognised as important method to control weeds, diseases, soil insects including nematodes. This practice involves exposure of soil to solar heat and desiccation, killing substantial proportion of insect-pest, pathogen and weeds and nematodes (Raghavan, 1964). In order to further enhance the efficacy of solar heating for nematode management polyethylene mulching of moist soil during hottest period of the year is being advocated and termed as summer solarization. Summer ploughing of nematode infested filed during May- June has been found to be effective in managing root-knot nematodes (Jain and Bhatti, 1987; Jain and Gupta, 1997).

The selection of nematode free healthy transplants and destruction of infected crop residue of previous crop needs to emphasis, since these if not cared can become major source of infection in scucessive crops. The use of organic amendments and phytotherapeutic substances implies the addition of various organic substances to soil to bring about physical, chemical or biological changes adverse to the survival and multiplication of nematodes. Due to intense microbial activity during decomposition of organic amendments, these amendments have proved very useful in nematode management in the tropical countries. There is a need for better understanding of the mechanisms involved in this type of control to gaurantee its success.

(d) Biological Control : There are several biological agents which parasitise on nematodes and reduce their populations. Some of these are fungi, bacteria, viruses, nematodes, insects and mites.

(i) Predacious Fungi : There are more than 50 species of predacious fungi which capture and kill nematodes in soil. Few important ones are *Arthrobotrys, Trichothecium, Dactylaria* and *Dactylella*. These belong to Hyphomycetes. Some species belong to Zoopagale. These fungi capture nematodes by traps, mechanical traps and constricting rings. Barrons (1977) considered only *Rhapalomyces elegans* as egg parasite. *Verticillium chlamydosporium* is a major pathogen of *H. schachtii* eggs. *Paecilomyces lilacinus* has been reported by Jatala *et al.*, (1979) infecting eggs of root-knot nematodes. This fungus also attacks eggs of *Globodera pallida*.

(ii) Bacteria : There are several reports of bacteria present inside the body of nematodes. However, *Pasteuria penetrans* has been described as potent biological agent against nematodes. The ability of *P. penetrans* to prevent reproduction and eventually kill the root-knot nematodes and several other species present one of the best opportunities for biological control of major nematode pest. The field application of *Pasteuria penetrans* for nematode management is still in progress. In addition certain rhizospheric bacteria like *Azotobacter chroococcum, Azospirillum lipoferum* and *Pseudomonas* sp. has been found to be promising in reducing the population of *H. avenae* infecting wheat (Bansal *et al.*, 1999).

(iii) Viruses : Root-knot nematode larvae infected with virus were observed to exhibit sluggishness. Aggregation of large number of individuals in aqueous solution has been considerd due to virus. In addition to above bioagents, protozoans, tardigrades, mites, turbellarians etc. has been reported as important nematode enemies. Further, certain predacious nematodes such as *Mononchus*, Diplogastrids, and *Seinura* etc. are known to feed on other nematodes. These genera has been found to be efficient under laboratory conditions. However, their efficacy under field conditions is doubtful. Although biological control offers a very attractive and promising area and many potential bioagents for nematode management are reported. But, one of the major constraints in their use is their mass production for field application.

(e) Host Plant Resistance : Growing of nematode resistant varieties in problem areas constitute an important approach for reducing nematode populations. It is an effective economical and environmentally safe means of reducing losses caused by phytoparasitic nematodes. The importance of use of nematode resistant varieties has gained further significance because of manufacturing restrictions imposed on DBCP due to its toxicity and pollution hazards. Resistance in host plant is normally dependent upon the physiological differences between varieties of host plants, which makes it difficult for the nematodes to meet its food requirements or in some cases in activates the salivary secretions released by nematodes, which play important role in establishment of compatible host-parasitic relationship. Once a resistant plant is developed, then it would normally be able to maintain its property of antibiosis against pathogen. A good account of different crop varieties against important plant parasitic nematodes has been presented by Bhatti & Jain (1994). Resistance breeding would continue to be one of the important methods of development of new crop varieties offering resistance to key nematode pest. A few newer crop varieties where resistance has been incorporated through resistance breeding in tomato againsat root knot nemnatode has been reported (Nandpuri *et al.*, 1988, Jain *et al.*, 1991). However, development of new races and biotypes among species of plant parasitic nematodes due to the selection pressure exerted by cultivation of new resistant genotypes play an important role in selection and development of resistant cultivars. The information on the mechanism of resistance to nematodes presented by Gommers, 1981; Premchandran & Dasgupta, 1983, helps in understanding the concept of resistance against nematodes. Advances in biotechnology, tissue culture, protoplast fusion technique and somaclonal variations etc. has opened up new dimensions in the area of development of resistant cultivars against plant pathogens including nematodes.

(f) Physical Methods : The use of physical methods needs special facilities which are sometimes beyond the reach of common farmers under Indian conditions. The most documented physical methods of nematode control include use of hot water treatment, soil/seed treatment by means of solar heating or by electrical means, radiation and seed cleaning etc. Hot water treatment of seed tuber, bulbs and root cuttings is most helpful in reducing nematode spread and damage. Hot water treatment of rice seeds against *Aphelenchoides besseii*, potato against root-knot nematode, rooted cuttings of citrus against *Tylenchulus semipenetrans*, grapevine and roses cutting against *Meloidogyne* spp. and *Rotylenchulus reniformis* and banana suckers for *Radopholus similis* are quite effective. Cleaning wheat seeds infested with earcockle nematode (*Anguina tritici*) or by water floatation or in 5-10 per cent brine solution is quite useful in separating cockles/galls from wheat seed.

3. INTEGRATED NEMATODE MANAGEMENT

The concept of integrated nematode control was proposed as early as 1933 by Tyler that the combination of two or more control strategies into an overall management programme is the only sound sustainable approach to control root-knot nematode effectively. The advantage of this approach include the use of partially effective practices and protection of highly effective ones vulnerable from nematode adaptation or environmental risk. The management options may be combined sequentially from season to season or applied simultaneously. Hence integrated nematode management (INM) has been defined as a system approach to reduce nematode populations below economic threshold level (ETL) through integration of various techniques such as :

(i) Use of biocontrol agents

(ii) Genetically resistant host cultivars

(iii) Modifications of natural environment or cultural practices

(iv) Judicious use of nematicides in a cost effective manner etc.

Implementation of INM programme is based on the principles of nematode exclusion, population modifications and tolerance (Bird, 1981). The aim of INM is

(a) To utilise several compatible control techniques in combination.

(b) To maximize natural environmental resistance to plant parasitic nematodes.

(c) To apply specific and drastic control measures only as and when necessary and

(d) To maximise profit of the grower with location and resource specific recommendations.

The requirements for integrated nematode management are the research, development, transfer of technology for implementation of combining different nematode management options in most effective manner (Bird, 1981). Hence, the components of INM system includes (a) biological and environmental monitoring (b) agricultural production system, (c) pest crop ecosystem models and system design and implementation (Bird, 1981).

4. CASE HISTORIES

Integrated management options for nematodes based upon the above mentioned components of INM have been documented and reported in a few crops viz., sugarbeet, carrot, apple and tomato etc. by Bird (1981). A few more cases under Indian conditions worked out by different researchers are mentioned below.

In wheat and barley, combined application of carbofuran with nitrogenous fertilizers and aldicarb + nitrogen gave 18.29 to 126.74 and 35.01 to 105.33 per cent increase in yield respectively in *Heterodera avenae* infected field (Handa, *et al.*, 1978). For integrated control of *H. avenae* it was seen that use of aldicarb @ 2 kg a.i./ha with early November sowing gave better yield and 81.6 per cent decrease in cyst populations (Bhatti *et al.*, 1980). Further Singh and Yadav (1986) recorded higher wheat yield and greater reduction in nematode population over control when organic amendments were combined with nitrogen fertilizer.

In vegetable crops for management of root-knot nematode, various effective unilateral practices (nursery bed treatment, seedling root-dip, spot application, seedling age at transplanting, summer solarization, use of biocontrol agents, resistant and antagonistic plants etc.) in various combinations have been tried (Jain, 1992). Application of subabool leaves (*Leucaena leucocephala*) @ 40 g/kg soil and allowed to decompose for 4 weeks before sowing + carbofuran @ 1 kg a.i./ha at sowing in okra resulted in minimum galling (35.4) in comparison to control (64.0 galls per plant) and better growth characters of okra (Paruthi *et al.*, 1987). In a field naturally infested with *Meloidogyne javanica*, age of tomato nursery (cv. HS 101) at transplanting in combination with bare root-dip treatment with phosphamidon at transplanting/field application of aldicarb or neem (*Azadirachta indica*) leaves were tried for control of root knot nematode. Spot treatment of aldicarb @ 1 kg a.i./ha at transplanting in tomato seedlings irrespective of the age gave significantly lower final nematode population (68.01/ 250 g soil) in comparison to (189.51/250 g) in control and higher yield (7.0 kg/plot) as compared to other treatment combinations with minimum yield (5.4 kg) in control. Bare-root-dip treatment with phosphamidon @ 1000 ppm for eight hours at transplanting, was effective but phytotoxic (Jain & Bhatti, 1989). Nursery bed treatment with carbofuran (@ 0.4 g a.i./m^2) or aldicarb (@ 0.4 g a.i./m^2) supplemented with carbofuran application @ 1 kg a.i. /ha at transplanting in tomato showed 64 per cent increase in yield and effectively controlled *M. javanica* (Jain *et al.*, 1988; Jain 1990). Okra seed treatment with U 54229 (@ 3 per cent a.i./w/w) integrated with thimet applied @ 1 kg a.i./ ha at sowing showed 61.5 per cent increases in yield (Gupta *et al.*, 1986).

Use of three summer ploughings during May-June at 10 days interval integrated with application of aldicarb @ 1 kg a.i./ha in healthy tomato seedlings of higher age (8-week-old) grown under *M. javanica* infected conditions gave 42.5 per cent increase in yield over check (Three summer ploughings + infected tomato seedlings (Jain & Bhatti, 1989).

In another trial two methods of ploughing i.e. normal (10 cm deep) and deep ploughing (20 cm deep) during the month of June (ambient temperature ranging between (42-47°C) were integrated with other effective practices such as nursery treatment with aldicarb (@ 0.4 g a.i./m^2) in tomato/seed treatment with carbosulfone @ 3 per cent a.i. (w/w) in okra and aldicarb application @ 1 kg a.i./ha in the main field at transplanting/sowing or no aldicarb treatment. In tomato, reduction in root-knot nematode population ranged between 78.2 to 92.3 per cent owing to different methods of ploughing. Treatment combination of deep ploughing + aldicarb treated nursery + aldicarb application @ 1 kg a.i./ha at transplanting of tomato gave maximum yield (7.8 kg/plot) with minimum root knot index (1.3) and final soil population (175 1/250 g soil) in comparison to 3.8 kg fruit yield per plot in check (normal ploughing + untreated seedlings + no aldicarb application). In okra, normal and deep ploughing led to 79.2 and 76.2 per cent reduction on *M. javanica* population, respectively. Further, aldicarb used @ a.i./ha irrespective of type of ploughing yielded significantly higher than untreated plots. Treated seeds had lower root-knot index in comparison to untreated seeds (Jain and Gupta, 1990). In brinjal for integrated control of *M. javanica*, maximum reduction in nematode population (94.9 per cent) was recorded in M_2 method of ploughing (root knot nematode infested field summer ploughed, covered with transparent polythene sheet and exposed to sun for 2 week). This was followed by 87.0 per cent reduction in M_1 method (*M. javanica* infected field ploughed and exposed to sun). Brinjal yield per plot (3.5 x 2.5 m) in M_1 (8.9 kg) was statistically at par with M_2 (8.2 kg). However, it was significantly higher than M_3 method (non-ploughed plots). Brinjal nursery treated with carbofuran @ 0.3 g a.i./m^2 had lower root knot index than the untreated seedlings (Jain & Gupta, 1991). Integration of castor leaves treatment with different doses of inorganic fertilizers in tomato against *M. javanica* showed that castor alone or in conjugation with different doses of NPK enhanced plant growth and reduced *M. javanica* infestation. Maximum reduction per g root (94.7%) and eggmasses (77.6%) was achieved with castor coupled with N (75 kg/ha) in tomato (Zaki & Bhatti, 1989).

In Okra, for management of reniform nematode (*Rotylenchulus reniformis*) deep ploughing (20 cm) followed by fallowing for one month or fallowing for one month after weeding or integration of aldicarb application @ 0.8 kg a.i./ha at sowing after either of cultural practices or deep ploughing together with carbofuran or aldicarb seed treatment proved to be most effective (Reddy, 1983).

A well known integration of methods can control potato cyst nematode (*Globodera rostochiensis*) in potato in a two year rotation combining soil fumigation, cultivation of resistant potato varieties and cultivation of susceptible varieties in a systematic programme.

Parvatha Reddy *et al.*, (1992) has developed in integrated management of the citrus nematode, (*Tylenchulus semipenetrans*) infecting acid lime based upon application of the parasitic fungus, *Paecilomyces lilacinus* @ 4 g per plant alongwith carbofuran application @ 1 kg a.i./ha. Another best combination was application of neem/castor cake @ 10 g/plant alongwith *P. lilacinus* @ 4 g per plant.

5. FUTURISTIC APPROACHES

Therefore, in view of the available information, our future thrust should centre on conducting preliminary researches in following areas for development of ecologically sound integrated nematode management system.

(1) To conduct researches and build up data on nematode population crop yield relations, role of biotic and abiotic factors and to work out the economic of crop protection system.

(2) To study the supplementary methods of nematode control available for use in a particular system and to harmonise these into the existing system to produce optimal crop production relative to nematode infestation.

(3) To study the comapatibility and integration of nematode management methods with control of other concomitant pests, diseases and weeds etc.

(4) Development of ecologically sound INM.

REFERENCES

Bansal, R.K.; Dahiya, R.S.; Lakshminarayana, K; Suneja, S.; Anand, R.C. and Narula, N. 1999. Effect of rhizospheric bacteria on plant growth of wheat infected with *Heterodera avenae. Nematol. Medit.*, 27 : 311-314.

Barron, G.L. 1977. The nematode destroying fungi. Guelph, Ontario; Canadian Biological publications, Limited, 140 pp.

Bhatti, D.S. and Jain, R.K. 1984. Chemical control of root-knot nematode in India. A review. *Haryana J. Hort. Sci.*, 13 : 62-68.

Bhatti, D.S.; Dalal, M.R.; Dahiya, R.S. and Dhawan, S.C. 1980. Integrated control of *Heterodera avenae* in wheat. *J. Nematol*, 12 : 214-215.

Bird, G.W. 1981. Integrated nematode management for plant production. In Plant Parasitic Nematodes vol. III (Eds. B.M. Zuckerman and R.A. Rhode) Academic Press. U.S.A. 1981, 355-372 pp.

Castro, C.E. and Thomson, I.J. 1971. Mode of action of nematicides. In plant Parasitic Nematodes vol. II (Eds. B.M. Zuckerman, W.F. Mai and R.A. Rhode). Academic Press. U.S.A. 1971. 289-296 pp.

Evan, A.A.F. 1974. Mode of action of nematicides. *Ann. Appl. Biol.* 75 : 469-473.

Gommers, F.J. 1981. Biochemical interaction between nematodes and plants and their relevance to control. *Helminth. Abstr. Ser.*, B. 50 : 9-24.

Gupta, D.C.; Jain, R.K. and Paruthi, I.J. 1986. Chemical control of root-knot nematode (*Meloidogyne javanica*) infecting okra." Bac. National Conf. on Short term and Long term hazards of pesticides for their safe use" held in Delhi, Feb. 1986, Abst. of papers pp. 68-69.

Handa, D.K., Mathur, R.L. and Mathur, B.N. 1978. Studies on the integrated control of molya disease of wheat and barley caused by *Heterodera avenae* in Rajasthan. *Indian J. Mycol. and Pl. Path.* 8 : 70.

Jain, R.K. 1990. Efficacy of carbofuran for the control of root-knot nematode (*Meloidogyne javanica*) in tomato and okra. *Int. Nematol. Netw. Newsl.*, 7 : 11-12.

Jain, R.K. 1992. Performance of root knot nematode (*Meloidogyne javanica*) resistant tomato genotypes in different types of infested soil. *Haryana J. Hort. Science* 21 : 106-109.

Jain, R.K. and Bhatti, D.S 1983. A note on seed treatment with systemics for the control of *Meloidogyne javanica* in okra. *Haryana J. Hort. Science* 12 : 143-144.

Jain, R.K. and Bhatti, D.S. 1987. Population development of root knot nematode (*Meloidogyne javanica*) and tomato yield as influenced by summer ploughings. *Tropical Pest Management.* 33 : 122-125.

Jain, R.K. and Bhatti, D.S. 1989. An integrated approach for the management of root-knot nematode (*Meloidogyne javanica*) in tomato. *Indian J. Nematol.* 19 : 159-161.

Jain, R.K. and Gupta, D.C. 1990. Integrated management of root-knot nematode (*Meloidogyne javanica*) infecting tomato and okra. *Indian J. Nematol.* 20 : 14-18.

Jain, R.K. and Gupta, D.C. 1991. Integrated management of root-knot nematode, *Meloidogyne javanica* infecting brinjal (*Solanum melongena*) *Afro. Asian. J. Nematol.*

1 : 159-160.

Jain, R.K. and Gupta, D.C. 1997. Solarization as nursery bed treatment in the management of root knot nematode (*Meloidogyne javanica*) infecting tomato. *Indian J. Nematol.* 27 : 261-263.

Jain, R.K., Bhutani, R.D., Kalloo, Bhatti, D.S. and Singh, K. 1991. Do not fear root-knot nematode, sow tomato hisar-Lalit. *Indian Horticulture* 36 : 24-28.

Jain, R.K., Paruthi, I.J. and Gupta, D.C. 1988. Control of root knot nematode (*Meloidogyne javanica*) in tomato through nursery bed treatment alone and in combination with application of carbofuran at transplanting. *Indian J. Nematol.* 18 : 340-341.

Jatala, P., Kaltenbach, R. and Boxangel, M. 1979. Biological control of *Meloidogyne incognita acrita* and *Globodera pallida* on potatoes. *J. Nematol.* 11 : 103.

Kuhn, 1881. Die Ergebnisse der versuche zur Ermittelung der ursache der Ruben mudigkeit und zur Erforschung der Natur der Nematode. *Ber physiol. Lab. Univ. Halle* 3 : 1-53.

Nandpuri, K.S., Singh, S. and Mahajan, R. 1988. Punjab NR-7, a new rootknot nematode resistant tomato variety. *Indian J. Nematol.* 18 : 121-122.

Paruthi, I.J., Jain, R.K. and Gupta, D.C. 1987. Effect of different periods of degradation of subabool leaves alone and in combination with nematicides in okra. *Indian J. Nematol.* 17 : 30-32.

Premchandran, D. and Dasgupta, D.R. 1983. Purification and some properties of ribonucleases from tomato plants susceptible and resistant to root-knot nematode (*Meloidogyne incognita*). *Indian J. Nematol.* 13 : 199-208.

Raghavan, D. 1964. Agricultur in ancient India. Indian Council of Agricultural Research, New Delhi, India.

Reddy, P.P. 1983. Plant Nematology. Published by Agricole Publishing Academy, 1983. pp. 287.

Reddy, P.P., Khan, R.M. and Rao, M.S. 1992. Integrated management of the citrus nematode using nematicides and biocontrol agent. Proc. Seminar on Current Trends in Diagnosis and management of Plant Diseases. Bangalore, 10 pp.

Sasser, J.M. and Freckman, D.W. 1987. A world perspective on nematology. The role of the society. In Vistas on nematology (Eds. J.A. Veech and D.W. Dickson) Published by Society of Nematologists, USA, 1987 pp. 7-14.

Sethi, C.L. and Gaur, H.S. 1986. Nematode Management : An overview. In Plant Parasitic Nematodes Problems and Progress (Eds. G. Swarup and D.R. Dasgupta). ICAR, New Delhi 1986. pp. 424-445.

Singh, H. and Yadav, B.S. 1986. Effect of integrated control of wheat yield and population of *Heterodera avenae. Proc. Nat. Conf. Plant Parasitic Nematodes of India Problems and Progress*, IARI, New Delhi, 1986. pp. 97.

Vangundy, S.D. and McKenry, M.V. 1977. Plant Diseases : An advanced treatise (Eds. J.G. Horsfall and E.B. Cowling) pp. 264.

Wright, D.J. 1981. Nematicides : Mode of action and new approaches to chemical control. In plant Parasitic Nematodes Vol. III (eds. B.M. Zuckerman & R.H. Phole) Academic Press, New York, USA. pp. 421-445.

Zaki, F. and Bhatti, D.S. 1989. Effect of castor (*Ricinus communis*) leaves in combination with different fertilizer doses on *Meloidogyne javanica* infecting tomato. *Indian J. Nematol.* 19 : 171-176.

CHAPTER - 18

APPLICATION OF VAM FUNGI IN THE MANAGEMENT OF PLANT PARASITIC NEMATODES

P.C. Trivedi

INTRODUCTION

Nematodes are invertebrate and roundworm organisms, they occur in enormous numbers, in all possible types of habitats, including soil and fresh water, seas and oceans and in the animal systems also. They belong to phylum-Nematoda and are free living bacteriophagous, myceophagous or predacious, or parasites of plants and animals including, both vertebrates and invertebrates. These microscopic-organisms occur in millions in one square metre area of soil. Approximately 3 per cent of the expected nemic-biodiversity have been identified. Nematode causes severe losses, not only in the form of reduced plant growth and yield but also in the marketable quality of the produce. Feeding on plant cells by nematodes may deprive the host of nutrients and water for its growth. The association of plant parasitic nematodes with fungi or bacteria on disease complexes often increase the severity of the incidence. It also acts as a vector for transmission of soil-borne viruses. There are some beneficial roles played by the nematodes, these includes : (i) participation in the cycling of minerals and nutrients essential for a soil ecosystem, (ii) decomposition of organic wastes including biodegradation of toxic compounds, (iii) acting as indicator of soil and water quality and (iv) application of biological control of insect pests.

There are numerous estimates of the economic importance of nematodes in crop production on a world-wide and individual on country basis, but precise values cannot be determined. It was because of the nematode's "small size and hidden way of life" and lack of definite information on it's occurrence and pathogenicity. In a world-wide survey conducted by Sasser (1989), the 10 most important genera of plant parasitic nematodes revealed, were : *Meloidogyne, Pratylenchus, Heterodera,*

Ditylenchus, Globodera, Tylenchulus, Xiphinema, Radopholus, Rotylenchulus and *Helicotylenchus*. This order of important various genera was found to be fairly representative for most regions of the world. The estimated overall annual yield loss of the world's major crops due to damage by plant parasitic nematodes is 12.3 per cent.

It is very important to protect our crops and increase the country's food supply while, simultaneously safe-guarding our environment by reducing the use of chemical pesticides. The biological plant protection, through a subject of immense complexity, is a fascinating and challenging area of study for plant pathologist. Serious efforts are required to discover new, dynamic and potentially important strains of antagonists and to study their ecology to fully understand how they can be best utilized in biological control. Among the various kinds of organisms, engaged in natural control of nematodes, Vesicular-Arbuscular Mycorrhizae (VAM) is now used not only to control nematode pest but usefully applied as bio-fertilizer.

MYCORRHIZA

The term mycorrhiza (literally 'fungus root') was first used by Frank (1885) to describe the long lived association between roots and fungal mycelium. It is a mutualistic symbiosis of co-existing partners, with fungal partner being obligate survival on plant partners. The association so developed, result in enhanced survival, nutrient acquisition, reproduction and growth for the component organisms.

Types of mycorrhizas

Based on morphology and anatomy of the host plant, five broad groupings of mycorrhizas have been suggested. These groups are :

1. The ectomycorrhizas
2. The vesicular - arbuscular mycorrhizas
3. The ericaceous mycorrhizas
4. The ectondomycorrhizas
5. The orchidaceous mycorrhizas

These groupings share the commonality of a host plant having a fungal associated in intimates, mutualistic symbiosis, where both partners are living and exchange of the substances can occur in both directions i.e., a biotrophic association (Lewis, 1973).

1. *Vesicular-arbuscular mycorrhizas* (VAM)

Vesicular-arbuscular mycorrhizas are commonly known as VAM, which has a more appropriate meaning. It refers to the presence of intracellular structures, vesicles and arbuscles - that are formed inside the root during different developmental stages. VAM fungi are aseptate phycomycetes which belong to the family Endogonaceae and include genera such as *Glomus, Gigaspora, Acaulospora, Sclerocystis, Endogene* and *Modicella*. Genera of VAM fungi showed characteristic taxonomic features Table 1. VAM occurs in roots of most angiosperms, gymnosperms, pteridophytes and bryophytes. From fossil records it was observed that this association existed for millions of years.

2. IMPORTANCE OF VAM

VAM is playing a very important role in soil. It provides organic link between the root and the bulk of soil. Although, the fungus provides to the plant with minerals, especially phosphorus, the plant supplies to the fungus with photosynthetic sugars (Norris *et al.*, 1991, 1992). The bi-directional transport leads to enhanced plant growth and completion of the fungal life cycle. For these reasons mycorrhizal

Table 1. Important taxonomic features of genera of Endogonaceae

	Genera	Fruiting	Kind of spores	Spore germination	Type of mycorn
1	*Acaulospora*	Single spores	Azygospores	Through wall	Vesicular-arbuscular
2	*Endogone*	Sporocarp	Zygospores	Unknown	Ectomycorrhiza or unknown
3	*Glaziella*	Sporocarps	Chalmydospores	Unknown	Unknown
4	*Glomus*	Sporocarps & single spores	Chalmydospores	Regrowth of attached hyphae	Vesicular-arbuscular
5	*Gigaspora*	Single spore	Azygospores	Through wall	Arbuscular
6	*Modicella*	Sporocarps	Sporangiophores	Through wall	Unknown
7	*Sclerocystis*	Sporocarps	Chalmydospores	Regrowth of attached hyphae	Vesicular-arbuscular

symbiosis are attractive systems in agriculture, flori- and horticulture, arboriculture viticulture and forest management to enhance crop and wood production in the sense of sustainable agriculture and restoring soil fertility.

These fungi are potential biofertilizers and bioprotectors to enhance plant growth yield, act as soil conditioners and play a vital role in sustainability of the rapidly degrading environment (Norris *et al.*, 1994). Mycorrhizal symbionts produce extra-radical hyphae that may extend several centimeters into the soil and exude organic materials that are substrates for other soil microbes. These hyphae associates frequently, produce sticky materials that cause soil particles to adhere and help to create small aeration and water percolation as well as stability (Tisdall, 1991). Functions of VAM fungi are described below—

(i) *Nutrient uptake*

Rate of nutrient uptake by the root will depend on the demand by the absorbing organ, the size and distribution of absorbing system and the mobility of the nutrient in soil (Nye, 1979). Biochemical mineralization and chelation by the fungal components of mycorrhizas could be important in solubilizing certain nutrients in the rhizospheres. The surface acid phosphatases present in VAM are useful in utilizing complex phosphates (Barlett and Lewis, 1973). Kaur *et al.* (1997) have reported enhanced alkali/acid phosphatase activity in mycorrhizal roots of soybean and gram. It has been estimated that nearly 25-30% of phosphatic fertilizer can be saved through inoculation with efficient VAM fungi (Bagyaraj, 1992). Besides P, the role of VAM in uptake and transport of S, Zn, Cu, N, K and Sr has been reported.

(ii) *Water uptake*

The increased surface area provided by the extra-material hyphal system can provide a direct pathway for water uptake to the root. Graham and Syvertsen (1994) suggested that hyphal transport alone is probably responsible for an increase in water uptake by VAM infected plants. The mycorrhizal plants are better adapted to survive in water scarce soils.

(iii) *Root pathogen control*

Root structure and morphology are changed by the infection with VAM fungi. Rhizosphere is a complex environment where different microorganisms live and interact with each other. VAM fungi play an important role in root pathogen control.

The pathogen control may be due to —

— competition for actual site of interaction in and on the root;

— selection of favourable microflora by changes in root exudate products;

— production of inhibitory or toxic substances in the rhizosphere; and

— change in physiology of host root.

(iv) *Interaction with nitrogen fixing microorganisms*

Dual inoculation of nitrogen fixing and VAM fungi may help the plant to acquire both nitrogen and potash which will in turn improve plant growth, nodulation and nitrogen fixation. VAM increases nitrogen nutrition by enhancing direct uptake of combined nitrogen by VAM, facilitated nitrogen transfer and enzyme activity involved in nitrogen fixation and metabolism. VAM also improves nitrogen fixation by enhancing availability of P and hence more production of ATP which in turn helps in nitrogen fixation process which is a highly energy requiring process. Dual inoculation of VAM fungi and Rhizobium, improved nodulation and nitrogen fixation in chickpea and soybean (Bagyaraj, 1984, Kaur *et al.* 1997).

(v) *Production of growth hormones*

Numerous mycorrhizal fungi liberate Indole acetic acid (IAA) and other indole compounds (IBA and IPA). According to Ek *et al.* (1993), the tendency to produce IAA varies greatly from one fungal species to another and even from one isolate to another. Besides, mycorrhizal fungi also produce cytokinins (Ng *et al.* 1982).

3. LIFE CYCLE OF VAM FUNGI

Life cycle of VAM fungi is not completely known. Chlamydospores produced by mycorrhizal roots of the previous crops are present in soil in dermant stage. Infection develops either from chlamydospore in the soil or via mycelium originating from infected roots.

Development and establishment of VAM fungi in roots involves a preinfection phase of germination, hyphal growth and appresorium formation followed by growth of fungus within the root (Smith and Bowen, 1979). Fungus infection within 2-5 days leads to first arbuscles (tree like structure filling infected plant cells that live for 7-10 days) formation and later to vesicles (swollen ended, thick walled hyphae filled with reserve lipids) within the cortical cells (Bonfanto, 1984). During stress situation these reserves are utilised and vesicles then degenerate. *Gigaspora* and *Scutellospora* never form vesicles. Vesicles later develop into thick walled chlamydospores which function as reproductive structure and can survive under unfavourable conditions. Newly formed spores get released into soil in the vicinity of roots. These spores may germinate during favourable conditions and infect the roots, causing rapid colonization. The proportion of mycorrhizal to non-mycorrhizal roots rarely exceeds 80% and this may be related to physiological changes in host roots.

4. NEMATODE - MYCORRHIZA INTERACTIONS

A number of reviews and numerous research papers have appeared on Nematode-VAM interaction using variety of crop plants as test host in the recent past decades from different parts of the world (Caron 1989; Carling *et al.*, 1989; Dehne, 1982; Jain & Hasan, 1998; Hussey & Roncadori, 1982; Ingham, 1989; Jain, 1994; Schonbeck, 1987; Smith, 1987; Trivedi, 1995, 1998).

(a) Sequence and Level of VAM inoculation

Prior establishment of mycorrhiza mitigated the deleterious effects of nematode on plant growth characters to a considerable extent (Suresh and Bagyaraj, 1984; Hussey and Roncadori, 1982; Sharma, 2001; Sharma & Trivedi, 1994; Sharma & Trivedi, 2001). It seems logical that a minimum level of VAM fungal root colonization should be required before nematode activities are affected. VAM fungal inoculum potential should be sufficient to achieve maximum root colonisation levels within a given time frame.

(b) Effect on Nematode Population

Baltruschat *et al.* (1973) were the first to show that plants preinoculated with VAM, *Glomus mosseae* were less susceptible to root-knot nematode infection. The potential of VAM fungi to alleviate nematode-induced plant stress has been widely investigated because of their ability to increase root growth and nutrient absorption (Smith, 1987). Many workers have demonstrated a reduction in nematode population or disease incidence (Bagyaraj *et al.*, 1979; Cooper and Grandison, 1986; Grandison and Cooper, 1986; Jain & Hasan, 1988; Kasaab & Taha, 1990; Sikora & Schonbeck, 1975; Nagesh & Reddy, 1999; Reddy *et al.*, 1977; Rao *et al.* 1996; Sitaramaiah & Sikora, 1982; Sharma & Trivedi, 1994; 2001); but cases where nematode population remains unaffected (O'Bannon *et al.*, 1974; Cason *et al.*, 1983; Hasan & Jain, 1987, 1988), or even increase under the influence of mycorrhiza (Atilano *et al.*, 1981; Kasaab & Taha, 1990) are not uncommon.

Schenck & Kinloch (1974) reported low endomycorrhizal population when associated with high population of root-knot nematode in soybean. Sikora (1978) and Priestel (1980) reported reduced penetration of *Meloidogyne incognita* in presence of *Glomus mosseae* in tomato. Similarly, Sikora and Sitaramaiah (1980, 1981) reported a significant reduction in penetration and development of *Rotylenchulus reniformis* in cotton, tomato, bushbean, cucumber and muskmelon plants preinoculated with *G. mosseae*.

Sikora & Schoenbeck (1975) revealed a 75 per cent decrease in the number of *M. incognita* juveniles that developed into adults on tobacco in presence of *G. mosseae*. However, the ability of *M. incognita* to suppress mycorrhizal development has been observed by several investigators (Bird *et al.*, 1974; Rich and Bird, 1974; Schenck *et al.*, 1975).

Jain and Sethi (1987, 1988) revealed the influence of endomycorrhizal fungi (*Glomus fasciculatum* and *G. epigaeus* on pathogenicity, penetration and development of *Heterodera cajani* on cowpea. Though some recent studies have been shown that VAM fungi may effect host response and nematode activities differently from P nutrition (Cooper & Grandison, 1986; Grandison & Copper, 1986).

Sedentary nematode like *Meloidogyne* species induce the formation of hypertrophic tissues, the formation of galls. Inside this more or less meristematic gall, the host reacts to nematode establishment with the formation of giant cells, which seem to function as transfer cells for the nutrient flow towards the parasite. Pathogenic fungi, which may influence nematodes, often parasitize these cells, destroy them, and thus inhibit nematode development. Direct parasitization of galls by mycorrhizal fungi is not apparent. Mycorrhizal fungi are unable to infect meristematic tissue and the centre of the nematode induced gall has to be regarded as meristematic (Shamra, 1992; Sharma,

1994). Usually the endophyte can be found in the unaltered, opposite site of the root (Dehne, 1982; Kellan & Schenk, 1980; Priestel, 1980; Sharma, 1992). The mycorrhizal fungi are competitors of sedentary nematodes, the parasite it committed to remain in the gall and be associated with a specific giant cell system, but the endophyte is not so limited and can continuously infect new appropriate root sites (Sharma, 1994).

The survey of literature reveals that the *Glomus* species are the most predominant genera of VAM fungi in India and one or the other species especially *G. fasciculatum* has shown great promise in the management of nematodes in several experimental trials (Hasan & Jain, 1987; Jain & Hasan, 1986; Jalali & Chand, 1990; Sharma & Trivedi, 2001).

Sharma & Trivedi (2001) studied the concomitant effect of VAM and Neem products on *Heterodera avenae* infected wheat. They reported that establishment of *Glomus fasciculatum* & *G. mosseae* in the root system of wheat, resulted in decrease in the final nematode population and significant plant growth enhancement. The plants treated with various neem products were at par with VAM treated plants, Application of VAM integrated with Nimin, Achook, Neemax, Neem leaf and Neem cake outbreaked in increase in length, fresh and dry weight of root, shoot and ears as compared to their individual inoculation.

Saleh & Sikora (1984) emphasized that each Nematode-VAM-Host combination is unique and not a subject of generalization, therefore, require studies of each combination. The nature of interaction varied to antagonistic, neutral or synergistic (Hussey & Roncadori, 1982) Table 2.

(c) Factors Affecting VAM Symbiosis in Field

Initial studies have shown that temperature, moisture, pH, amendments and other microorganisms, besides some management practices e.g. fallowing, tillage, cropping pattern, fertilizers and pesticides may all affect the conduciveness of soil to mycorrhizal symbiosis (Brumdret, 1991; Wani & Lee, 1995). There is much evidence that different mycorrhizal fungi are adapted to different temperature regimes and that even different isolate of the same *Glomus* species can differ,

Table 2. Possible effects on interactions between plant parasitic Nematodes and VA Mycorrhizae (After Hussey and Roncadori, 1982)

Type of interaction	Component	Effect on component
Neutral	Fungus	Root infection or sporulation not altered.
	Host	Mycorrhizal stimulation of vegetative growth or yield not altered nematode suppression of vegetative growth or yield not offset.
	Nematode	Attraction to roots, penetration, development and reproduction not altered.
Positive	Fungus	Root infection or sporulation increase.
	Host	Nematode suppression of vegetative growth or yield offset.
	Nematode	Attraction to roots, penetration, development and reproduction suppressed.
Negative	Fungus	Root infection or sporulation suppressed.
	Host	Vegetative growth or yield response to mycorrhizae suppressed.
	Nematode	Attraction to roots, penetration development and reproduction increased.

therefore, further investigations of this is highly desirable. However, in general agreement that excess water, low oxygen levels and reducing conditions inhibit mycorrhizal formation but the precise mechanisms are not clear. Apart from a consensus that *Glomus* species favour neutral and alkaline soil, whereas, *Acaulospora* and *Gigaspora* speices are more tolerant to acidity, no generalization seem possible.

Wani & Lee (1995) has reviewed and examined the effects of different management practices such as fallowing, tillage, cropping pattern, fertilizers and pesticides on mycorrhizal association with crop plants. The present literature appears to suggest varying view on effects of such management practices for harnessing VAM benefits for increasing crop production. However, in general, long fallowing, excessive secondry tillage, high inputs of fertilizers and pesticides especially broad spectrum fumigants, substituted aromatic hydrocarbons and benzemidazoles leads to a decline in the inoculum in soil and in VAM colonization of the succeeding crop. Further, maintaining a plant cover on soil for as long as possible during a year helps in building up the inoculum of VAM in soil. Moreover, since crop species differ in association with VAM fungi, selection of appropriate crop/culture in the cropping sequence can be expected to help in raising the population of VAM in soil.

(d) Culture and Multiplication of VAM fungi

VAM has very broad specificity towards plants including various species of field, fruit and forest plant species, but ability to produce VAM in bulk quantities is a major bottleneck. VAM based biofertilizers are currently recommended only for transplanted and nursery raised crops because of difficulty in inoculum production. Since VAM fungi are obligate symbionts, the axenic culture in an important chellange from both scientific and practical point of view. The inability of culture to VAM fungi has been the major limiting factor in their application in agriculture.

Among the various host plants tested Rhode grass *Chloris gayana* was found to be best host of *G. fasciculatum* on the basis of root colonization (98%), spore production and infective propagules (Jain & Hasan, 1994). Further Perlite : Soilrite mixture (1 : 1) was found to be the best substrate (Singh, 1990; Sreenivasa and Bagyaraj, 1988a, b). Moreover work at MRCP at BAIF, Pune on Freeze drying technology in 30 ml glass vial or lyphalization of VAM culture has given encouraging results in prolonging the viability and economizing transportation cost. The freeze dried material has also been successfully used to inoculate some field grown plants (Singh, 1990).

(e) Axenic multiplication of VAM

In the recent years, culture of isolated roots have exhibited great potential for growing VAM fungi *in vitro*. Different forms of inocula have been used to associate VAM with root organ culture. Sporocarpic chalmydospores of *Glomus mosseae*, segments of mycorrhizal roots, internal vesicles extracted from infected roots, and non-sporocarpic azygospores have been used. Root culture method has been successfully used for *in vitro* production of different species of *Glomus* viz. *G. fasciculatum, G. intraradices, G. mosseae, G. aggregatum, G. caledomium* and *G. deserticola* (Mathur & Vyas, 1995). The culture and subculture of VAM fungi with transformed roots are suitable for an *in vitro* collection of fungal isolates and for increasing the fungal biomass. Based on these *in vitro* studies life cycle and the mode of survival of growth has been proposed (Strullu *et al.*, 1997). Gosal *et al.* (2000)

reviewed the work on VAM - *in vitro* multiplication and application to micropropagated plants.

(f) Mechanism of VAM Action

Many mechanism operate simultaneously during interaction of VAM and plant parasitic nematode on a particular host. A possible mechanism of interaction is given in Fig. 1. Following factors play major role in reducing nematode population and in enhancing plant growth.

(1) VAM fungi may increase root growth, expand the absorptive capacity of the root system for nutrients and water and enhance cellular processes in roots (Hayman, 1982).

(2) VAM fungi improve plant nutrition and by doing so may aid the host in compensating for damage caused by parasitic nematodes, thereby increasing plant tolerance to these pathogens.

(3) Mycorrhizal root colonization has been shown to affect root exudation (Gerdemann, 1968; Harley & Smith, 1983; Hayman, 1982). These changes could alter chemotactic attraction of nematodes to roots or directly retarded nematode development within root tissues.

(4) VAM fungi have reduced nematode infection and development on several hosts in spite of larger root systems on mycorrhizal plants (Cooper and Grandison, 1986; Grandison & Cooper, 1986; Smith *et al.* 1986; Sharma, 1992, 1993, 1994, 2001). It may be due to unfavourable conditions for nematodes developed on root system due to VAM colonization. Changes in hormones,

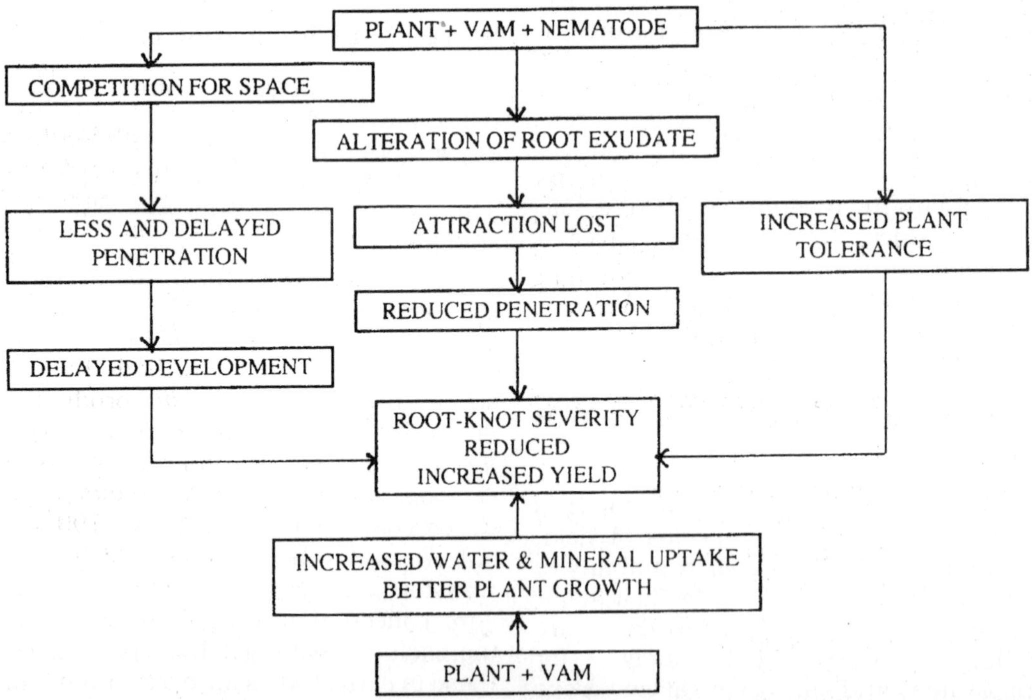

Fig. 1. A diagrammatic scheme of the mechanism of interaction of root-knot nematode and VAM.

amino acids and cell permeability in roots have been attributed to mycorrhizal symbiosis (Hayman, 1982).

(5) VAM fungi may cause change in the post-infection nematode host interaction by altering nematode reproduction and development.

(6) Direct competition for space may account for reduced nematode infection on mycorrhizal root system since endoparasitic nematodes occupy similar root tissues as VAM fungi. However, results of several studies do not support this hypothesis.

In addition to above given factors, increased tolerance or resistance to nematodes by VAM fungi may be due to competition for host photosynthates, production of nematistatic compound and parasitism of eggs. It may be concluded from observations that VAM fungi do not directly interact with plant parasitic nematodes inspite of their proximity in root tissue. More likely, VAM fungi alter the host either physically or physiologically, and thus indirectly affect the host-nematode relationship (Smith, 1987).

5. CONCLUSION

Environmental and economic concerns are causing scientist to consider low input sustainable agricultures as an alternative to conventional, high input agricultures. VAM fungi in need to be considered as part of such concept of minimal inputs in crop production.

The potential role of VAM fungi as biocontrol agent for nematode eradication has received considerable attention. It has been observed that the plants colonised with VAM fungi, generally grow well inspite of the presence of damaging levels of plant parasitic nematodes. Following aspects need special attention for the future work on VAM-Nematode interaction study -

(1) Considerable efforts are needed to select native efficient VAM fungi for each crop for getting maximum benefits out of these fungi.

(2) Positive results of greenhouse experiment should be verified under field conditions to assess potential utilization of VAM as biocontrol agent.

(3) Because of obligate nature of VAM fungi, these cannot be cultured on large scale. Studies on priority basis should be undertaken to standardized technique for mass scale multiplication and application.

(4) It would be valuable to establish germplasm bank within a country, perhaps at one University and one large research institute to help streamline the choice of endophytes for field trials.

(5) Efforts should also be made to enhance the field of VAM inoculum with least contaminants, longer shelf life and easy transportation.

(6) More technical improvements in inoculation especially by seed pelleting and artificially by seed pelleting and artificial culture of VAM are needed for large scale commercial applications.

(7) Field studies also confirm the importance of natural mycorrhizas in stimulating plant growth relative to adjacent soils lacking effective mycorrhiza due to fumigation, fallowing or non-mycorrhizal cover crops. More field experiments are needed before valid judgement can be made for the potential value of field inoculations.

The existence of variability in the interrelationship between VAM fungi and plant parasitic nematodes, in different combinations, is unique and an understanding of these interactions need intensive study. Studies should determine whether, VAM fungi can be of sufficient value to warrant the development of technology for using these organisms. VAM fungi should be used with other methods for nematode control as a part of integrated pest management strategies.

REFERENCES

Atilano, R.A., J.A. Menge and S.S. Vangundi 1981. Interaction on *Meloidogyne arenaria* and *Glomus fasciculatum* in Grape. *J. Nematol.* 13 : 52-57.

Bagyaraj, D.J. 1984. Biological interaction with VA mycorrhizal fungi. In : Powell, C.L.I. and Bagyaraj, D.J. (Eds.) VA Mycorrhiza, CRC Press, Boca Raton, FL, pp. 131-153.

Bagyaraj, D.J. 1992. Vesicular-arbuscular mycorrhiza- Application in agriculture. Methods in Microbiology. 24 : 359- 373.

Bagyaraj, D.J., A. Manjunath and D.D.R. Reddy 1979. Interaction of vesicular-arbuscular mycorrhizae with root-knot nematode in tomato. *Plant Soil.* 51 : 397-407.

Baltruschat, H. and Schonbeek, F. 1975. The influence of endotrophic mycorrhiza on the infestation of tobacco by *Thielaviopsis basicola. Phytopathology Z.* 84 : 172- 188.

Barlett, E.M. and Lewis, D.H. 1973. Surface phosphate activity of mycorrhizal roots of beech. *Soil Biol. Biochem.* 5 : 249-257.

Bonfante, F.P. 1984. Anatomy and morphology of VA mycorrhizal. In : Powell, C.L., Bagyaraj, D.G. (Eds.) VA Mycorrhizae pp. 5-32, CRC Press, Boca Raton, Jla.

Brundret, M. 1991. Mycorrhiza in natural ecosystems. In : Ecological Research (Eds. Begon *et al.*) Vol. 21, Academic Press, London, New York, pp. 171-313.

Bird, C.W., Rich, J.R. and Glover, S.U. 1974. Increased endomycorrhizae of cotton roots in soil treated with nematicide. *Phytopathology* 64 : 48-51.

Carling, D.E., Roncadori, R.W. and Hussey, R.S. 1989. Interaction of VAM fungi, root-knot nematode and phosphorus fertilization on soybean. *Plant Dis.* 73 : 730-733.

Caron, M. 1989. Potential use of mycorrhizae in control of soil borne disease. *Can. J. Pl. Path.* 11 : 177-179.

Cason, T.K.M., Hussey, R.S. and Roncadori, R.W. 1983. Interaction of vesicular-arbuscular mycorrhizal fungi and phosphorus with *Meloidogyne incognita* on tomato. *J. Nematol.* 15 : 410-417.

Cooper, K.M. and Grandison, G.S. 1986. Interaction of vesicular arbuscular mycorrhizal fungi and root-knot nematode on cultivars of tomato and white clover susceptible to *Meloidogyne hapla. Annals of Applied Biology* 108 : 555-556.

Dehne, H.W. 1982. Interaction between vesicular arbuscular mycorrhizal fungi and plant pathogens. *Phytopathology* 72 : 1115-1119.

Ek, M., Ljungquist, P.O. and Stenstrom, E. 1983. Indole-3 acetic acid production by mycorrhizal fungi determined by gas chromatography-mass spectrometry. *New Phytol.* 94 : 401- 403.

Frank, A.B. 1885. Uber die an Wurzely mliose, berunhende, Ernabrung, gewisser Baume durch unterirdische. Pilze. Ber. Dt. Bot. Ges. 3 : 128-145.

Gerdemann, J.W. 1968. Vesicular-arbuscular mycorrhiza and plant growth. *A. Rev. Phytopath.* 6 : 397-418.

Graham, J.H. and Syvestsen, J.P. 1984. Influence of vesicular arbuscular mycorrhiza on the hydraulic conductivity of roots of two citrus root stocks. *New Phytol.* 97 : 227-284.

Grandison, G.S. and K.M. Cooper 1986. Interactions of vesicular arbuscular mycorrhizae and cultivars of alfalfa susceptible and resistant to *Meloidogyne hapla. J. Nematol.* 18 : 141-149.

Gosal, S.K., Gupta, R.P. and Gosal, S.S. 2000. Vesicular arbuscular mycorrhiza *in vitro* multiplication and application to micropropagated plants. *In*: Plant Biotechnology - Recent Advances. (Eds. P.C. Trivedi), Panima Publisher, New Delhi, PP. 435-450.

Harley, J.L. and Smith, J.F. 1983. Interactions of vesicular arbuscular mycorrhizae and cultivars of alfalfa susceptible and resistant to *Meloidogyne hapla J. Nematol.* 18 : 141-149.

Hayman, D.S. 1982. Practical aspects of vesicular arbuscular mycorrhiza. PP. 325-373. In : Subba Rao, N.S., Advances in Agricultural Microbiology. Oxford and IBH Publishing, Co.

Hasan, N. and Jain, R.K. 1987. Parasitic nematodes and vesicular-arbuscular mycorrhizal (VAM) fungi associated with berseem (*Trifolium alexandrium* L.) in Bundelkhand region. *Indian J. Nematol.* 17 : 184-188.

Hussey, R.S. and Roncadori, R.W. 1982. Vesicular-arbuscular mycorrhizae may limit nematode activity and improve plant growth. *Pl. Dis.* 66 : 9-14.

Ingham, R.E. 1989. Interaction between nematode and vesicular- arbuscular mycorrhiza. *Agriculture Ecosystems and Environment.* 24 : 169-182.

Jain, R.K. and Hasan, N. 1986. Association of vesicular- arbuscular mycorrhizal (VAM) fungi and plant parasitic nematodes with forage sorgham (*Sorgham bicolor* L.) *Sorghum Newslett.* 29 : 89.

Jain, R.K. and Hasan, N. 1988. Role of VAM fungi and nematode activities in forage production. *Acta Bot. Indica* 16 : 84- 88.

Jain, R.K. and Hasan, N. 1994. Vesicular-Arbuscular mycorrhizal (VAM) Fungi. A potential biocontrol agent for nematodes. (In: Vistas in Seed Biology- Ed. Singh, T. & Trivedi, P.C. 1994). Printwell Publishers, Jaipur, P. 49-60.

Jain, R.K. and Hasan, N. 1998. Vesicular arbuscular mycorrhizal (VAM) fungi : Nematode management and growth responses in forage crops. *In*: Recent Advances in Plant Nematology (Ed. P.C. Trivedi) CBS Publishers & Distributors, N. Delhi, PP. 201-215.

Jain, R.K. and Sethi, C.L. 1987. Pathogenicity of *Heterodera cajani* on cowpea as influenced by the presence of VAM fungi, *Glomus fasciculatum* or *G. epigaeus. Indian J. Nematol.* 17 : 165-170.

Jain, R.K. and Sethi, C.L. 1988. Interaction between vesicular arbuscular mycorrhiza and *Meloidogyne incognita, Heterodera cajani* on cowpea as influenced by time of inoculation. *Indian J. Nematol.* 18 : 263-268.

Jalali, B.L. and Chand, H. 1990. Current trends in mycorrhizal research. Proce. of the National Conference on Mycorrhiza, HAU, Hissar, Haryana, India. pp. 210.

Kasaab, A.S. and Taha, A.H.Y. 1990. Interaction between plant parasitic nematodes, vesicular - arbuscular mycorrhiza, rhizobia and nematicide on Egyptian Clover. *Ann. Agric. Sci.* 35 : 509-520.

Kellam, M.K. and Schenk, N.C. 1980. Interactions between a vesicular arbuscular mycorrhizal fungus and root-knot nematodes on soybean. *Phytopathology.* 20 : 293-296.

Kaur, J., Gupta, R.P., Pandher, M.S., Holly, R.S. and Gupta, S.K. 1997. Synergistic effect of VAM and *Rhizobium* on soybean. Proc. Third Agricultural Science Congress. P. 47. Punjab Agricultural University, Ludhiana.

Mathur, N. and Vyas, A. 1995. *In vitro* production of *Glomus deserticola* in association with *Zizyphus nummularia. Plant Cell. Reports.* 14 : 735-737.

Nagesh, M., Reddy, Parvatha and Rao, M.S. 1999. Comparative efficacy of VAM fungi in combination with neem cake against *Meloidogyne incognita* on *Crossanelra undulaefolia. Mycorrhiza News.* 11 : 11.

Ng, P.P., Cole, A.L.J., Jameson, P.E. and Newha, J.A. 1982. Cytokinin production by ectomycorrizal fungi. *New Phytol.* 91 : 57-62.

Norris, J.R., Read, D.J. and Varma, A.K. 1991. Methods in Microbiology 23 : Academic Press, London.

Norris, J.R., Read, D.J. and Varma, A.K. 1992. Methods in Microbiology 24 : Academic Press, London.

Norris, J.R., Read, D.J. and Varma, A.K. 1994. Techniques in Mycorrhizal Research. Academic Press, London.

Nye, P.H. 1979. Soil properties controlling the supply of nutrients to the root surface. In : J.L. Harley & R.S. Russel (Eds.). The Soil-Root interface. Academic Press, London, PP. 39- 49.

O'Bannon, J.H., Inserra, R.N., Nemec, S. and Vovlas, N. 1979. The influence of *Glomus mosseae* on *Tylenchulus semipenetrans* infected and uninfected citrus limon seedlings. *J. Nematol.* 11 : 247-250.

Priestel, G. 1990. Wechselbeziehung Zwischen der endotrophen Mycorrhiza and dem Wurelgallen nematoden *Meloidogyne incognita* on Gurke Dissertation, University, Hannover, W. Germany.

Rao, M.S., Reddy, Parvatha & Mohandas, S. 1996. Effect of integration of *Calotropis procera* leaf and *Glomus fasciculatum* on the management of *Meloidogyne incognita* infesting tomato. *Nematologica Mediterranea* 24(1): 59.

Rich, J.R. and Bird, G.W. 1974. Association of early season vesicular arbuscular mycorrhizae with increased growth and development of cotton. *Phytopathology* 64 : 1421-1425.

Saleh, H. and Sikora, R.A. 1984. Relationship between *Glomus fasciculatums* root colonization on cotton and its effect on *Meloidogyne incognita. Nematologica* 30 : 230-237.

Schenck, N.C. and Kinlock, R.A. 1974. Pathogenic fungi, parasitic nematodes and endomycorrhizal fungi associated with soybean roots in Florida. *Pl. Dis. Reptr.* 58 : 169-173.

Schenck, N.C., Kinlock, R.A. and Dickson, D.W. 1975. Interaction of endomycorrhizal fungi and root-knot nematode on soybean. PP. 606-617. In : Endomycorrhizas (Eds. Sanders, F.E., Mosse, B. & Tinker, P.). Academic Press, London.

Schonbeck, F. 1979. Endomycorrhiza in relation to plant disease. In : Soil-borne plant pathogens. Int. Symp. Soil-borne Pl. Pathogens. 4th Munick, 1978, New York. Academic Press, PP. 271-280.

Sharma, I.R. 1994. Biology, biochemistry and biocontrol of barley cyst nematode, *Heterodera avenae.* Ph.D. Thesis, University of Rajasthan, Jaipur.

Sharma, Renu 1993. Biological control of root-knot nematode, *Meloidogyne* sp., Ph.D. Thesis, University of Rajasthan, Jaipur.

Sharma, Ameeta 2001. Application of VAM Fungi in the management of *Heterodera avenae* on wheat. Ph.D. Thesis, University of Rajasthan, Jaipur.

Sharma, Ameeta and Trivedi, P.C. 2001. Concomitant effect of VAM and Neem Products on *Heterodera avenae* infected Wheat. *J. Indian Bot. Soc.* 80 : 9-13.

Sharma, Renu and Trivedi, P.C. 1994. Interaction of root-knot nematode, *Meloidogyne incognita* and VA mycorrhizae *Glomus fasciculatum* and *Glomus mosseae* on brinjal (*Solanum melongena* L.) *J. Indian Bot. Soc.* 73 : 221-224.

Sharma, Widhi 1992. Studies on biochemical alterations resulting *Melodogyne* sp. infection in *Abelmoschus esculentus* and their control. Ph.D. Thesis, University of Rajasthan, Jaipur.

Sharma, Widhi and Trivedi, P.C. 1994. Interaction of root-knot nematode with VAM on *Abelmoschus esculentus* and their effect on net yield. *In : Vistas in seed Biology, Vol. II (Eds. Singh, T. & Trivedi, P.C. 1994). Printwel Publishers, Jaipur.* pp. 61-81.

Sikora, R.A. 1978. Einflub der endotrophen mykorrhiza (*Glomus mossae*) In: Root knot nematode, *Meloidogyne* spp. Systematic Biology and Control (Eds. Lamberti, R. & C.E. Taylor). Academic Press, London, PP. 399-404.

Sikora, R.A. and Schonbeck, F. 1975. Mycorrhiza (*Endogone mosseae*) on the population dynamics of the root-knot nematode (*Meloidogyne incognita* & *M. hapla*) 8th Int. Gen. Pl. Protec. 5 : 80-86.

Sikora, R.A. and Sitaramaiah, K. 1980. Antagonistic interaction between the endotrophic mycorrhizal fungus *Glomus mosseae* and *Rotylenchulus reniformis* on cotton. *Nematropica.* 10 : 72-73.

Sikora, R.A. and Sitaramaiah, K. 1981. Influence of the endomycorrhizal fungus, *Glomus mosseae* on *Rotylenchulus reniformis* penetration and development of bushbean, cucumber and muskmelon. Mededelingen Van de Faculteit. Landbouwwetens chappen Rijksumbesiteit Gent. 45 : 695-702.

Singh, S.N. 1990. Biotechnology for mass production of VA Mycorrhiza inocula. In : Current trends in mycorrhizal reserch. Proc. Natin. Conf. on Mycorrhiza (Eds. Jalali, B.L. and H. Chand) HAU, Hissar, India Pp. 80-86.

Smith, S.E. and Bowen, G.E. 1979. Soil temperature mycorrhizal infection and nodulation of *Medicago truicutula* and *Trifolium subterraneum. Soil Biol. & Biochem.* 11 : 469-473.

Smith, G.S., Hussey, R.S. and Roncadori, R.W. 1986. Penetration and post infection development of *Meloidogyne incognita* on cotton as affected by *Glomus intradices* and phosphorus. *J. Nematol.* 18 : 429-435.

Sreenivasan, M.N. and Bagyaraj, D.J. 1988a. *Chloris gayana* (Rhode grass) a better host for mass production of *Glomus fasciculatm* inoculum. *Pl. & Soil* 106 : 289- 290.

Sreenivasan, M.N. and Bagyaraj, D.J. 1988b. Selection of a suitable substrate for mass multiplicaton of *Glomus fasciculetum Pl. & Soil.* 109 : 125-127.

Suresh, C.K. and Bagyaraj, D.J. 1984. Interaction between a vesicular arbuscular mycorrhiza and a root-knot nematode and its affect on growth and chemical composition of tomato. *Nematologica Medit.* 12 : 31-39.

Trivedi, P.C. 1995. Biocontrol potential of VAM fungi in the management of plant parasitic nematodes. *J. Indian bot. Soc.* 74 : 165-171.

Trivedi, P.C. 1998. Potential of mycorrhizal fungi in the management of nematode- An overview. In : Ecofriendly Pest Suppression (Eds. R.D. Gautam and D. Prasad), Westvill Publ. House, N. Delhi, pp. 153-169.

Trivedi, P.C. 1998. Editor. Nematode diseases in plants. CBS Publisher & Dist., N. Delhi, PP. 414.

Trivedi, P.C. 1998. Editor. Plant Nematode Management. CBS Publ. & Dist. N. Delhi. pp. 280.

Trivedi, P.C. 1998. Editor. Recent Advances in Plant Nematology, CBS Publi. & Dist., New Delhi, pp. 277.

Trivedi, P.C. 1998. Phytonematology in India. CBS Publi. & Dist., New Delhi, pp. 205.

Trivedi, P.C. 2000. Plant Diseases, Pointer Publishers, Jaipur, pp. 452.

Trivedi, P.C. 2001. Plant Pathology, Pointer Publishers, Jaipur, pp. 422.

Trivedi, P.C. 2002. Plant Pest Management, Aavishkar Publ. & Dist. Jaipur, pp. 416.

Wani, S.P. and Lee, K.K. 1995. Exploiting Vesicular arbuscular mycorrhizae through crop and soil management practices. *Mycorrhiza News* 6 : 1-7.

INDEX